D1260103

On Geography

Europe supported by Africa and America
From J. G. Stedman, *Narrative of a five years' expedition, against the revolted Negroes of Surinam, in Guiana, on the wild coast of South America: from the year 1772, to 1777, elucidating the history of that country, and describing its productions, viz. quadrupeds, birds, fishes, reptiles, trees, shrubs, fruits and roots; with an account of the Indians of Guiana, and Negroes of Guiana* (London: J. Johnson, 2 volumes, 1796), volume 2, p. 394.

D. R. Stoddart

On Geography
and its History

Basil Blackwell

First published 1986

Basil Blackwell Ltd
108 Cowley Road, Oxford OX4 1JF, UK

Basil Blackwell Inc.
432 Park Avenue South, Suite 1505,
New York, NY 10016, USA

British Library Cataloguing in Publication Data

Stoddart, D. R.
On geography: and its history.
1. Geography
I. Title
910 G116
ISBN 0–631–13488–3

Library of Congress Cataloging in Publication Data

Stoddart, D. R. (David Ross)
On geography: and its history.
Includes bibliographical references and index.
1. Geography—History. I. Title.
G80.S785 1985 910.9 85–11252
ISBN 0–631–13488–3

Typeset by Oxford Publishing Services
Printed in Great Britain by
The Bath Press, Avon

Contents

Preface

I am glad to offer this book to those who have made it possible. First, Colin Nichols and Ken Stott, who helped turn me to geography. Second, Alfred Steers and Benny Farmer, who set the tone in both university and college for the development of the most liberal as well as the finest department of geography in the country. Third, my colleagues and students, who have lived through the ideas presented here. And finally my family – all practical geographers on islands around the world.

Acknowledgements

I am indebted to the following for permission to reprint material from essays previously published: the director and secretary of the Royal Geographical Society and the editor of the *Geographical Journal* for material in chapters 3 (*Geographical Journal*, 147 (1981), 287–297), 4 (*ibid.*, 146 (1980), 190–202) and 5 (*ibid.*, 141 (1975), 216–239); the honorary secretary of the Institute of British Geographers and the editor of its *Transactions* for chapter 9 (*Transactions of the Institute of British Geographers*, 66 (1975), 17–40) and the Institute and the editor of *Area* for material in chapter 6 (*Area*, 7 (1975), 188–190); the honorary secretary of the Geographical Association and the editor of *Geography* for material in chapter 2 (*Geography*, 67 (1982), 289–296; the secretary of the Association of American Geographers and the editor of its *Annals* for chapter 8 (*Annals of the Association of American Geographers*, 56 (1966), 683–698); the editor of the *Bulletin of Marine Science* for chapter 10 (*Bulletin of Marine Science*, 33 (1983), 521–527); Edward Arnold (Publishers) and the editors of *Progress in Human Geography* for material in chapter 6 (*Progress in Human Geography*, 5 (1981), 119–124); Basil Blackwell Ltd. for material in chapter 1 from D. R. Stoddart, ed., *Geography, ideology and social concern* (Oxford: Basil Blackwell, 1981), 1–7 and 70–80; Associated Book Publishers for chapter 11 from R. J. Chorley and P. Haggett, eds, *Models in geography* (London: Methuen, 1967), 511–548; and Edward Arnold (Publishers) for material in chapter 12 from *Progress in Physical Geography*, 1 (1977), 537–543, 2 (1978), 514–528, 5 (1981), 575–590, and 7 (1983), 256–264.

I am grateful to the Hakluyt Society and Cambridge University Press for permission to reproduce plate 3; to the Syndics of Cambridge University Library for plates 5, 6, 7 and 8; to the Department of Geography at Cambridge University for plates 9 and 10; to Professor Denys Brunsden and the Department of Geography at King's College, London, for plate 16; to Mr P. Titheradge and the Royal College of Surgeons of England, Down House, for plate 17; and to Sir Andrew Huxley for plate 18.

I thank Professor Brunsden for sharing with me his archive of Wooldridgeana, and Dr I. R. Bishop, of the Department of Zoology, British Museum (Natural History), for recondite information on the eating habits of zoologists.

The whole book has been typed by Maria Constantinou in the Department of Geography at Cambridge: I trust she found it less deadly than most of the typing that comes her way. And the project has been urged on in a variety of Chinese, Middle Eastern and even French restaurants by John Davey. The indexes have been compiled in California, Arizona, and the District of Columbia (in all of which there are many more exciting things to do) by June, Aldabra and Michael Stoddart, and myself.

Plates appear between pages 148 and 149.

Introduction

I have spent most of my life in the study of coral reefs and islands – in the Caribbean, the Indian Ocean and the central and west Pacific. Colleagues on my expeditions have often expressed surprise that a student of geomorphology, sedimentology, botany and zoology should work in a department of geography. I share no such feeling: I am a geographer, and my work lies squarely in a great tradition.

But I recognize – and regret – the fact that for many people the geographer has long since appeared to have surrendered to other specialists his catholic concern for the diversity of the natural world. We have too long accepted the artificial constructions of the bookmen about what geography is, what it should be concerned with and how it should be done. I confess to a feeling of unreality about much of the literature on the philosophy, methodology and even history of the subject, much of it written by people who signally fail to practise what they preach. Meanwhile so many retreat into increasingly restrictive and esoteric specialities, where they protect themselves with secret languages and erudite techniques.

The view of geography offered here – in essays that have been written on various occasions, some as formal papers, others as lectures and addresses, over the past twenty years – has grown directly from my own research. On uninhabited Pacific atolls, sailing along the barrier reefs of Australia and Belize, in the mangrove swamps of Bangladesh, on English coastal marshes, I have been concerned with making sense of nature. It is a tradition which in this book I trace back especially to Forster, though it is doubtless of great antiquity. How the field of study it represents came to be labelled 'geography' is, of course, another matter, and it is a story I seek to tell of the growing organization and expansion of scholarly activity over the last 150 years. Not surprisingly, my heroes are not the usual ones – the Ritters, Ratzels, Hettners, entombed by conventional wisdom. My geography springs from Forster, Darwin, Huxley: and it works.

Not for me the doubt and despair of many of my contemporaries on the human and quantitative sides of the subject. I read with astonishment statements by apparent leaders in these fields of the 'vacuum' at the heart of the subject, that it is a 'formless discipline', 'unproductive and disillusioning', 'lacking any rigour'.[1] There is a simple remedy: do some *real* geography. I cannot be surprised that 'hours of thoughtless number crunching' on things like the distribution of gasworks in Lancashire or service stations on motorways lead to this kind of moroseness. After all, Wooldridge long ago made plain his view – I share it – that there is vastly more to geography than 'careful studies of how bootlaces are made, or where Ovaltine comes from'.[2]

There is a different world, of great beauty and diversity, waiting for exploration. Its geography brings deep satisfaction and immense pleasure: it is also fun. Over the years it has attracted a remarkable collection of frequently bizarre and occasionally outrageous people, some of whom appear in this book.[3] Dull they rarely are, and the reader will forgive me if from time to time I record anecdotes about them of perhaps only marginal relevance to my argument.

This book is thus a very personal one. It supplies the context and the justification for years of active research in the tropical seas. It claims for geography a central role among the sciences of nature. It underlines the essential need to have faith in and enthusiasm for the subject we profess.

[1] I do not exaggerate. Take this, for example, from a leader of our profession: 'To be quite honest, I no longer care very much about geography, with its smug self-satisfaction and nauseating, narrow-minded chauvinism. . . . Most geography is inconsequential claptrap.' This and the phrases quoted in the text (which come from the most indefatigable authors of instruction manuals for students on how to do geography) are charitably left anonymous, but the inquisitive reader will find them (and many more) in various contributions to *Recollections of a revolution: geography as spatial science*, edited by M. Billinge, D. J. Gregory and R. L. Martin (London: Macmillan, 1984).

[2] S. W. Wooldridge, 'On taking the ge- out of geography', *Geography*, 34 (1949), 9–18; reference on p. 15. This paper is reprinted in Wooldridge's *The geographer as scientist: essays on the scope and nature of geography* (London: Thomas Nelson, 1956), pp. 7–25. I do not deny that quite respectable studies might be made of such things, though even the economic geographers seem to be ashamed of them nowadays. The rot goes back to Rodwell Jones, in his inaugural lecture at (of all places) the London School of Economics, where after noting that 'Northampton was noted for boots and shoes, Kidderminster for carpets, and that Nottingham was notorious for lace curtains', he concluded: 'That sort of stuff is no longer geography.' Perhaps this accounts for the evident unhappiness of the human geographers today. See Ll. Rodwell Jones, 'Geography and the university', *Scottish Geographical Magazine*, 42 (1926), 65–79; reference on p. 69.

[3] Alas, I have restricted myself to those no longer with us: I recall that one of my teachers at Cambridge lost a large sum in the High Court, having injudiciously neglected this precaution (*The Times* Law Report, 11 March 1965). But my files suggest that things have not changed greatly in more recent years.

Above all, it is a book written from the field and not from the armchair, for a very good reason grasped 300 years ago by Robert Cresswell:[4]

> Though most Geographers have the good hap
> To travel in a safe expencelesse Map,
> And while the World to us they represent,
> No further yet then Pilgrim *Purchas* went,
> Past *Dovers* dreadful cliffe afraid to go,
> And took the lands end for the worlds end too;
> Spand Countreys at the fingers end at ease,
> Crack'd with their nail all *France*, turn'd blots to seas;
> Of whom this strong line we may ridling say,
> They travel not, but sit still a great way.
> I must applaud whither they choise, or lot
> Which hath beyond their lazie knowledge got,
> Who onely in the Globe do crosse the Line,
> There raise the Pole, and draw whole Maps in wine
> Spil'd on the Table; measure Seas and Lands
> By scale of miles wherein the Compasse stands.
> But you the truths eye-witnesse have not been.

[4] In Edward Terry, *A voyage to East-India. Wherein some things are taken notice of in our passage thither, but many more in our abode there, within that rich and most spacious Empire of the Great Mogul. Mix't with some parallel observations and inferences upon the storie, to profit as well as delight the Reader* (London: J. Martin & J. Allstraye, 1655).

1

Geography and its History

There are many books about the history of geography. This one is different. To understand not only why this is so, but also why it is demanded, we must first explore the nature of historical understanding of intellectual and scientific endeavour, and examine the ways in which that understanding has been organized.

THE HISTORY OF GEOGRAPHY

In its simplest form, history is but a chronology of events. The main organizing device is narrative of change over time, and the archetype of such history in the field of geography is that of exploration. Baker,[1] to take but one example, provides a narrative listing over centuries of the progressive discovery of the lands and seas of the world. The story told is necessarily one of cumulative advance in knowledge towards the present, and insofar as there is any coherence to the story it is provided by the sequence of events themselves. Little attempt is made to seek explanations of the incidents described, or to recognize the deeper structures or meanings of the exploratory process.[2]

It is a commonplace that under the influence of Kantian ideas, this methodology was readily transferred to the history of geography as a whole. Classic examples of works similarly organized are given by the recent books of Dickinson.[3] Such histories share common characteristics.[4] First, there is

[1] J. N. L. Baker, *A history of geographical discovery and exploration* (London: Harrap, 1931; 2nd edition, 1937).

[2] But see J. K. Wright, 'Terrae incognitae: the place of the imagination in geography', *Annals of the Association of American Geographers*, 37 (1947), 1–15; J. D. Overton, 'A theory of exploration', *Journal of Historical Geography*, 7 (1981), 53–70; and also chapter 7.

[3] R. E. Dickinson, *The makers of modern geography* (London: Routledge & Kegan Paul, 1969), and *Regional concept: the Anglo-American leaders* (London: Routledge & Kegan Paul, 1976). For analyses of the scope of the history of geography within this framework, see H. Beck, 'Methoden und Aufgaben der Geschichte der Geographie', *Erdkunde*, 8 (1954), 51–57, and 'Das Problemfeld der Geschichte der Geographie', *Erdkunde*, 31 (1977), 81–85.

[4] These were first analysed in the brilliantly polemical book of Joseph Agassi, *Towards an historiography of science* (*History and Theory*, Beiheft 2, 1963).

an emphasis on chronology, cumulation and continuity. Scientific advance is seen as pressing relentlessly forwards towards the present, and issues such as priority in discovery or publication loom large as criteria of significance. Second, such inductivist history is markedly internalist. Written from the perspectives of the present, such studies identify a continuous series of men and ideas, linked together in chronology and content. Thus Hartshorne[5] and subsequently many others have traced a development within geography from Kant through Humboldt and Ritter to Richthofen and Hettner. Set squarely within what Schaefer[6] termed the 'exceptionalist' tradition, it is hence not surprising that the names of Darwin, Marx and Freud are absent from Hartshorne's and similar works; that they give little or no attention to philosophical or epistemological issues; and that the history traced remains unrelated to social, economic and political conditions.[7] Fourth, the identification of such a mainstream of progress involves value judgements about the past from the standpoint of what is clearly an evanescent present. Indeed Hartshorne devotes a section of his book to 'deviations from the course of historical development'.

Hartshorne, of course, was writing a methodological tract rather than making an historical enquiry alone, but his methods had profound consequences for historical writing in the subject and for the general perception of what was significant in our history. For him, history is

[5] R. Hartshorne, 'The nature of geography: a critical survey of current thought in the light of the past', *Annals of the Association of American Geographers*, 29 (1939), 171–658; reprinted as *The nature of geography* (Lancaster, Pa.: Association of American Geographers, 1939 and later printings), to which subsequent citation is made.

[6] F. K. Schaefer, 'Exceptionalism in geography: a methodological examination', *Annals of the Association of American Geographers*, 43 (1953), 226–249. Schaefer attacked the notion that geography in its aims and procedures was fundamentally different from other branches of knowledge. His paper came to provide a philosophical justification for geography as a positive spatial science, using especially the methods of mathematics and geometry which are the common tools of enquiry in all the sciences. But in a broader sense Schaefer was arguing for the reintegration of geography and geographers, isolated far too long, into a wider intellectual tradition. It is one of the paradoxes of our recent history that the lead which he gave propelled the subject into a quarter of a century of obsession with spatial analysis, the limitations and restrictions of which have only belatedly been recognized. See, diversely, L. Guelke, *Historical understanding in geography: an idealist approach* (Cambridge: Cambridge University Press, 1982), and D. J. Gregory, 'Suspended animation: the stasis of diffusion theory', in D. J. Gregory and J. Urry, eds, *Social relations and spatial structures* (London: Macmillan, 1985).

[7] This is the more remarkable when we consider that Hartshorne wrote the final version of his book in Vienna in 1939, at a time when the political situation rendered fieldwork on the borders of Germany, Poland, Austria, Hungary and Czechoslovakia impossible. See R. Hartshorne, 'Notes toward a bibliobiography of *The Nature of Geography*', *Annals of the Association of American Geographers*, 69 (1979), 63–76.

normative: 'If we wish to keep on the track – or return to the proper track . . . we must first look back of us to see in what direction that track has led.'[8] Furthermore, the actors in the history are readily characterized into those who followed the track (and who were therefore right) and those who blundered off (and were hence wrong). The former heroes include men such as Varenius, Vidal de la Blache, Humboldt and Ritter. But such reputations are particularly vulnerable to changes in one's vantage point. Depending on the track one discerns, for example, one can see such a figure as Carl Sauer either as hero or as villain: indeed the ambiguity in Sauer's position and the ambivalence with which he is often regarded stem largely from the difficulty with which he can be fitted into the standard unilinear pattern.

Finally, such narrative history says nothing of the processes of change. Ideas, insofar as they are discussed at all, are linked only by the simplest mechanisms of cause and effect, largely inferred from temporal succession. There is little consideration of individuals and their intellectual milieux; and such reconstructions hardly reflect the contingencies and pressures which all scholars recognize in their own intellectual development.

But there is another view of the history of the subject, recognized more than 50 years ago by John K. Wright, when he wrote that 'the history of geography . . . is the history of geographical ideas.'[9] Sadly, Wright's words were scarcely heeded. Such a history will emphasize the development of problems and theories and the social and intellectual context of their protagonists, rather than the cataloguing of people, institutions and publications. The analysis of the history of ideas is necessarily contextual, as Berdoulay has convincingly argued.[10] The truth or falsity of any particular proposition is no longer the prime criterion of its importance, whether from an absolute or a relative standpoint: indeed, questions must arise about the status of such categories themselves. Rather, we seek to understand how geographers as individual scholars recognized and grappled with intellectual issues in their time, in specific intellectual, social and economic environments. Can the history of geography in Europe between 1870 and 1914 begin to be understood without reference to the Franco-Prussian and first world wars? – to the educational expansion taking place in schools and universities? – to the Darwinian ferment, to which the French alone

[8] Hartshorne, *op. cit.* (note 5), p. 31.

[9] J. K. Wright, 'A plea for the history of geography', *Isis*, 8 (1926), 477–491; reprinted in J. K. Wright, *Human nature in geography* (Cambridge: Harvard University Press, 1966), 11–23, which is cited subsequently.

[10] V. Berdoulay, 'The contextual approach', in D. R. Stoddart, ed. *Geography, ideology and social concern* (Oxford, Basil Blackwell, 1981), 8–16.

remained largely immune? – or to the emergence of the social and the human sciences in the eighties and nineties?[11] Such a history simply cannot be written from the pages of the leading geographical journals of the time.

It follows that individuals themselves can no longer be categorized as 'good' or 'bad', and neither can their views. Wright in the same prescient paper realized this clearly: 'Is not the history of error, folly, and emotion often as enlightening as the history of wisdom?'[12] Agassi similarly in 1963 drew attention to the importance for scientific advance of 'clever mistakes' and 'stimulating error', but in geography only James has begun to explore this field.[13]

My emphasis thus is on context and contingency as they affect the development of geographical thought. In the history of geography there are surprisingly few analyses of particular concepts. There are important papers by Plewe on comparative geography,[14] by Bartels on 'harmony',[15] and by Hard on the concept of *Landschaft*.[16] Others have explored the ways in which geographical ideas are embedded in and respond to a wider intellectual context. Lehmann in 1937 provided a remarkable but wholly overlooked consideration of the implications for geographical reasoning of the new statistical mechanics and atomic physics of Ernst Mach and Max Planck (though later commentators have perhaps wrongly claimed a greater influence for Heisenberg).[17] There is, too, a growing literature on the impact of Darwin's ideas on the organization of knowledge and the methodology of explanation.[18] Lukermann in a brief, allusive paper has

[11] N. Broc, 'La géographie française face à la science allemande (1870–1914)', *Annals de Géographie*, 86 (1977), 71–94.

[12] J. K. Wright, *op. cit.* (note 9), p. 17.

[13] J. Agassi, *op. cit.* (note 4); P. E. James, 'On the origin and persistence of error in geography', *Annals of the Association of American Geographers*, 57 (1967), 124; J. Leighly, 'Error in geography', in J. Jastrow, ed., *The story of human error* (New York: Appleton-Century, 1936), 89–119.

[14] E. Plewe, 'Untersuchung über die Begriff der "Vergleichenden" Erdkunde und seine Anwendung in der neueren Geographie', *Zeitschrift der Gesellschaft für Erdkunde zu Berlin*, Ergänzungsheft 4 (1932), 1–92.

[15] D. Bartels, 'Der Harmoniebegriff in der Geographie', *Die Erde*, 100 (1969), 124–137.

[16] G. Hard, 'Die "Landschaft" der Sprache und die "Landschaft" der Geographen. Semantische und forschungslogische Studien zu einiger zentralen Denkfiguren in der deutsche geographischen Literatur', *Colloquium Geographicum*, 11 (1970), 1–278.

[17] O. Lehmann, *Der Zerfall der Kausalität und die Geographie* (Zurich: Im Selbstverlage der Verfassers, 1937).

[18] See chapters 8 and 9; and also D. R. Stoddart, 'Darwin's influence on the development of geography in the United States, 1859–1914', in B. W. Blouet, ed., *The origins of academic geography in the United States* (Hamden: Archon Books, 1981), 265–278.

linked ideas in French geography at the turn of the century with the probabilistic work of Henri Poincaré and other mathematicians. But in general this is a field which remains largely unexplored.[19]

PURPOSE

Hartshorne's justification for an interest in the history of geography was essentially that it should be a guide to what one ought to do. The justification for contextual history is somewhat different. Can we recognize homologies in problems and responses between past and present? Can analysis of past dilemmas help us to cope with those which now beset us? But more, can we so deepen our understanding of the past that we cease to view its practitioners as two-dimensional figures, and recognize throughout the development of our subject the complexity and subtlety of the intellectual endeavour? Inductivist and internalist history bears a heavy responsibility for reducing these activities to such simplistic terms in our standard histories that their intellectual content has been demeaned, and our leading practitioners, while identified as the great men of the past, have been paradoxically reduced in stature. And it is precisely in revealing the complexity of the past that we can throw light on the processes of historical change.

We could, of course, go further, and argue that these historiographical problems are basically epistemological: that they depend ultimately on what we understand knowledge to be. There is some discussion of these issues by Lowenthal[20] and Kirk,[21] but only Gregory[22] has made any sustained attempt to incorporate such issues into the fabric of the subject. There is undoubtedly an archaeology of geography, as Foucault has argued of knowledge in general, but it would be premature to attempt it here.[23] Rather, we need to discuss the analytical structures and procedures which could lead us to deeper understanding of both geography and its history. Of these, the concept of paradigm change requires particular critical discussion because of its popularity; and in its place I wish to propose an

[19] F. Lukermann, 'The "calcul des probabilités" and the Ecole française de géographie', *Canadian Geographer*, 9 (1965), 128–137.

[20] D. Lowenthal, 'Geography, experience and imagination: towards a geographical epistemology', *Annals of the Association of American Geographers*, 51 (1961), 241–260.

[21] W. Kirk, 'Problems of geography', *Geography*, 48 (1963), 357–371.

[22] D. J. Gregory, *Ideology, science and human geography* (London: Hutchinson, 1978).

[23] Nevertheless, see P. Claval, 'Epistemology and the history of geographical thought', *Progress in Human Geography*, 4 (1980), 371–384.

alternative emphasis on social and political relationships between individuals and within institutions.

KUHN AND PARADIGMS

It is almost 20 years since T. S. Kuhn introduced the notion of the paradigm in his analysis of *The structure of scientific revolutions.*[24] Since then both term and concept have been widely adopted in philosophical and historiographical discussion, especially in the social sciences, and after some hesitation the paradigm idea is suddenly common currency in geographical writings too. We need to consider the ways in which the term has been applied, its value in geographical historiography, and especially the reasons why it has had such sudden and widespread popularity.

Kuhn used the term paradigm to denote a generally accepted set of assumptions and procedures which served to define both subjects and methods of scientific enquiry. For him, 'normal science' was carried out within the context of a prevailing paradigm, which itself both defined the importance of questions for study and set criteria for the acceptability of solutions. Within this framework, much of 'normal science' was of a 'puzzle-solving' kind, seeking laws, constants, coefficients and other relationships within the context of the paradigm. Such work, while necessarily restricted in its scope, was nevertheless highly focused and usually productive.

But from time to time the goals and procedures within which scientists operated came to appear less satisfactory, and the prevailing paradigm would be replaced by a new. This process of paradigm change thus supplied a key to the interpretation of historical development in the sciences: change came to be seen as episodic, or indeed 'revolutionary'. With a change in paradigm, old problems lost their significance, old methods their relevance, and the focus of research moved abruptly to new areas. 'To desert the paradigm is to cease practising the science it defines.'[25]

Kuhn's concept, used as a key to the understanding both of the formal structure of investigation and of the interpretation of change in the history of science, was first explicitly introduced into geography by Haggett and

[24] T. S. Kuhn, *The structure of scientific revolutions* (Chicago: University of Chicago Press, 1962).

[25] *Ibid.*, p. 34.

Chorley.[26] They defined paradigms both operationally as 'stable patterns of scientific activity' and more formally as 'large-scale models'; and Kuhn's views were then used somewhat confusingly to argue for a 'model-based paradigm' for geography itself. This polemical use of Kuhn's ideas as a means of promoting particular views in geography was accepted implicitly by many commentators at the time,[27] and subsequently Harvey in his influential *Explanation in geography* used the paradigm as an organizing framework.[28] Thereafter the term passed into common use, and has recently been used to structure a history of Anglo-American human geography since 1945 (though with a postscript disavowal).[29]

GEOGRAPHICAL PARADIGMS?

The only existing geographical paradigm discussed by Haggett and Chorley[30] was a classificatory or regional one, exemplified by Berry's proposed data matrix[31] and Grigg's logic of regional systems.[32] Against this they set their own proposal for a model-based paradigm. In their discussion, this latter is clearly an inclusive category, defined in terms of characteristics rather than of content. A new paradigm, they suggest, 'must be able to solve at least some of the problems that have brought the old to crisis point'; it

[26] P. Haggett and R. J. Chorley, 'Models, paradigms and the new geography', in R. J. Chorley and P. Haggett, eds, *Models in geography* (London: Methuen, 1965), 19–41. Kuhn's views are central to the structure of this essay, but for a reason which, though comparatively trivial, illustrates perfectly the contingency at the heart of scientific advance. I happened to read Kuhn's book early in 1965, was greatly excited by it, and loaned it at once to Haggett, who at that time was working on the floor above. Immediately it became a subject for student seminars and supervisions at Cambridge. This was only a matter of weeks before the Madingley Lectures at which the Haggett and Chorley paper was delivered. I myself (somewhat uncritically) used Kuhn's idea in my own contribution to the same series (see chapter 13 of that book). Yet only one other contributor to *Models in geography* mentioned Kuhn, which indicates that at that time his ideas were far from common currency.

[27] For example, P. Saey, 'Toward a new paradigm? Methodological appraisal of integrated models in geography', *Bulletin of the Ghana Geographical Association*, 13 (1968, published 1973), 51–60.

[28] D. W. Harvey, *Explanation in geography* (London: Edward Arnold, 1969).

[29] R. J. Johnston, *Geography and geographers: Anglo-American human geography since 1945* (London: Edward Arnold, 1979).

[30] P. Haggett and R. J. Chorley, *op. cit.* (note 26).

[31] B. J. L. Berry, 'Approaches to regional analysis: a synthesis', *Annals of the Association of American Geographers*, 54 (1964), 2–11.

[32] D. Grigg, 'The logic of regional systems', *Annals of the Association of American Geographers*, 55 (1965), 465–491.

must be 'elegant, appropriate and simple'; and it must contain 'potential for expansion'.[33] The samples they cite (derived from work on migration, point patterns, search theory, network analysis and diffusion studies) scarcely compared in scale with Kuhn's own examples of paradigm change (which hinge on the work of Copernicus, Newton and Einstein).

Haggett had, of course, already suggested the existence of what he then termed simply 'widely held views' about the nature of geography: as the study of areal differentiation of the earth's surface, as the study of landscape, as the study of the relationship between earth and man, and as the study of distributions or location;[34] and possibly it is on this scale, rather than that of specific techniques or theories, that the application of the paradigm idea should be sought. Perhaps in Haggett's third category (though he himself treats it in his second) the work of the Berkeley school under the powerful and inspiring influence of Carl Sauer must come closest to the main criteria set by Kuhn for paradigm recognition. The extraordinary dominance exercised by Sauer, the loyalty he commanded and the affection and regard in which he was held, are made very clear by Leighly[35] and Parsons.[36] But the Berkeley school coexisted with other very different research traditions, not only in Europe but also in the United States; and its distinctiveness did not prevent Sauer from twice serving as president of the Association of American Geographers.

Apart from the Berkeley School, the paradigm idea has been applied in geography in three main areas. The first and most popular is that of continental drift and plate tectonics.[37] Here commentators have been fascinated by the way in which Alfred Wegener's basically correct conclusions were powerfully opposed for years by physicists such as Harold Jeffreys, because of the difficulty raised over a suitable mechanism for drift, until the discovery of the significance of magnetic lineations on the ocean

[33] P. Haggett and R. J. Chorley, *op cit.* (note 26), 37–38.

[34] P. Haggett, *Locational analysis in human geography* (London: Edward Arnold, 1965), 10–13.

[35] J. Leighly, 'Drifting into geography in the twenties', *Annals of the Association of American Geographers*, 69 (1969), 4–9.

[36] J. J. Parsons, 'The later Sauer years', *Annals of the Association of American Geographers*, 69 (1979), 9–15.

[37] See, for example, H. Frankel, 'Arthur Holmes and continental drift', *British Journal for the History of Science*, 11 (1978), 130–150; A. Hallam, *A revolution in the earth sciences: from continent drift to plate tectonics* (Oxford: Clarendon Press, 1973); D. B. Kitts 'Continental drift and scientific revolution', *Bulletin of the Association of American Petroleum Geologists*, 58 (1974), 2490–2496; I. Moffatt, 'Paradigm development in geology', *University of Newcastle-upon-Tyne, Department of Geography, Seminar Paper* 33 (1977), 1–34; F. J. Vine, 'The continental drift debate', *Nature, Lond.*, 266 (1977), 19–22.

floor by Vine and Matthews demonstrated the reality of major continental movements. Second, Davisian geomorphology is often seen as establishing a paradigm for the historical analysis of landforms (denudation chronology) which dominated geomorphic work from the end of the nineteenth century until after Davis's death in 1935 and its final overturning by a new concern with process and system studies introduced by the work of Horton.[38] Third, Haggett's own book on *Locational analysis in human Geography* is often taken to make a convenient break between an older classificatory and descriptive mode of study and a newer analytical and quantitative one.[39]

Recently, however, the paradigm idea has been used on a variety of scales. Berry proposed a new paradigm for geography involving action and change,[40] and he has also discussed the history of environmental determinism in paradigm terms.[41] Meyer[42] and Herbert and Johnston[43] apply the idea to urban locational analysis. Garrison refers in the same paper both to 'causal paradigms' and to 'paradigms for the study of urban areas, transportation, and regional science'.[44] The list could be expanded, and also readily duplicated for neighbouring fields such as economics and sociology. In each case the paradigm terminology has been used either to illuminate the establishment of views of which a commentator approved or to advocate the rejection of those he did not.

APPLICABILITY?

This ready acceptance of Kuhn's vocabulary has occurred without any close attention to Kuhn's own statements or to the critical literature on them in the history and philosophy of science. We must first enquire whether the

[38] R. J. Chorley, 'A re-evaluation of the geomorphic system of W. M. Davis', in R. J. Chorley and P. Haggett, eds, *Frontiers in geographical teaching* (London: Methuen, 1965), 21-38.

[39] P. Haggett, *op cit.* (note 34).

[40] B. J. L. Berry, 'A paradigm for modern geography', in R. J. Chorley, ed. *Directions in geography* (London: Methuen, 1973), 3–22.

[41] B. J. L. Berry, 'Geographical theories of social change', in B. J. L. Berry, ed., *The nature of change in geographical ideas* (DeKalb: Northern Illinois University Press, 1978), 18–35.

[42] D. Meyer, 'Urban locational analysis: a paradigm in need of revolution', *Proceedings of the Association of American Geographers*, 5 (1973), 169–173.

[43] D. T. Herbert and R. J. Johnston, 'Geography and the urban environment', in D. T. Herbert and R. J. Johnston, eds., *Geography and the urban environment: progress in research and applications* (Chichester: John Wiley, 1979), vol. 1, 1–33.

[44] W. L. Garrison, 'Playing with ideas', *Annals of the Association of American Geographers*, 69 (1979), 118–120.

paradigm idea is useful in understanding the processes of change in geography on other than a superficial level.

The confusion implicit in Haggett and Chorley's initial usage and the plasticity with which the concept has subsequently been applied mirrors the multiplicity of meanings attached to the idea at different times by Kuhn himself: Masterman identifies 21 discrete definitions in his 1962 book alone.[45] But if we leave this to one side, and accept the view of a paradigm as a consensus of aims and methods which defines, until replaced, the pursuit of normal science, then the closer the analysis the less apposite the concept appears in geography.

Thus, for example, whereas it is true that Davisian geomorphology was codified in textbooks and widely adopted by both researchers and pedagogues during the first half of the twentieth century, Davis's views were received with scepticism by many, even in his own lifetime. In Britain H. R. Mill and J. E. Marr, in the United States Fenneman and Chamberlin, and in Germany Albrecht Penck and Alfred Hettner all declined to be persuaded. Alternative geomorphic systems, notably that of Walther Penck, were not only actively used in research (for example in Sauer's paper on the tectonic landforms of Chiricahua[46]) but also given textbook authority, as in Von Engeln's influential *Geomorphology*.[47] The Berkeley Department of Geography virtually ignored Davisian methods.[48] Kuhn's formulation scarcely makes allowances for coexisting paradigms (especially coexisting over nearly half a century), and if Davisian geomorphology is to be described as a paradigm the meaning of the term requires revision. Only by analysing the effects of Davis's views on non-scientists (such as the historian Walter Prescott Webb)[49] and on educationalists is it possible to maintain the value

[45] M. Masterman, 'The nature of a paradigm', in I. Lakatos and A. Musgrave, eds, *Criticism and the growth of knowledge* (Cambridge: Cambridge University Press, 1970), 59–89. See also D. G. Cedarbaum, 'Paradigms', *Studies in the History and Philosophy of Science*, 14 (1983), 173–213.

[46] C. O. Sauer, 'Basin and range forms in the Chiricahua area', *University of California Publications in Geography*, 3 (1930), 339–414.

[47] O. D. von Engeln, *Geomorphology systematic and regional* (New York: Macmillan, 1942), especially chapter 13, 'The Walther Penck geomorphic system', 256–268.

[48] J. Leighly, *op. cit.* (note 35).

[49] W. P. Webb, 'Geographical–historical concepts in American history', *Annals of the Association of American Geographers*, 50 (1960), 85–93. Webb's account of the impact that Davisian physiography had on him illustrates perfectly the appeal of Davis's formulation and the uncritical acceptance by a non-professional audience of such ideas. Webb attempted to recall the origin of his interest in geography: 'Probably it dates back to my boyhood when I roamed the pastures with my dog and unconsciously absorbed some knowledge of the land forms, noted that the grass grew in the valleys and the rocks on the high hills. My real

of the paradigm, but then on essentially social rather than scientific grounds.

A similar analysis could be made of the 'widely held views' – not universally held views – listed by Haggett [50] as well as of some of the other recent candidates for paradigmacy: 'relevant' geography, behavioural geography, and the wide range of attitudes discussed by Gregory and others.[51] None has been uniquely accepted as a geographic paradigm in Kuhn's original sense, all have a long history within the subject, and all have coexisted with many other divergent schools of thought. Johnston has indeed concluded from his 'failure to match Kuhn's model to recent events in human geography' that 'the model is irrelevant to this social science, and perhaps to social science in general.'[52] In a similar way, Tribe has concluded that Kuhn's proposals 'are of no use in considering the problems faced by the history of economics'.[53]

If the concept of the paradigm has been used so loosely, on a variety of levels and without specific reference to Kuhn's own criteria, it is clearly not possible to infer from its use the operation in each case of the processes of scientific change which Kuhn described. But since, even when loosely used, the term carries with it connotations of revolutionary change, it is necessary

---

awakening, however, came when I began teaching a country school, and found that one of the courses I had to teach was called physical geography. There was a red clothbound text, entitled *Physical Geography*, by a man named W. E. Davis [*sic*]. I had never seen the book or heard of the subject before, and so it became necessary for me to read it pretty closely in order to stay ahead of the class. This book differed from Wentworth's arithmetic and Reed and Kellogg's grammar in that it dealt with things I knew about. It differed from political geography where I had to memorize all the states in the union and their capitals, memorize the names of five oceans which I knew only by reputation, and learn that Belize was the capital of British Honduras. It differed from spelling because I did not have to do puzzles with letters, getting them in exact order to form a word which often was pronounced as if spelled in another way. But here in Davis's *Physical Geography* was something that made sense from the first, something I had seen when hunting horses in the early morning. From the book I went back to the pasture, and sure enough *there* things were going on just like the author said they should. It was a wonderful experience, and I have never gotten over it' (p. 85).

[50] P. Haggett, *op. cit.* (note 34).

[51] D. J. Gregory, *op. cit.* (note 22); M. E. Harvey and B. P. Holly, eds, *Themes in geographic thought* (London: Croom Helm, 1981); R. J. Johnston, *Philosophy and human geography: an introduction to contemporary approaches* (London: Edward Arnold, 1983).

[52] R. J. Johnston, 'Paradigms and revolution or evolution? Observations on human geography since the Second World War', *Progress in Human Geography*, 2 (1978), 189–206; see p. 201.

[53] K. Tribe, *Land, labour and economic discourse* (London: Routledge & Kegan Paul, 1978), p. 15.

to consider briefly how change has occurred in some of the geography cases cited as paradigms.

PARADIGM CHANGE

The importance of the paradigm idea, in Kuhn's terms, lies not only in the way it apparently supplies an interpretative framework for historical studies, but in the implications it carries for why and in what manner change occurs. When change is viewed as paradigm replacement or paradigm shift, large and often unrecognized assumptions are made about the processes involved. Simplistic notions of change imply simplistic assumptions about the behaviour of individual scientists, assumptions which require consideration before and not after the paradigm idea is accepted. As more is understood of changes in geographical thought over the past hundred years, and particularly of the subtle interrelationships of geographers themselves, the less appropriate and useful does the notion of revolutionary change appear.

The main area of controversy has been that between Kuhn and Popper over the criteria for paradigm rejection.[54] Essentially Popper emphasizes the importance of methodological procedures, especially those leading to the rejection of predictions rather than to their confirmation, as the only sound criterion in science; whereas Kuhn's argument is concerned more with the changing attitudes and values of groups of paradigm adherents, not necessarily resulting from any demonstration of error in scientific terms. The problem is particularly intractable even in the more 'scientific' kinds of geographical work, such as quantitative human geography, where low-level predictions (called 'forecasts') are rarely used to test theory, and where the relationship between theory and reality often differs in fundamental respects from that in the physical sciences.

Consider the manner of paradigm change. In the Kuhnian view, this is by a revolutionary process, often expressed in military metaphor: old views are attacked and suddenly 'overthrown'. The introduction of quantification into British geography in the 1960s is seen by Taylor in this way.[55] In

[54] I. Lakatos, 'Falsification and the methodology of scientific research programmes', in I. Lakatos and A. Musgrave, eds, *op. cit.* (note 45), 91–195; M. Blaug, 'Kuhn versus Lakatos or Paradigm versus research programmes in the history of economics', in S. J. Latsis, ed., *Method and appraisal in economics* (Cambridge: Cambridge University Press, 1976), 149–180.

[55] P. J. Taylor, 'An interpretation of the quantification debate in British geography', *Transactions of the Institute of British Geographers*, n.s. 1 (1976), 129–142.

analysing the progress of the revolution, Taylor suggests that the day was carried because mathematics was used by the insurgents as a specialist secret language inaccessible to its opponents, and as intellectual camouflage intended to impress, bolstered by frequent impressive references to Newton, Einstein, Planck, Heisenberg and 'the scientific method', to establish intellectual respectability. The revolution was over, at least according to Burton,[56] by the early 1960s: the old paradigm had been vanquished and a new one had taken over.

Two obvious questions arise. First, what of the quantifiers who had worked happily enough under the old regime? How can their continuing activities be reconciled with this simplistic view of change? I refer not simply to people like Spottiswoode and Cayley, who managed to publish in spite of the admirals and explorers in the journals of the Royal Geographical Society,[57] but to more central figures such as Christaller and to his many predecessors.[58] Those who say that Christaller was ignored – perhaps even suppressed – in the dark days before 1960 should read their Dickinson.[59] To give another example, there is the clearest foreshadowing of the structure of Haggett's *Locational analysis* in a remarkable paper published by P. E. James, a leader of the old paradigm.[60] Here James considered the problem of pattern, process and scale, and attempted to reconcile them with the prevailing theme of geography as the study of areal differentiation. And it is not difficult to show how James's views developed over time from his first involvement in technical debates in the *Annals* of the Association of American Geographers in the early 1930s.[61] In these particular cases the revolution, if there was one, had long and respectable antecedents: the precursors themselves were central figures of the old paradigm. These

[56] I. Burton, 'The quantitative revolution and theoretical geography', *Canadian Geographer*, 7 (1963), 151–162.

[57] G. Cayley, 'On the colouring of maps', *Proceedings of the Royal Geographical Society*, n.s. 1 (1879), 259–261; W. Spottiswoode, 'On typical mountain ranges: an application of the calculus of probabilities to physical geography', *Journal of the Royal Geographical Society*, 31 (1861), 149–154.

[58] C. F. Müller-Wille, 'The forgotten heritage: Christaller's antecedents', in B. J. L. Berry, ed., *op. cit.* (note 41), 38–64.

[59] R. E. Dickinson, *City region and regionalism: a geographical contribution to human ecology* (London: Kegan Paul, Trench, Trubner, & Co., 1947).

[60] P. E. James, 'Toward a further understanding of the regional concept', *Annals of the Association of American Geographers*, 42 (1952), 195–222.

[61] See, for example, P. E. James, 'The terminology of regional description', *Annals of the Association of American Geographers*, 24 (1934), 93-107; 'On the treatment of surface features in regional studies', *ibid.*, 27 (1937), 213–218; 'Formulating objectives of geographic research', *ibid.*, 38 (1948), 271–276.

revolutions were thus processes rather than events, involving a shift in emphasis rather than the wholesale replacement of one set of attitudes by another.

Second, if the revolution succeeded as rapidly as Burton suggests,[62] why do we find its leading practitioners in disarray, lamenting their 'confusion and doubt',[63] the 'growing isolation of those involved in theoretical and quantitative work',[64] and the alleged lack of effect of the changes on university admissions policies and curricula?[65] I raise these points simply to indicate that the revolutionary model of change is self-evidently inappropriate to anything as complex as scholarly endeavour.

But more crucially, the revolutionary model says nothing of the ways in which change is effected: why some views appeal to particular individuals and others do not, why some workers in some localities are attracted by, adopt and transmit new ideas. There are very few studies of the processes and contexts of change in geographical thought. Consider the introduction of the 'New Geography' in Britain in the 1880s, said to have been brought about by Mackinder's paper 'On the scope and methods of geography' in 1887.[66] Examination shows that this paper was very far from being an unheralded frontal assault on the entrenched forces of exploration, which won the day by the force of its intellectual argument. Mackinder's argument had been almost wholly anticipated by others, and was indeed common currency in the Royal Geographical Society in the later eighties.[67] More to the point, however, its content reflects social and economic as well as intellectual tensions not only in geography but also in neighbouring subjects: the 'New Geography' was simply part of a general readjustment of roles and subject matter in the earth sciences at a time of widespread educational reform.[68] The more this complexity is understood, the less revolutionary the process seems and the less dominant a figure Mackinder appears.[69]

[62] I. Burton, *op. cit.* (note 56).

[63] B. J. L. Berry, *op. cit.* (note 40), p. 3.

[64] L. J. King, 'The seventies: disillusionment and disillusion', *Annals of the Association of American Geographers*, 69 (1979), 155–164; see p. 157.

[65] P. Gould, 'Geography 1957–1977: the Augean period', *Annals of the Association of American Geographers*, 69 (1979), 139–151; see pp. 149–150.

[66] J. F. Unstead, 'H. J. Mackinder and the new geography', *Geographical Journal*, 113 (1949), 47–57.

[67] See chapter 4.

[68] See chapter 3.

[69] D. R. Stoddart, 'Mackinder: myth and reality in the establishment of British geography', XXIII International Geographical Congress, Leningrad, Symposium K5, *Abstracts* (1976), 4 pp.

Such an analysis has yet to be made for the progress of the quantitative revolution, though Duncan[70] has outlined its progress in diffusion terms, and Pred[71] has described the constant contingencies involved in his own intellectual development. What led Pred to Göteborg in 1960–1961 (and ultimately to his association with Hägerstrand), and Harvey to Uppsala (and his association with Olsson) in the same year? The processes of change are concerned with matters such as these, which are simply not considered in a revolutionary view. It is true that the contingencies were somewhat reduced in the emergence of quantitative geography with the organization of summer schools, symposia and new journals in the 1960s, but these are the central devices of 'normal science', not of revolutions.

It follows from these examples that the adoption of Kuhn's terminology, far from clarifying history, actively distorts it, largely by reducing the participants to caricature figures. Some very clearly become heroes. This is demonstrated to the point of absurdity by Bunge's treatment of Schaefer and Christaller.[72] Schaefer's personal history and political convictions were unknown (as indeed was Schaefer) to almost every reader of the only paper of any consequence he ever published. Whether Christaller was or was not a fascist is in terms of present attitudes to his contribution of no significance.

Supporters of the old paradigm readily become fools, if not knaves. W. M. Davis, for example, came readily to be considered simply as 'an old duffer with a butterfly-catcher's sort of interest in scenery', as Mackin describes it.[73] Taylor[74] notes that much of the opposition to quantitative geography came from older men, such as L. Dudley Stamp (who injudiciously compared the quantifiers to communists)[75] and a succession of presidents of the Institute of British Geographers. But a reading of their presidential addresses gives an impression of cautious sympathy rather than outright hostility. Bunge again especially identifies Hartshorne as an 'enemy on a personal scale' and a 'life-long protagonist' (presumably meaning antagonist), without any consideration of the magnitude of his scholarly

[70] S. S. Duncan, 'The isolation of scientific discovery: indifference and resistance to a new idea', *Science Studies*, 4 (1974), 109–134.

[71] A. Pred, 'The academic past through a time-geographic looking glass', *Annals of the Association of American Geographers*, 69 (1979), 175–180.

[72] W. Bunge, 'Fred K. Schaefer and the science of geography', *Harvard Papers in Theoretical Geography, Special Paper Series*, A (1968), 1–21; 'Walter Christaller was not a fascist', *Ontario Geographer*, 11 (1977), 84–86.

[73] J. H. Mackin, 'Rational and empirical methods of investigation in geology', in C. C. Albritton, ed., *The fabric of geology* (Reading: Addison-Wesley, 1963), 135–165; see p. 136.

[74] P. J. Taylor, *op. cit.* (note 55), pp. 138–139.

[75] L. D. Stamp, 'Ten years on', *Transactions of the Institute of British Geographers*, 40 (1966), 11–20.

contribution.[76] It is perhaps worth noting that the 'Planck principle' (that new ideas do not convince opponents, who simply die and are replaced by younger and more receptive men),[77] while superficially attractive, has been disproved on the only occasion on which it has been formally tested.[78]

This is not to say that generalization is impossible about the nature and progress of scientific change and the characteristics of those who are involved in it. Many such generalizations predate Kuhn's formulation and are independent of it. Thus Huxley long ago suggested a general reaction to any new proposal.[79]

T. H. Huxley's four stages of public opinion

I (*Just after publication*)
The Novelty is absurd and subversive of Religion and Morality. The propounder both fool and knave.

II (*20 years later*)
The Novelty is absolute Truth and will yield a full and satisfactory explanation of things in general. The propounder a man of sublime genius and perfect virtue.

III (*40 years later*)
The Novelty will not explain things in general after all and therefore is a wretched failure. The propounder a very ordinary person advertised by a clique.

IV (*100 years later*)
The Novelty a mixture of truth and error. Explains as much as could reasonably be expected. The propounder worthy of all honour in spite of his share of human frailties, as one who has added to the permanent possessions of science.

Beveridge repeated much the same idea: 'The reception of an original contribution to knowledge may be divided into three phases: during the first it is ridiculed as not true, impossible or useless; during the second, people say there may be something in it but it would never be of any practical use; and in the third and final phase, when the discovery has received general

[76] W. Bunge, 'Fred K. Schaefer and the science of geography', *op. cit.* (note 72).

[77] M. Planck, *Wissenschaftliche Selbstbiographie. Mit einem Bildnis und der von Max von Laue gehältenen Traueransprache* (Leipzig: J. A. Barth, 1948).

[78] D. L. Hull, P. D. Tessner and A. M. Diamond, 'Planck's principle', *Science, N.Y.* 202 (1978), 717–723.

[79] C. Bibby, *T. H. Huxley: scientist, humanist and educator* (London: Watts, 1959), p. 77.

recognition, there are usually people who say that it is not original and has been anticipated by others.'[80] And there are also interesting generalizations about the innovators themselves.[81] Rogers, in one of the founding documents of diffusion theory, quotes Linton's view that innovators in general are 'very frequently misfits in their societies, handicapped by atypical personalities', and Barnett's that 'the disgruntled, the maladjusted, the frustrated, or the incompetent are pre-eminently the acceptors of culture innovation and change.'[82] I quote these views in illustration of the fact that the common characterization of the innovator as hero is by no means axiomatically true.

Why, if this interpretation is valid, is the paradigm idea so popular, not only in human geography (and other social sciences) but in physical geography too? I suggest that a major reason lies in the way in which the concept of revolution bolsters the heroic self-image of those who see themselves as innovators and who use the term paradigm in a polemical manner, coupled with the fact that Kuhn's terminology supplies an apparently 'scientific' justification for the advocacy of change on social rather than strictly scientific grounds.

In its simplest formulation, the paradigm idea suggests the replacement rather than the testing of ideas, and by extension the replacement of practitioners also: in this lies the heart of Popper's criticism of Kuhn's thesis.[83] In this sense there is room for sociological enquiry into the extent to which the concept has been used in recent years as a slogan in interactions between different age-groups, schools of thought, and centres of learning, rather than as a useful heuristic model of how and in what manner science is structured and change occurs. We might usefully analyse why some geographers choose to identify themselves as paradigm-changers at the present day, and whether, by their actions since 1960, they have so simplified our perceptions of the processes of change that the paradigm idea comes to appear analytically useful. In other words, those who propound the Kuhnian interpretation have done so in ways which tend to make it self-fulfilling. It is this, rather than its value as a framework for studying historical change, that makes the paradigm idea of interest to the historian

[80] W. I. B. Beveridge, *The art of scientific investigation* (New York: Vantage Books, 1953), pp. 151–152.

[81] B. Barbour, 'Resistance by scientists to scientific discovery', *Science, N.Y.* 134 (1961), 596–602.

[82] E. M. Rogers, *Diffusion of innovations* (New York: Free Press, 1962), p. 194; see also A. Roe, 'Personal problems and science', in C. W. Taylor and F. Barron, eds, *Scientific creativity: its recognition and development* (London: Wiley, 1963), 132–138.

[83] I. Lakatos, *op. cit.* (note 54).

of science: as itself an object of study, rather than as a means of understanding the complexities of change.

AN ALTERNATIVE FRAMEWORK

If the paradigm idea is so manifestly inadequate as a guide to the history of geography, what alternatives are available? Much of this book deals with the growing institutionalization and professionalization which characterized the emergence of geography as a discipline in the nineteenth century. It is thus necessarily concerned with the structure of social relations, between both individuals and groups. Gramsci long ago introduced the concept of hegemony to describe the maintenance of cultural supremacy by particular groups in society, a concept involving the reciprocal components of dominance and subordination, control and deference.[84] In introducing the idea to the history of science, Morris Berman suggested that 'the evolution of the English scientific community becomes understandable only when seen within the framework of the cultural imprint of the ruling class,'[85] and he subsequently explored its implications in a study of the early years of the Royal Institution.[86] Can such an analysis be extended to geography?

Central to the institutionalization of the subject in Britain in the nineteenth century is the role of the Royal Geographical Society. Let us examine some aspects of its foundation, organization and operation as a test of the usefulness in interpretation of Gramsci's thesis.

The first point of importance is that the R.G.S. was not founded until 1830, more than 60 years after Cook first entered the Pacific. The reason for the delay is not far to seek. Sir Joseph Banks was elected president of the Royal Society in 1778, a position he held on to tenaciously, though crippled and immobilized by gout,[87] until his death at the age of 77 in 1820. An overbearing and imperious man, obsessed with public honour and ceremonial, he jealously guarded the Royal Society's monopoly of metropolitan science. 'The Autocrat of the Philosophers' – 'a man self-sufficient

[84] G. Williams, 'The concept of "egemonia" in the thought of Antonio Gramsci: more notes on interpretation', *Journal of the History of Ideas*, 21 (1960), 586–599.

[85] M. Berman, '"Hegemony" and the amateur tradition in British science', *Journal of Social History*, 8 (1975), 30–50.

[86] M. Berman, *Social change and scientific organization: the Royal Institution, 1799–1844* (London: Heinemann Educational Books, 1978).

[87] Holland describes him 'wheeled about in his armchair, his limbs hopelessly knotted with gouty tumours'. H. Holland, *Recollections of past life* (London: Longmans, Green, 1872), p. 216.

and intolerant of advice, haughty, vainglorious and not a little puffed up by the honours he received from a King with whom he made it his business to stand well'[88] – Banks personified for four decades hegemony in practice.[89] He himself played a leading role in promoting exploration, and with Banks in charge there was clearly no need for any more formal organization for this purpose. He was prepared to sanction dining clubs and other less substantial bodies, and looked benevolently, for example, on the African Association (founded for 'ladies of distinction as well as noblemen and gentlemen'),[90] but as soon as they became more organized and ambitious, Banks showed his disfavour.[91] He opposed the formation of the Geological Society of London in 1807,[92] and as a result the growing specialization of science did not begin to be reflected in its public organization until the foundation of the Royal Astronomical Society in the year of Banks's death.

The details of the actual foundation of the Royal Geographical Society, while confused, also lend themselves to Gramsci's interpretation. It seems clear that during 1828 and 1829 the idea of such a Society was being

[88] H. C. Cameron, *Sir Joseph Banks: the autocrat of the philosophers 1744–1820* (London: Batchworth Press, 1952), p. 277.

[89] Banks, increasingly gross and ponderous, habitually appeared at the Royal Society in court dress with decorations: 'To question his rulings or dispute his arguments, to encounter the formidable lowering of his massive eyebrows, with every year that passed, become a more alarming proposition seldom adventured': Cameron, *op. cit.* (note 88), p. 158.

[90] H. R. Mill, *The record of the Royal Geographical Society 1830–1930* (London: Royal Geographical Society, 1930), p. 7.

[91] An exception was the founding of the Linnean in 1788, but the circumstances were rather special: Banks had been instrumental in the purchase of Linnaeus' collections by J. E. Smith; was immediately elected as honorary fellow; and after his death the Society moved into his house. Banks also presided over the foundation of the Horticultural Society, which perhaps did not count as scientific.

[92] The Geological was founded in November 1807; Humphrey Davy called it 'a little talking Geological Dinner Club'. Banks joined in January 1808 but resigned 13 months later. In March 1809, doubtless at Banks's behest, the new Society's patron, the Rt Hon. Charles Greville, proposed that there be two classes of membership; one comprising fellows of the Royal Society, the other the rest, called 'assistant members'. The Society would be called 'the Assistant Society to the Royal Society, for the advancement of Geology', and the Royal Society would have first choice in the selection of papers for publication in its own journal. The Geological Society not surprisingly roundly rejected 'any proposition tending to render this Society dependent upon, or subservient to, any other Society', whereupon Greville too resigned, together with Humphrey Davy, the Royal Society's secretary. Banks's counterattack in this case failed, but no one was left in any doubt of his magisterial disapprobation. See H. B. Woodward, *The history of the Geological Society of London* (London: Geological Society, 1907), pp. 10–29: also M. J. S. Rudwick, 'The foundation of the Geological Society of London: its scheme for co-operative research and its struggle for independence', *British Journal for the History of Science*, 1 (1963), 326–355.

discussed by William Jerdan (editor of the *Literary Gazette*), William Huttmann (a clerk at India House) and John Britton. They met more formally on 12 April 1830 at the home of Francis Baily, the first president of the new Royal Astronomical Society, together with Captain W. H. Smyth, R.N. The involvement of Smyth was crucial, for he in turn involved the Hydrographer of the Navy, Captain Beaufort, and the permanent secretary to the Admiralty, John Barrow. Clearly, now that the idea of a new society had come to the stage of formal meetings and printed prospectuses, it was too important to be left to the likes of Jerdan and Britton, and a quite different group of people moved swiftly to take it over.[93]

Barrow, as a member of the Raleigh Club (a dining club for travellers), convened a meeting at which he proposed the formation of a new society; the details were worked out in a series of discussions held in the Admiralty; and Barrow himself chaired the inaugural meeting held on 16 July 1830. Britton, naturally displeased at the turn of events, nevertheless achieved a place on the council. It is not surprising that he found it dominated by naval men (Captain Beaufort himself, 'whom it was rather hazardous, in any affair, to contradict';[94] Captain Basil Hall; Captain Lord Prudhoe; Captain Smyth), soldiers (Major the Hon. G. Keppel; Colonel Sir Augustus Frazer, R.A.; Colonel Jones, R.E.; Colonel Monteith, E.I.C.S.), and gentlemen of substance and position (John Cam Hobhouse; Sir Arthur de Capell Brooke; the former Resident at Poona, the Hon. Mountstuart Elphinstone; the Rt Hon. Sir George Murray; the Dean of Wells). The first president was Viscount Goderich, former Prime Minister and future Earl of Ripon, and the Chancellor of the Exchequer, Lord Althorp, also joined the council. Scholarship was represented by the orientalist G.C. Renouard, the botanist Robert Brown and the geologist Roderick Murchison, shortly to begin his first term as president of the Geological Society.[95]

Poor Britton, who began life as 'a sort of pot-boy in a public-house' and then became a journalist,[96] can scarcely have felt at home or at ease in such company, either on a social or on a personal level. Even his friend Jerdan recalled that Britton's 'personal appearance was not imposing. . . . Neither was his character of mind of a high or solid cast. The soul befitted the body. He was easily offended, and, when provoked or prejudiced, not very measured in his terms of resentment or reprobation.'[97] Britton's disapproval of Goderich as 'a mere amiable peer' is significant, for it demonstrates that

93 W. Jerdan, *Men I have known* (London: G. Routledge, 1866), p. 44.
94 *Ibid.*
95 H. R. Mill, *op. cit.* (note 90), p. 39.
96 W. Jerdan, *op. cit.* (note 93), p. 39.
97 *Ibid.*, p. 38.

he did not understand Barrow's strategy at all: 'a working geographer', whom Britton wanted, simply could not have supplied the status the Society required.[98] Feeling 'browbeaten' by his fellow councillors, Britton stayed for only two years and then resigned. Thus was the original idea for a geographical society transformed into a substantial social institution, exercising from the first a notable influence in London scientific society.

Of the 29 members of the inaugural council (including officers), 18 were also fellows of the Royal Society. But this at that time was a social rather than a scientific label; 1830 was after all the year of Babbage's famous attack on the Royal Society,[99] which had become increasingly dilettante and amateur through Bank's declining years. Indeed, it was not until 1858 that its president was unfailingly a scientific man, and not until 1860 that scientists actually formed a majority of its fellows.

Once it had come into being the Royal Geographical Society confirmed its hegemony in a variety of ways. The first and most obvious was to secure the patronage of the new King, William IV, the fount of privilege and honour, through the intercession of Sir Robert Peel. Not content with that alone, the King's brother, the Duke of Sussex, was enrolled as vice-patron. This was a particularly significant step, for it was Sussex who defeated Herschel in the election for the presidency of the Royal Society in the year R.G.S. was founded, a step generally interpreted as 'representing the continuation of rule by gentlemen' rather than by scientists.[100]

Second, as if to emphasize its exclusiveness, the Society organized its inner circle into a private dining club, just as did the Zoological, the Geological and indeed the Royal itself.[101] The Geographical Club was itself the

[98] [J. Britton], 'The autobiography of Sir John Barrow. 3', *United Service Magazine* (1847), 241–255; see p. 255.

[99] C. Babbage, *Reflections on the decline of science in England* (London: B. Fellowes, 1830). See also M. B. Hall, *All scientists now: the Royal Society in the nineteenth century* (Cambridge: Cambridge University Press, 1984), especially chapter 3.

[100] W. F. Cannon, *Science in culture: the early Victorian period* (London: Dawson, 1978), p. 170.

[101] A. Geikie, *The annals of the Royal Society Club: the record of a London dining-club in the eighteenth and nineteenth centuries* (London: Macmillan, 1917). Then, as now, the advantages of belonging to such clubs was not solely of the mind. I cannot resist quoting the experience of Faujas de St Fond at a dinner presided over by Banks: 'The dishes were of the solid kind, such as roast beef, boiled beef and mutton prepared in various ways, with abundance of potatoes and other vegetables. . . . The beefsteaks and the roast beef were at first drenched with copious bumpers of strong beer, called porter, drunk out of cylindrical pewter pots, which are much preferred to glasses because one can swallow a whole pint at a draught. This prelude being finished, the cloth was removed and a handsome and well polished table was covered, as if by magic, with a number of fine crystal decanters filled with the best port,

successor to the Raleigh Club (which Barrow later claimed to be the direct progenitor of the R.G.S.[102] The Raleigh had been founded as a dining society by Arthur de Capell Brooke, as an offshoot of the Travellers', formed in May 1819 by 'a few noble and distinguished Travellers and

---

madeira and claret; this last is the wine of Bordeaux. Several glasses, as brilliant in lustre as fine in shape, were distributed to each person and the libation began on a grand scale, in the midst of different kinds of cheese, which, rolling in mahogany boxes from one end of the table to the other, provoked the thirst of the drinkers. To give more liveliness to the scene, the President proposed the health of the Prince of Wales: . . . we then drank to the Elector Palatine. . . . The same compliment was next paid to us foreigners of whom there were five present. The members of the Club afterwards saluted each other, one by one, with a glass of wine. According to this custom, one must drink as many times as there are guests. . . . A few bottles of champagne completed the enlivenment of everyone. . . . Brandy, rum and some other strong liqueur closed this philosophic banquet, which terminated at half past seven, as we had to be at a meeting of the Royal Society summoned for eight o'clock. . . . We were all pretty much enlivend but our gaiety was decorous.' Quoted by H. C. Cameron, *Sir Joseph Banks K.B., P.R.S.: autocrat of the philosophers* (London: Batchworth, 1952), pp. 168–170.

The menu in the Club's minute books reads as follows:

Soals	Fruit Pie
Chickens Boil'd	Sallad
Pye	A Lambs head and minced
Bacon and Greens	Colly flower
Cold ribs of Lamb	Chine of Mutton Rt.
Veal Cutlets	Pye
Potatoes	Soals
Rabbits and Onion	

Sir James Marshall-Cornwall gives details of a number of more sophisticated menus at the Geographical Club in his *History of the Geographical Club* (London: privately printed for the Club, 1976). Both were doubtless preferable to the Zoological, where for years they were in the habit of eating whatever had recently died in the Gardens. Thus on 10 January 1867 members dined on Spotted Cavy 'without any recorded ill effect'; on 5 February 1889 on Wapiti (a North American ungulate), the superintendent of the Gardens declining to partake; and in 1975 the 1000th dinner was marked by the consumption of Chinese Water Deer. Members have also demolished kudu, zebra, yak, bear and canned rattlesnake. It is recorded that 'the club's emblem – a massive penis bone of a walrus – was placed before the Chairman . . . who was expected to elevate this before dinner was served.' Thank God one is a geographer. See W. P. C. Tenison, *The Zoological Club 1866–1948* (London: privately printed for the Club, 1948), p. 5, and J. R. Norman, *Squire: memories of Charles Davies Sherborn* (London: Harrap, 1944), p. 129. I am indebted to Ian Bishop for these details.

[102] W. Jerdan, *op. cit.* (note 93), p. 254, was particularly incensed by the idea in Barrow's autobiography that the R.G.S. owed its origins to 'the patronizing condescension of a Dining Club'.

Diplomatists';[103] and its members included a remarkable collection of active explorers, many of them naval and military men. It continued as a dining club after the foundation of the R.G.S., but in 1854 at Murchison's behest it changed its name to the Geographical Club, and Murchison became its president for virtually the rest of his life. The Club continued as a bastion of exclusive male privilege until 1972, when it was resolved to admit lady fellows of the R.G.S. – 59 years after they had first been admitted to the Society.[104] A junior dining club, the Kosmos, was founded in 1859 for those who failed to get elected to the Geographical, thus cementing social distinctions, and it survived until 1958.[105] The existence of such clubs served to emphasize the dominance of the metropolis, for with the exception of Oxford and Cambridge it was obviously impracticable for more distant country members to attend at all regularly; and the process of election to membership was such that the social network was automatically preserved.

Third, from the outset the Society codified and publicized its dispensation of patronage through the distribution of medals, a technique pioneered (at Sir Robert Peel's initiative) by the Royal Society in 1826.[106] The King himself made funds available to the R.G.S. for royal medals (the Founder's and the Patron's). During the period 1832–1979 280 were awarded, of which perhaps 43 could be considered to be for academic achievement and perhaps 25 for non-exploratory geographical work. Over the first 50 years of the Society's existence the number of academics so honoured was minute: Carl Ritter (at the age of 66) in 1845, John Arrowsmith in 1863, August Petermann in 1868, Mary Somerville in 1869, and Keith Johnston in 1871. Overwhelmingly, public recognition was given primarily to explorers and secondarily to surveyors, many of them military men.[107] Only after 1890 did scholars and academics begin to feature at all regularly; even so the next 50 years included but a handful of names of professional geographers: Reclus, Murray, Grandidier, Albrecht Penck, Davis, Cvijic. Not until the

[103] J. Marshall-Cornwall, *op. cit.* (note 101), and 'An early Scandinavian traveller' [Brooke], *Geographical Journal*, 144 (1978), 250–253.

[104] J. Marshall-Cornwall, *op. cit.* (note 101, pp. 43–44. The Geographical was not alone in its reluctance to admit women. I recall dining on several occasions at the Zoological in the 1960s when any lady unfortunate enough to be present was hidden behind a curtain so that the sight of her would not disturb the gentlemen; apparently the ladies did not object. Nor was the reform in the Geographical unanimously approved: the Arabian explorer Wilfred Thesiger resigned on the issue.

[105] *Ibid.*, appendix II, p. 49.

[106] R. M. McLeod, 'Of medals and men: a reward system in Victorian science 1826–1914', *Notes and Records of the Royal Society of London*, 26 (1971), 81–105.

[107] See the lists in I. Cameron, *To the farthest ends of the earth: the history of the Royal Geographical Society 1830–1980* (London: Macdonald, 1980), pp. 262–275.

1940s were the professionals honoured in any numbers, with royal medals for Bowman, Lattimore, Mackinder, Stamp, J. K. Wright and L. C. King. Not that the Society ignored more intellectual claims: but rather than dilute the roll of royal medallists it chose to institute a separate award, the Victoria medal, especially for academic work. Established in 1902, 56 awards had been made by 1979, and simply to read the list of names of the recipients, almost all professional men in the universities, is to grasp the profound gulf in social standing between them and the royal medallists.

Finally, the Society extended its patronage directly through the scientific community: by financial support and approval for expeditions; by meeting a major part of the cost of geographical instruction in the older universities; and by a system of gold medals for boys at selected public schools. Doubtless its influence also operated in many other ways, for as Billinge reminds us, 'hegemony is not mechanical, nor are its forms of extension primarily overt.'[108]

It is a central tenet of Gramsci's thesis that the subordinates are constantly attempting to subvert the dominant, and the latter react by quelling or subsuming the unruly. The R.G.S. hegemony did not go unchallenged, not least from within. To some extent the internal criticism, which centred round the exploration/education dichotomy, could be controlled, partly by placing representatives of the dissidents on the council, partly by giving them Victoria medals, partly by paying the older universities to employ them. But from time to time the strategy failed, often in highly symbolic ways. One such was the open conflict in 1892 over the admission of women, when the future Viceroy of India, Lord Curzon, chose to overturn a decision of the Council. Freshfield, the secretary, resigned and went off to found the Geographical Association, thus formalizing a clear divergence in interest and authority between the R.G.S. and newly emergent professional groups.[109] Almost inevitably it followed that a further group of dissident red-brick academics should in due course also take themselves off to found the Institute of British Geographers, to distance themselves from the explorers and to establish their own hegemony over the already rigidly stratified sub-population of university geographers.[110]

[108] M. D. Billinge, 'Hegemony, class and power in late Georgian and early Victorian England: towards a cultural geography', in A. R. H. Baker and D. J. Gregory, eds, *Explorations in historical geography: interpretative essays* (Cambridge: Cambridge University Press, 1984), 28–67, 203–217; reference on p. 30.

[109] H. R. Mill, *op. cit.* (note 90), pp. 107–112; see also chapter 4.

[110] D. R. Stoddart, 'Progress in geography: the record of the I.B.G.', in D. R. Stoddart, ed., *The Institute of British Geographers 1933–1983*, *Transactions of the Institute of British Geographers*, n.s. 9 (1983), 1–124; R. W. Steel, *The Institute of British Geographers: the first fifty years* (London: Institute of British Geographers, 1984).

It is not difficult to discern that these formal moves in the organization of British geography reflected profound social distinctions which in large degree determined and explain the behaviour of those involved. The point can readily be emphasized by a simple comparison of the holders of the presidencies of the Royal Geographical Society and the Institute of British Geographers.[111] Of the 48 presidents in the first 150 years of the R.G.S., 40 were titled (including seven earls, a marquess and seven other peers), and only one (Stamp) was an academic. Of the 37 presidents of the Institute of British Geographers, all except one were holders of university chairs (the exception was a reader), only three came from Oxford or Cambridge, and only one (again, Stamp) received a title in his lifetime (though many years after his tenure of office in the I.B.G. and ironically enough during his presidency of the R.G.S.).

DIFFERENT APPROACHES

I do not wish by this brief sketch to suggest that such an approach to our history, through the analysis of the social and political relations between individuals within institutions, is necessarily the most appropriate: merely that it merits attention. There are other possibilities, some of which I have explored elsewhere, others in this book.

At a fairly elementary level we can study the external characteristics of the development of the subject – increasing numbers of practitioners, departments, degrees, publications – and be intrigued by the elegant simplicity of the functions which describe them.[112] Likewise we can approach the internal coherence of the discipline through analysis of citation structures, with similar consequences.[113] And we can measure the great men themselves, listing their publications, counting their medals and honours, ranking their achievements.[114] Such activities give much innocent pleasure

[111] Lists are given by I. Cameron, *op. cit.* (note 107), pp. 260–261, and R. W. Steel, *op. cit.* (note 110), pp. 152–154.

[112] D. R. Stoddart, 'Growth and structure of geography', *Transactions of the Institute of British Geographers*, 41 (1967), 1–19.

[113] *Ibid.* Also K. M. Clayton, 'Publication and communication in geography', *Geographia Polonica*, 18 (1970), 13–20. For examples of the interpretative use of such techniques in geography, see S. S. Duncan, *op. cit.* (note 70) and A. C. Gatrell, 'Describing the structure of a research literature: spatial diffusion modelling in geography', *Environment and Planning*, B 11 (1984), 29–45.

[114] D. R. Stoddart, 'Some statistical characteristics of eminent geographers', *University of Bristol Department of Geography Seminar Paper Series*, A 5 (1967), 1–18.

as well as some enlightenment.[115]

Alternatively one can adopt a more specifically biographical approach. Experience in this field is less encouraging. Those of our number who have produced autobiographies – H. R. Mill, E. A. Reeves, Griffith Taylor – sadly tell us little other than that their subjects rushed from A to B and from time to time did this and that.[116] It is a tradition unfortunately followed in the few full biographies of geographers so far published – of Jefferson, Huntington, Bowman, J. Russell Smith.[117] A barren catalogue of activities tells us little about life or thought. Of course it need not be thus, as Anne Buttimer has vigorously urged. There is much to intrigue in recent collections of autobiographical recollections of individual geographical experience: some, such as those by Glacken, Mead and Leighly, are gems in our literature.[118] But how to make sense of the detail, how to generalize and to assess significance, other than in a purely aggregative manner? It is not enough to let such accounts stand by themselves, and hope the connections will be obvious.

Lastly, there is the history of ideas and of themes. This may be highly specific, as in Wright's initial bibliobiography of Ellen Churchill Semple's *Influences of geographic environment*[119] or the study of Huxley's *Physio-*

[115] I recall with pleasure my (predictable) discovery that one-third of the membership of the Institute of British Geographers had published nothing at all and thus fell into the class of 'non-producing deviates': D. R. Stoddart, *op. cit.* (note 112), p. 8.

[116] H. R. Mill, *An autobiography* (London: Longmans, Green, 1951); E. A. Reeves, *Recollections of a geographer* (London: Seeley, Service, n.d.); T. G. Taylor, *Journeyman Taylor: the education of a scientist* [*sic*] (London: Robert Hale, 1958). I have not seen the unpublished autobiography of L. Dudley Stamp, however.

[117] These criticisms, alas, apply to the most sustained biographical attempts in recent years: G. J. Martin, *Mark Jefferson, Geographer* (Ypsilanti: Eastern Michigan University Press, 1968), *Ellsworth Huntington: his life and thought* (Hamden: Archon Books, 1973), and *The life and thought of Isaiah Bowman* (Hamden: Archon Books, 1980), as well as to V. M. Rowley, *J. Russell Smith: geographer, educator and conservationist* (Philadelphia: University of Pennsylvania Press, 1964). Fortunately others achieve more, notably David Lowenthal, *George Perkins Marsh: versatile Vermonter* (New York: Columbia University Press, 1958), G. S. Dunbar, *Elisée Reclus, historian of nature* (Hamden: Archon Books, 1978), W. H. Parker, *Mackinder: geography as an aid to statecraft* (Oxford: Clarendon Press, 1982), and David Livingstone's forthcoming book on Shaler. Walter Freeman's great achievement, *Geographers: biobibliographical studies* (1977–), has different aims.

[118] A. Buttimer, ed., *The practice of geography* (London: Longman, 1983), especially C. J. Glacken, 'A late arrival in academia', 20–34, W. R. Mead, 'Autobiographical reflections in a geographical context', 44–61, and J. B. Leighly, 'Memory as mirror', 80–89; M. D. Billinge, D. Gregory and R. Martin, *Recollections of a revolution: geography as spatial science* (London: Macmillan, 1984), on which I have commented in the Introduction.

[119] J. K. Wright, 'Miss Semple's "Influences of geographic environment": notes towards a bibliobiography', *Geographical Review*, 52 (1962), 346–361, reprinted in his *Human nature in geography*, *op. cit.* (note 9 [1966]), 188–204.

graphy in chapter 9; or it may treat wider issues, as in the survey of integrative models in chapter 11. There is great scope for this kind of study, and it is not my purpose to suggest that the kind of social and institutional analysis advocated here should in any sense replace it.

Finally, there are studies of institutions themselves and their histories. The recent diverse anniversaries of the Royal Geographical Society (150th), Royal Scottish Geographical Society and Geographical Society of Berlin (100th) and the Institute of British Geographers (50th), and their attendant retrospective accounts, give an index of both the potential and the limitations of such studies.[120]

Each of these approaches has its own validity and its own strengths. I claim here only that the history of geography is an infinitely richer and more varied landscape than acquaintance with the standard works on the subject would suggest. This book represents some initial explorations in these *terrae plerumque incognitae*.

[120] T. W. Freeman, 'The Royal Geographical Society and the development of geography', in E. H. Brown, ed., *Geography yesterday and tomorrow* (London: Oxford University Press, 1980), 1–99: D. R. Stoddart, *op. cit.* (note 110); R. W. Steel, *op. cit.* (note 110). See also A. D. Grady, 'The role of geographical societies in the development of geography in Britain from 1900–1914' (University of London Ph.D. thesis, 1972).

2

Geography – A European Science

It is now a commonplace to think of the geography of the last 25 years as a discipline different in fundamental respects from that of previous times. For some, the differences are so profound that an understanding of the recent development of the subject need be rooted no further back than 1945, and be largely confined to England and to two or three North American universities.[1] I shall argue here for a very different view: that the central core of our discipline – why we call ourselves geographers and why what we do has so much in common – has been essentially European in its derivation and characteristics, and that it developed over a remarkably short space of time at the end of the eighteenth and the beginning of the nineteenth centuries. By implication I shall suggest that these characteristics have persisted as enduring concerns through the series of 'revolutions' to which the discipline is said to have been subjected over the past quarter of a century.

Geography as a body of knowledge is of course of great antiquity. Our standard histories speak of the work of Strabo and Eratosthenes, Varenius, Hakluyt, Purchas. But all of these figures seem to us remote. Their contributions have meaning in the contexts only of their own time, not of ours. Their significance to us is as precursors, its study largely antiquarian. We need to ask how geography emerged as a method or activity – an intellectual structure – rather than simply as a body of information, and what the past has to say to the present both in the problems we recognize and the way we tackle them.

One might well think that this is not a particularly fruitful line of enquiry to explore. After all, we know our history. It is set down in magisterial form

[1] R. J. Johnston, *Geography and geographers: Anglo-American human geography since 1945* (London: Edward Arnold, 1979).

by a succession of scholars: Bunbury[2] and Beazley[3] for the ancient and medieval worlds, Eva Taylor[4] for the sixteenth and seventeenth centuries, Dickinson[5] and Freeman[6] for more modern times; above all, of course, in that 'leviathan of learning', as it has been called, Hartshorne's *Nature of geography*[7] – methodology justified by history, or history given meaning by methodology, depending on your point of view.

But I am not alone in thinking that whatever these works do – and they document in encyclopaedic detail the achievements of those who have called themselves geographers – they tell us very little indeed about *why* men turned to geography, *what* they did when they became geographers, *how* they did it, rather than becoming, for example, historians, archaeologists, economists, sociologists, anthropologists, geologists. To use these labels is itself to indicate the recency of our conventional sub–divisions of knowledge, since many of them (like Whewell's term 'scientist', coined in 1840) are little more than a century or two old.

So I shall argue that what distinguishes geography as an intellectual activity from these other branches of knowledge is a set of attitudes, methods, techniques and questions, all of them developed in Europe towards the end of the eighteenth century. I am less concerned with the minutiae of the historical record – of who wrote this or who discovered that – than with context, arguments, questions; with social structures; and with individuals as human beings in their time. In short, for our history to speak to us today, it cannot simply be narrative, cumulative, internalist, unilinear, but must rather be social and contextual.[8] We must appreciate what our founders did in order to know what we ourselves are doing.

[2] E. H. Bunbury, *A history of ancient geography among the Greeks and Romans from the earliest times to the fall of the Roman Empire* (London: John Murray, 2 volumes, 1883).

[3] C. R. Beazley, *The dawn of modern geography* (Oxford: Clarendon Press, 3 volumes, 1897–1906).

[4] E. G. R. Taylor, *Tudor geography 1485–1583* (London: Methuen, 1930), *Late Tudor and early Stuart geography 1583–1650* (London: Methuen, 1934).

[5] R. E. Dickinson, *The makers of modern geography* (London: Routledge & Kegan Paul, 1969).

[6] T. W. Freeman, *A hundred years of geography* (London: Duckworth, 1961).

[7] R. Hartshorne, 'The nature of geography: a critical survey of current thought in the light of the past', *Annals of the Association of American Geographers*, 29 (1939), 171–658; reprinted as *The nature of geography* (Lancaster, Pa.: Association of American Geographers, 1939 and later printings).

[8] D. R. Stoddart, 'Ideas and interpretation in the history of geography', in D. R. Stoddart, ed. *Geography, ideology and social concern* (Oxford: Basil Blackwell, 1981), pp. 1–7; see also chapter 1.

How to begin? What are our organizing themes? Clarence Glacken[9] attacked this problem in his *Traces on the Rhodian shore*, when he distinguished three major ideas as a structure for his analysis: of an earth designed for man, with a place for everything and everything in its place; of the influence of environment on man, as traditionally expounded in the works of Hippocrates and Montesquieu; and of the increasing effect of man on environment. But Glacken's great work ends at the close of the eighteenth century, just when, as I shall argue, the story becomes interesting.

It seems to me that there are three prerequisites for the development of modern geography, and all of them emerged in Europe. The first, perhaps paradoxically, is our recognition of the immensity of time, for on this depends our understanding of processes and rates. The sixteenth century was concerned with calendars, a subject which greatly exercised the popes, and the seventeenth and eighteenth centuries with clocks, culminating in Harrison's chronometer of 1761.[10] This was a period of technical developments to do with measurement and calibration, though the practical significance of time, and in particular of synchrony, awaited the needs of the railway and the telegraph in the 1830s and later.[11] But time also involves profound concepts with wide-ranging implications.[12] James Hutton ended his first paper on his 'theory of the earth' in 1785 with the statement that he could find 'no vestige of a beginning, – no prospect of an end',[13] and thus broke through the barrier of the date of 4004 B.C. for the creation of the world derived from the Bible by Archbishop Ussher in 1650–1654. Hutton's concept of time is contained in one of his most famous statements: 'Time, which measures every thing in our idea and is often deficient to our schemes, is to nature endless and as nothing: it cannot limit that by which alone it had existence.'[14] Time, for Hutton, was defined by processes,

[9] C. J. Glacken, *Traces on the Rhodian shore: nature and culture in western thought from ancient times to the end of the eighteenth century* (Berkeley: University of California Press, 1967).

[10] S. Gaye and H. Michel, *Time and space: measuring instruments from the 15th to the 19th century* (London: Pall Mall Press, 1971); R. T. Gould, *The marine chronometer: its history and development* (London: Holland Press, 1960 [reprint of the 1923 edition]).

[11] N. J. Thrift, 'The diffusion of Greenwich Mean Time in Great Britain: an essay on a neglected aspect of social and economic history', *University of Leeds Department of Geography Working Paper*, 188 (1977), 1–12; D. Parkes and N. J. Thrift, *Times, spaces and places: a chronogeographic perspective* (Chichester: Wiley, 1980).

[12] F. C. Haber, *The age of world: Moses to Darwin* (Baltimore: Johns Hopkins University Press, 1959).

[13] J. Hutton, 'Theory of the earth', *Transactions of the Royal Society of Edinburgh*, 1 (1788), 209–304; reference on p. 304.

[14] J. Hutton, *Theory of the earth with proofs and illustrations* (Edinburgh: William Creech, 2 volumes, 1795), reference in volume 1, p. 15.

rather than applied to them as an external measuring rod, just as today we define and measure time through radioactive decay processes. And once the limitation on the age of the earth was abandoned, it made possible new attitudes to the antiquity of man himself and his prehistory, especially following the exploration of the Somme gravels in the 1820s.

The second theme is the importance of space and of scale. Again, instrumental developments made possible conceptual developments. The invention of the microscope in 1593 (and especially its imaginative use by Robert Hooke in the seventeenth century) and the invention of the telescope in 1608 helped to define the narrow band of the cosmos in which man lived, and suggested the operation of causes operating at scales hitherto not accessible to direct observation. The need for standardization and codification of measurements led in 1790 to the definition of the metre by the Paris Academy as one ten-millionth of the earth's quadrant.[15] Surveying techniques were revitalized, not least by the use of the barometer, pioneered by Boyle and Halley, for the measurement of altitude.[16] Pascal first used it to measure the height of the Puy de Dome in 1648; Scheuchzer took it into the Alps in 1709; but a great advance occurred when Tournefort employed it to determine the heights of the vegetation zones of Mount Ararat in 1730. The effect of these developments on the European imagination was profound – witness Swift's *Gulliver's travels* in 1726 – and readily transmitted to a rapidly expanding knowledge of the earth itself. And it is a theme, of course, which has powerfully re-emerged in the concern with scale and process during the 1960s.

Third, there was an increasing recognition of man's capacity to interpret and to modify his environment, and so to influence his position on the planet. The first large-scale directed environmental modification (as opposed to the piecemeal clearing of the woodlands of earlier centuries) came with the drainage and engineering works, canals and bridges of the seventeenth and eighteenth centuries: artificial landscapes were created in the Fens and in Holland.[17] It was also the time of the first concern about environmental quality, with the seventeenth-century forest ordinances, Evelyn's concern with air pollution in London in 1661, and even the first

[15] B. W. Petley, 'New definition of the metre', *Nature, Lond.*, 303 (1983), 373–376.

[16] E. Halley, 'A discourse of the rule of the decrease in the height of the mercury in the barometer, according as places are elevated above the surface of the Earth, with an attempt to discover the true reason of the rising and falling of the mercury, upon change of weather', *Philosophical Transactions of the Royal Society of London*, 16 (1686–1687 [1688]), 104–116.

[17] G. P. Marsh, *Man and nature; or, physical geography as modified by human action* (New York: Charles Scribner, 1864); H. C. Darby, *The draining of the Fens* (Cambridge: Cambridge University Press, 1940).

measures to protect the rapidly disappearing Green Turtle in far-away Bermuda in 1620.[18]

These were the contexts of change, both technical and conceptual, in which man related anew to his environment at the dawn of modern times, when Newton was still alive and when the Royal Society had but recently been founded. But we need to go one step further. Time, scale, the effects of man: all these can be intuitively, even emotionally perceived. For us, the crucial point is: when did *truth* become our central criterion? When did geography become an objective science?

It is instructive to compare the European reaction to the discovery of America in the sixteenth century with the reaction to that of Australia and the Pacific in the eighteenth.[19] In the first case the questions raised were primarily theological and philosophical rather than scientific, with the rare exceptions of the natural histories of Oviedo and Acosta. It proved impossible to subsume the cultures of the Andes and the Amazon within accepted European models: in consequence European accounts of the New World are overwhelmingly records of conquest and destruction. Indeed, it was only with the greatest reluctance that the Church conceded, in the bull *Sublimis deus* in 1537, that the American Indians were indeed people like ourselves.

Myth and imagination (as in the idea of El Dorado) were difficult to separate from fact in the geographical literature of the time. Many geographical accounts, even of extraordinary verisimilitude, were fictional, written by people who had never visited the lands they described: some were so lifelike that it is difficult now to know whether they are true or false.[20] There was, however, a spectrum in such works, from idealistic tracts such as More's *Utopia* (1518), Bacon's *Nova Atlantis* (1627) and Campanella's *Civitas solis* (1637), to overtly satirical and allegorical works such as *Gulliver's travels*.[21] Books like *Robinson Crusoe* (1719) emphasized the

[18] J. J. Parsons, *The Green Turtle and man* (Gainesville: University of Florida Press, 1952), p. 24.

[19] J. H. Elliott, *The Old World and the New 1492–1650* (Cambridge: Cambridge University Press, 1970).

[20] Geoffroy Atkinson, *The extraordinary voyage in French literature before 1700* (New York: Columbia University Press, 1920); *The extraordinary voyage in French literature from 1700 to 1720* (Paris: Libraire Ancienne Honoré Champion, 1922). For an example of the difficulty in separating truth from fiction in the literature of that time, see my 'Pinnipeds or sirenians at western Indian Ocean islands?', *Journal of Zoology*, 167 (1972), 207–217.

[21] E. L. Tuveson, *Milennium and utopia: a study in the background of the idea of progress* (Berkeley: California University Press, 1949). Many attempted to transform these fictions into fact: see W. H. G. Armytage, *Heavens below: utopian experiments in England, 1560–1960* (Toronto: Toronto University Press, 1961).

capability of man in coping with and modifying his own environment, based loosely on the story of a real castaway in the Juan Fernandez Islands. But the characteristic of all of these works is that none of them is based, whatever their aims, on *direct observation*.

I date the transformation to the year – 1769 – when Cook first entered the Pacific, and in that brief decade before his death in Hawaii in 1779 charted one-third of the coastlines of the world. Suddenly empirical science displaced the old concerns. Cook took with him scientists, illustrators and collectors – Banks and Parkinson, Johann Reinhold and Georg Forster – starting a great European naval expeditionary tradition, which has only begun to falter in Britain in the last few years. The elder Forster[22] stated his aims in his account of Cook's second voyage: 'My object was nature in its greatest extent: the Earth, the Sea, the Air, the organic and animated Creations, and more particularly that class of beings to which we ourselves belong'; and his title page gives his agenda for research:

> *Observations made during a voyage round the world, on Physical Geography, Natural History, and Ethic Philosophy. Especially on 1. The Earth and its Strata, 2. Water and the Ocean, 3. The Atmosphere, 4. The Changes of the Globe, 5. Organic Bodies, and 6. The Human Species.*

With Harrison's chronometer and the solving of the longitude problem in 1761, marine surveying had become accurate. Charting by sextant was rapid and simple. It was also carried out from a floating fortress which served too as a laboratory. One reason why the exploration of interior North America and Siberia lagged for a further century after Cook was partly because land surveying is inherently a much more tedious and slower business than nautical surveying, but also because one is greatly more exposed to danger in a tent than in a man-of-war. A naval surveyor must be either careless or unfortunate to succumb to sharks, tempests or natives; whereas in the interior of the great continents the possibility of being eaten by wolves, bears or people was an omnipresent hazard.[23]

[22] J. R. Forster, *Observations made during a voyage round the world* (London: G. Robinson, 1778), p. ii.

[23] See my 'Humboldt and the emergence of scientific geography', in P. Alter, ed. *Humboldt* (London: William Heinemann, 1985). It is interesting to compare the work of Pallas along the Volga and in Siberia at the same time that Cook was exploring the Pacific with Banks and the Forsters. See F. Wendland and K.-B. Jubitz, 'Der Beitrag von Peter Simon Pallas (1741–1811) zum Weltbild der Geowissenschaften. Eine Bestandsaufnahme', *Zeitschrift für geologische Wissenschaften*, 8 (1890), 119–133; and E. Stresemann, 'Leben und Werk von Peter Simon Pallas', *Quellen und Studien zur Geschichte Osteuropas*, 12 (1962), 247–257.

With the rigorous requirements of charting there was also a new concern for realism in illustration and description.[24] This applied of course to the 'nautical views' with which charts were illustrated – not as decorative cartouches but as precise guides for the survival of navigators, and on which men's lives could depend. The botanical illustrations of Parkinson,[25] and later the depiction of shells, birds and minerals with minute accuracy became characteristic. Words too were used with more precision, especially after Gilbert White depicted *The natural history of Selborne* in 1789. Places came to be seen as composed of objects which could be recorded and related to each other in an objective manner, rather than simply as triggers to mood and to expression.[26]

There were two great devices used to bring the huge diversity of nature revealed by this new wave of exploration within the bounds of reason and of comprehension. One was classification. Linnaeus, who died in 1778 and who was a close contemporary of Cook, provided the basis for action through the many editions of his *Systema naturae* (1735) and *Genera plantarum* (1737).[27] Specialists began to emerge who saw themselves as botanists, entomologists, conchologists, geologists, united not just by subject matter but by method and technique. The first use of the geological hammer in a scientific rather than a utilitarian sense dates from 1696; of the vasculum 1704; of the butterfly net 1711. It was the beginning of the specialization and professionalization of the field sciences.[28]

The second device was the invention of the comparative method. One sees it beautifully employed by Reinhold Forster, when he tries to explain the existence of uplifted coral reefs in the tropical Pacific in the same framework as the emergence of land from the sea in Scandinavia, as observed during the course of a human lifetime by Linnaeus himself, and which we now know to

[24] B. Smith, *European vision and the South Pacific 1768–1850: a study in the history of art and ideas* (Oxford: Clarendon Press, 1960); S. J. A. Stephens, 'Reflections in a Pacific mirror: English perceptions of mankind 1772–1798' (Trinity College Dublin M.Litt. thesis, 1983).

[25] D. J. Carr, ed., *Sydney Parkinson: artist of Cook's Endeavour voyage* (Canberrra: Australian National University Press, 1984).

[26] K. Clark, *Landscape into art* (New York: Harper & Row, new edition, 1976); B. Novak, *Nature and culture: American landscape and painting 1825–1875* (New York: Oxford University Press, 1980). For an illustration of the juxtaposition of incongruous styles of illustration, contrast the landscape drawings and the panoramic views by F. W. von Egloffstein in *Report upon the Colorado River of the West, explored in 1857 and 1858 by Lieutenant Joseph C. Ives* (Washington: Government Printing Office, 1861).

[27] W. Blunt, *The complete naturalist: a life of Linnaeus* (London: Collins, 1971); N. von Hofsten, 'Linnaeus's conception of nature', *Kungliga Vetenskaps-Societetens Årsbok* (Uppsala), 1957, 65–105.

[28] D. E. Allen, *The naturalist in Britain: a social history* (London: Allen Lane, 1976).

result from isostatic movements.[29] One sees it too in the attempts to connect the mechanism behind the great Lisbon earthquake of November 1755, which so shocked the European consciousness, with that in Java two days later.[30]

But above all one sees it in attitudes to man. It is surely the extension of scientific methods of observation, classification and comparison to peoples and societies that made our own subject possible.[31] The great European voyages began with a heavy weight of presupposition – the Garden of Eden, the Golden Age, Paradise on Earth – inherited from earlier centuries.[32] And these were strongly reinforced by the first idyllic contacts with the Great South Sea. After the appalling journey round Cape Horn from Europe, with scurvy, rotten food and stinking water, not to mention sodomy and the lash, islands appeared where breadfruit and coconuts grew abundantly and no one had to work.[33] Bougainville's arrival at Tahiti on 6 April 1768 was one of the first and most characteristic of such experiences. The *Boudeuse* was surrounded by canoes of naked women as it edged through the reef, and the marines had to be called out to keep order among the sailors. The cook, who disobeyed orders and reached the beach, was immediately dragged into the bushes by a band of women, stripped of his clothing, and instructed to perform publicly the act for which he had come ashore: he later told Bougainville that whatever his punishment might be it would not terrify him more than had the ladies on the beach.[34] The effect of such stories on the

[29] The range of Forster's interests extended far beyond the natural sciences. For example, see 'On the management of carp in Polish Prussia', *Philosophical Transactions of the Royal Society of London*, 61 (1771), 310–325, and 'Some account of the roots used by Indians, in the neighbourhood of Hudson's bay, to dye Porcupine Quills', *ibid.*, 62 (1772), 54–59. The importance of the elder Forster to geography was first signalled by J. Rittau, 'Johann Reinhold Forsters Bemerkungen auf seine Reise um die Welt gesammelt. Ein Beitrag zur Geschichte der Geographie' (Marburg-Lahn, dissertation, 1881). See also M. E. Hoare, *The tactless philosopher: Johann Reinhold Forster (1729–98)* (Melbourne: The Hawthorn Press, 1976).

[30] T. D. Kendrick, *The Lisbon earthquake* (London: Methuen, 1956); A. Kemmerer, 'Das Erdbeben von Lissabon in seiner Beziehung zum Problem des Ubels in der Welt' (Frankfurt-am-Main, thesis, 1958).

[31] See, for example, G. Chinard, *L'Amérique et le rêve exotique dans la littérature française au XVIIe et XVIIIe siècle* (Paris: Librairie Hachette, 1913); C. L. Sanford, *The quest for paradise: Europe and the American moral imagination* (Urbana: University of Illinois Press, 1961); and G. H. Williams, *Wilderness and paradise in Christian thought* (New York: Harper, 1962).

[32] See note 24.

[33] J. R. Forster, *op. cit.* (note 22), pp. 610–649. See also M. E. Hoare, ed., *The Resolution journal of Johann Reinhold Forster 1772–1775* (London: Hakluyt Society, 4 volumes, 1982).

[34] L. de Bougainville, *A voyage round the world. Performed by order of His Most Christian Majesty, in the years 1766, 1767, 1768, and 1769* (London: J. Nourse & T. Davies, 1772), reference on p. 219.

European imagination was enormous. Polynesians and Micronesians were brought back to European capitals – Ahutoru to Paris, Omai and Lee Boo to London – and introduced to royalty and society. *Omai* was the subject of the Covent Garden pantomime at Christmas 1785: it ran for 50 performances and was an outstanding success.[35]

Contrast the objectivity with which Darwin describes the Fuegians during the voyage of the *Beagle* in 1833, 70 years later:

> whilst going on shore, we pulled alongside a canoe with 6 Fuegians. I never saw more miserable creatures; stunted in their growth, their hideous faces bedaubed with white paint and quite naked. One full aged woman absolutely so, the rain and spray were dripping from her body. Their red skins filthy and greasy, their hair entangled, their voices discordant, their gesticulation violent and without dignity. Viewing such men, one can hardly make oneself believe that they are fellow creatures placed in the same world. I can scarcely imagine that there is any spectacle more interesting and worthy of reflection, than one of these unbroken savages.[36]

Darwin described the scene again for a different world in *The descent of man* in 1871, but added that 'the reflection at once rushed into my mind – such were our ancestors', adding that he would himself prefer to be descended from a little monkey or an old baboon than from such savages.[37] It was a new attitude to man and his works which was to lead before the end of the century to empirical recording of artefacts and ways of life on a global scale, as in Ratzel's *Völkerkunde* of 1885–1888.[38]

Humboldt was born in the year Cook first saw the Pacific, Ritter in the year he died there. Both men died in the year in which the *Origin of species* appeared. Their achievement was to seize the technical and conceptual advances of the Pacific voyages and so to organize and order knowledge as

[35] B. Smith, *op. cit.* (note 24), p. 80; M. Alexander, *Omai, noble savage* (London: Collins, 1977); T. B. Clark, *Omai, first Polynesian ambassador to England* (San Francisco: Colt Press, 1940); E. H. McCormick, *Omai, Pacific envoy* (Auckland: University of Auckland Press, 1977); R. Joppien, 'Philippe Jacques de Loutherbourg's pantomime "Omai; or, a Trip round the World" and the artists of Captain Cook's voyages', *British Museum Yearbook*, 3 (1979), 81–136.

[36] N. Barlow, ed., *Charles Darwin's diary of the voyage of H.M.S. 'Beagle'* (Cambridge: Cambridge University Press, 1933), p. 212.

[37] C. R. Darwin, *The descent of man, and selection in relation to sex* (London: John Murray, 2 volumes, 1871), reference in volume 1, p. 326.

[38] F. Ratzel, *Völkerkunde* (Leipzig: Bibliographisches Institut, 3 volumes, 1885–1888). The second edition (1894–1895) was translated as *The history of mankind* (London: Macmillan, 3 volumes, 1896–1898).

to show its coherence and significance, Humboldt ecologically, Ritter historically and regionally. Humboldt himself was a great critical observer. The list of his field equipment is astonishing – barometers, pedometers, galvanometers, compasses, levels, sextants (pocket 2-inch as well as standard 8-inch models), chronometers, cyanometers (to measure the blueness of the skies), thermometers, telescopes, microscopes, balances, magnetometers, and many others.[39] As an instrumentalist he was a measurer and quantifier. But he was also a comparer, looking for trends and regularities: the first man to carry out a barometric traverse to establish the profile of a mountain range (in this case, the Andes), the first man to make sense of temperature data by drawing isotherms in 1817.[40] And he was also a conceptualizer and theorist – the discoverer, for example, of the great fact of continentality of climates by the device of reducing temperatures to sea-level and plotting them on maps. The same kinds of methods he applied too to people and political economy as well as to the natural world, which is why his *Political essay on the island of Cuba* of 1811 reads so freshly still.

Neither Humboldt nor Ritter fully succeeded in their aims to demonstrate the fundamental congruity of man and environment – Humboldt caught as he was between the German romantic *Naturphilosophie* and the new mathematical physics of Laplace and Lagrange, Ritter unable to see beyond his simplistic teleological religion.[41] It was their misfortune never to know the key to nature which Darwin supplied, and which transformed nineteenth-century thought. Today, a century after Darwin's death on 19 April 1882, we *necessarily* see evolutionary order everywhere, with all things interrelated in time and space. It was a vision denied to the critical observers, classifiers, comparers of the previous decades. I examine elsewhere the ways in which Darwin's ideas swept through the sciences, not least our own,[42] and the extent to which evolution, ecology, selection and adaptation gave common explanations to diverse phenomena.

What was, however, remarkable was the way in which, in Europe, this scientific revolution coincided with the social and educational upheavals consequent on the industrial revolution.[43] The Education Act of 1870 made

[39] S. F. Cannon, 'Humboldtian science', in *Science in culture: the early Victorian period* (New York: Dawson, 1978), pp. 75–76. Cannon's early work appeared authored by W(alter); by the time this book was published he had become S(usan).

[40] D. Botting, *Humboldt and the cosmos* (London: Sphere Books, 1973); A. de Humboldt, 'Des lignes isothermes et de la distribution de la chaleur sur la globe', *Mémoires de Physique et de Chimie de lat Société d'Arcueil*, 3 (1817), 462–602.

[41] D. R. Stoddart, *op. cit.* (note 23).

[42] See chapter 7.

[43] See chapters 3 and 8.

elementary education compulsory in Britain, and it was paralleled by similar developments in France and Germany. Schools and universities proliferated; national examinations with national standards were organized; knowledge was suddenly divided into *subjects*, themselves soon constrained by buildings, staff and syllabuses; the 'new geography' began in new journals, especially the *Geographical Journal* and the *Annales de Géographie*; and the first appointment of the young Vidal de la Blache in Paris in 1877 might be taken to symbolize the new professionalism of those who called themselves geographers. These developments, in the brief span between 1870 and the end of the century, determined the future organization of our subject, just as the empirical recording and observation of earth's diversity over the preceding century had shaped its content.

But there was one further component of our subject already well developed in these formative years. It arose as people began to ask questions not just of the Polynesians or the Fuegians or the Aborigines, but of themselves. After the experience of Easter Island, Stonehenge could never be the same again. The Pyramids of Egypt were seen afresh in the light of the Aztecs and the Mayas (Napoleon's expedition to Egypt was contemporary with Humboldt's to the New World). Slowly it dawned on men that at the heart of European civilization there was savagery and barbarism not so far removed in either time or space from that of darkest Africa. The shock that was felt at the end of the nineteenth century when it was found that working men in the East End of London had a vocabulary smaller than that of Melanesian islanders in the Torres Straits brought to general acceptance a new cultural relativism which had been completely absent when the first European voyages set sail a century before.[44]

Two men realized the significance of these changes, and the potential role that geography could play. One was Elisée Reclus, the other Peter Kropotkin. They were not of course the first to have their consciences stirred by what they saw about them in the world. Humboldt himself had been moved to anger and shame – just as Darwin was 30 years later – by slavery in South America. But both were acutely aware of the need for social change and (to use a current phrase) social justice, and both devoted their lives to working for it.[45]

It is, of course, impossible in this brief discussion adequately to cover the rich diversity of European geography in its formative stages, or properly to

[43] G. Weber, 'Science and society in nineteenth century anthropology', *History of Science*, 12 (1974), 260–283.

[45] G. S. Dunbar, *Elisée Reclus: historian of nature* (Hamden: Archon Books, 1978); and also chapter 6.

explore its intellectual heritage. I have singled out the rise of objective methods in both the human as well as the natural sciences; the role of Darwin and the integration of knowledge to which his theory led; and the development of social awareness and social concern not only in distant lands but also in Europe itself, as hallmarks of the European geographical tradition. Other dominant figures of our European heritage – Marx (of whom perhaps enough has been said of late) and Freud (who still awaits a proper treatment) – clearly should also be considered in a comprehensive study of our intellectual heritage.

But I have attempted to show that in method and in concept geography as we know it today is overwhelmingly a European discipline. It emerged as Europe encountered the rest of the world, and indeed itself, with the tools of the new objective science, and all other geographical traditions are necessarily derivative and indeed imitative of it. Quantification, perception, social concern - all were dominant concerns of European geography in its formative period, just as they are today.

Forster himself spelled out what I believe must be our central theme, at the end of his *Observations* in 1778:

> mankind ought to be considered as the members of one great family; therefore let us not despise any of them, though they be our inferiors in regard to many improvements and points of civilisation; none of them is so despicable that he should not, in some one point or other, know more than the wisest man of the most polished nation. This knowledge may be easily obtained from them by friendliness, kindness, and gentleness; and if so bought is cheaply obtained. The second observation points out the necessity of sending out men versed in science, and the knowledge of nature on all occasions to remote parts of the world, in order to investigate the powers and qualities of natural objects; and it is not enough to send them out, but they ought likewise to be encouraged in their laborious task, liberally supported and generously enabled to make such enquiries as may prevent their fellow creatures in future times becoming sacrifices of their own ignorance.[46]

[46] J. R. Forster, *op. cit.* (note 22), pp. 648–649. Georg Forster, whose account of the voyage predated his father's by a year, made the same point with perhaps greater force. Stating that 'neither attachment nor aversion to particular nations have influenced my praise or censure,' he was able to look objectively and comparatively even on cannibalism in New Zealand: 'Though we are too much polished to be cannibals, we do not find it unnaturally and savagely cruel to take the field, and to cut one another's throats by the thousands, without a single motive, beside the ambition of a prince, or the caprice of his mistress' (J. G. A. Forster: *A voyage round the world in His Britannic Majesty's sloop, Resolution, commanded by Capt James Cook,*

It is these moral and ethical imperatives which give our subject meaning and direction. Recently Paterson told us that one of the functions of geography was to lead us 'to wonder, and to weep, over this world of man'.[47] One might add: to attempt to do something about it too. This is what European geography stands for, with its breadth of vision and catholicity of interest. To adopt Darwin's phrase in the famous last paragraph of *The origin of species*: 'There is grandeur in this view of life'[48] – and justification for all of us in what we seek to do.

---

during the years 1772, 3, 4 and 5 [London: B. White, J. Robson, P. Elmsby and G. Robinson, 2 volumes, 1777] references in volume 1, pp. xiii and 517). Georg's contribution to geography is perhaps less celebrated than his father's, though it was he who travelled down the Rhine with Humboldt and across to England in 1790, and it was Georg's descriptions of the Pacific islands in the *Voyage* that roused in Humboldt 'the first beginnings of an inextinguishable longing to visit the tropics' (A. von Humboldt, *Cosmos: a sketch of a physical description of the Universite*, translated by E. C. Otté [London: Henry Bohn, 2 volumes, 1848–1849], reference in volume 2, pp. 317–372). Idealist and reformist, Georg supported the French when they took Mainz in 1792 and was elected to the Convention. His wife (a 'faithless, frivolous, dissolute' woman, his father called her) left him, the Terror began, and Georg died of scurvy and pneumonia in a Paris garret, disowned by his father, in 1794. There is much of this tragedy in M. E. Hoare, *op. cit.* (note 29), pp. 282–306; see also L. Bodi, 'Georg Forster: the "Pacific expert" of eighteenth-century Germany', *Historical Studies of Australia and New Zealand*, 8 (1959), 345–363; E. Lange: 'Grundzüge der philosophischen Entwicklung Georg Forsters' (University of Jena, dissertation, 1961); and H. Fieldler, S. Scheibe and E. Germer, *Georg Forster, Naturförscher, Weltreisender, Humanist und Revolutionär* (Worlitz: Staatliche Schlosser und Garten Worlitz, 1970).

[47] J. Paterson, 'Some dimensions of geography', *Geography*, 64 (1979), 266–278; reference on p. 276.

[48] C. R. Darwin, *On the origin of species by means of natural selection, or the preservation of favoured races in the struggle for life* (London: John Murray, 1859), p. 490.

3

Geography, Education and Research

The Royal Geographical Society was formally constituted on 16 July 1830, under the patronage of the King, its 'sole object' the 'promotion and diffusion of that most important and entertaining branch of knowledge, GEOGRAPHY'. In July 1830, too, Charles Lyell published the first volume of hs *Principles of geology*, a book which was to establish both new procedures and new modes of explanation in the earth sciences and bring the study of the earth's surface within the bounds of observation and rational understanding. And Charles Darwin, still at Cambridge, was before long to join the five-year voyage round the world of H.M.S. *Beagle*, taking with him the volume of Lyell's *Principles* given him by FitzRoy. It was largely the combination of Lyell's *Principles* and the *Beagle* observations which ultimately led to *The origin of species* and all that that implied for the transformation of nineteenth-century thought.

As the nineteenth century progressed, the Royal Geographical Society mobilized its resources to persuade the universities of the worth of the subject as an academic discipline. In this chapter I shall show how after slow beginnings academic geography has flourished and prospered, and I shall trace the influence of the major figures in our discipline to whom we owe these developments. I shall consider what geography meant to these pioneers, in terms of education as well as of research, and I shall emphasize the importance of their views as a basis for what we need to do in the very changed circumstances of today.

When the R.G.S. was founded it was a vivid and exciting time to call oneself geographer, and the new Society vied with the Linnean, the Geological, the Zoological and many others in the intellectual life of the capital. At least in origin, the R.G.S. was primarily a club for gentlemen travellers, soldiers and sailors, though with a substantial number of natural scientists, especially geologists. As the century progressed, the Society's role in the progress of exploration proved its major occupation, as might be expected from its origins in the Raleigh Club and the African Association, and names such as those of Livingstone and Stanley, Burton and Speke, Scott and Shackleton, are part of its history. It is a story of great achievement, and one often told, most recently by Ian Cameron[1] and Walter Freeman.[2]

But there was another role to be played, in environments as unknown and as challenging to the founders of the 1830s as the heart of Africa or the polar wastes. In 1830, the population of England and Wales was 14 million, and that of the world 1,000 million. Perhaps only one Englishman in two lived in towns when the Society was founded, and only one in five earned his living from industry. By the end of the century the situation had been transformed. The country's population stood at 37 million and the world's at 1,600 million. The coming of the railway and the industrial revolution had changed not only the face of the land but also the social structure of the population.

As the urban middle classes and the poor multiplied, scholars such as Huxley saw the need for a radical restructuring of both the aims and the methods of education, in which, at least for him, geography would play a central role. Lyell himself led the way in denouncing the intellectual and educational decrepitude of the older universities, and his lectures at King's College, London, in 1832 may be thought to symbolize, together with Buckland's and Sedgwick's field trips at Oxford and Cambridge, a new attitude to instruction in British education. In this new climate, no other body pressed the case for geographical education, or pressed it so successfully, as did the R.G.S.

DEVELOPMENTS

Before considering how this came about, and what its consequences were, it is worth recalling how recently our university system has been established in Britain. With the exceptions of Oxford, Cambridge and the Scottish universities, the only institution in existence when the Society was founded was University College, London, established in 1828. King's College, London, followed in 1831, Durham in 1832, and the formal constitution of the University of London in 1836. Bedford College for Women followed in

[1] I. Cameron, *To the farthest ends of the earth* (London: Macdonald, 1980).

[2] T. W. Freeman, 'The Royal Geographical Society and the development of geography', in E. H. Brown, ed., *Geography yesterday and tomorrow* (Oxford: Oxford University Press, 1980), 1–99.

1849, and the first of the civic universities, Owen College, Manchester, in 1851. Southampton began in 1862, and the 1870s saw the foundation of the College of Natural Science at Newcastle (1871), the Yorkshire College of Science at Leeds, Firth College at Sheffield (1874), and the College of Science at Bristol (1876). Mason College, Birmingham, followed in 1880, university colleges at Liverpool and Nottingham in 1881, and university colleges at Reading and Exeter in 1892 and 1895, though some of these new foundations had to wait many years for their charters. A further great advance occurred in the 1950s and 1960s, and the total number of British universities now stands at 44.

Given these late developments, it is not surprising that the movement for the establishment of geography as a university discipline, which began in the 1870s, did not meet with immediate success. The problems began at Oxford and Cambridge. Here there was little support or understanding of the subject from within the colleges, with the notable exception of scattered and often idiosyncratic individuals such as the ornithologist Alfred Newton. Thus the Society had to establish a new concept of what geography was rather than simply argue a practical case for its adoption.[3] More to the point, the subject met active opposition from many geologists. The days of Lyell (who died in 1875) and his contemporaries, the great field interpreters, were over. Geologists were now largely professional men employed by the Geological Survey, and their confidence was deeply shaken when its professional staff was cut by half in the early 1880s. The university geologists had become specialized and technical – mineralogists, palaeontologists, petrologists – and the rigours of lecturing, examining and writing textbooks had replaced the stimulation and excitement of the field. When the geographers began to press their claims, particularly to the science of landforms, it was seen by geologists as a yet further assault on an already beleaguered position, and as a result was almost universally opposed. In this at least, the position in Britain differed radically from that in North America, where university departments of geography were largely created by geologists of the calibre of Chamberlin, Salisbury and Davis, and frequently originated as sub-departments within departments of geology.

That the Society succeeded at all in these rather dismal circumstances can be attributed to three factors. The first was the remarkable advocacy of John Scott Keltie, who later was to be its secretary for 30 years.[4] He was

[3] See chapter 5.

[4] H. R. Mill and D. W. Freshfield, 'Sir John Scott Keltie', *Geographical Journal*, 49 (1927), 281–287; G. G. Chisholm, 'Sir J. Scott Keltie, Ll.D.', *Scottish Geographical Magazine*, 43 (1927), 102–105.

appointed in 1884 to report on the state of geographical education, and his report included a wide-ranging survey of the state of the subject in continental and American as well as British universities and schools.[5] It was launched with a great fanfare of public lectures and exhibitions in 1885, and Keltie was at pains to show, especially from the work of Von Richthofen, that the widespread doubts among the natural scientists about the respectability of geography as a discipline were unfounded.

The second factor was the powerful if generally acerbic and occasionally ill-natured advocacy, often in competition with each other, of Francis Galton and Douglas Freshfield, supported (sometimes almost reluctantly) by such officers as Sir Clements Markham and Sir Richard Strachey. Galton and Freshfield, both wealthy men, were utterly different in outlook and interests. Freshfield – mountaineer, classical scholar, poet and traveller – rarely wrote on academic matters, whereas Galton's prodigious, original and eccentric output included the first newspaper weather map,[6] a map of the distribution of female beauty (classified as good, medium and bad), a notorious statistical study of the utility of prayer,[7] and a paper on the optimal method of cutting a fruit cake.[8] Yet it was their powerful advocacy within the council that enabled the Society to speak with a unitary voice in its dealings with the universities. And when the battle had been won in principle for the recognition of geography at Oxford and Cambridge, successive councils agreed, over a period of nearly half a century and from severely limited financial resources, actually to pay the salaries of the lecturers and readers appointed, since either from parsimony or obtuseness the universities themselves declined to do so. It is perhaps worth recalling, too, how in similar manner the Society's librarian, Hugh Robert Mill, personally organized the collection of rainfall data throughout the British Isles for the first 20 years of the present century, when the government declined to become financially involved in any operation of so wasteful and useless a kind.[9]

[5] J. S. Keltie, 'Geographical education – Report to the Council of the Royal Geographical Society', *Supplementary Paper of the Royal Geographical Society* , 1 (1886), 439–594.

[6] T. W. Freeman, 'Francis Galton: a Victorian geographer', in *The geographer's craft* (Manchester: Manchester University Press, 1967), 22–43. See also F. Galton, 'On the employment of meteorological statistics in determining the best course for a ship', *Proceedings of the Royal Society of London*, 21 (1873), 263–274; and, from a much later date, compare W. Warntz, 'Transatlantic flights and pressure patterns', *Geographical Review*, 51 (1961), 187–212.

[7] F. Galton, 'Statistical enquiries into the efficacy of prayer', *Fortnightly Review*, 12 (1872), 125–135.

[8] F. Galton, 'Cutting a round cake on scientific principles', *Nature, London*, 75 (1906), 173.

[9] H. R. Mill, *An autobiography* (London: Longmans, Green, 1950), pp. 109–134.

And thirdly, of course, there is the character of the geographers themselves, notably of Halford Mackinder, who was appointed to Oxford in 1887.[10] It was the Society's secretary, Henry Walter Bates, who asked Mackinder to prepare his famous paper, 'On the scope and methods of geography', delivered in January 1887, which remains one of the founding documents of our discipline.[11] His contemporaries at Cambridge, alas, could not match Mackinder's magic, and it says much for the pertinacity of the R.G.S. that they held the younger university to its commitments until the situation stabilized under Philip Lake and Frank Debenham – a period of uncertainty lasting nearly 30 years. In the civic universities, success came later. Often (as at Leeds, Nottingham and University College London) the geologists retained physical geography for themselves, and confined the geographers to the human aspects of the subject. When Lyde was appointed at University College London in 1903, for example, it was as Professor of Economic Geography, to supplement the work of the Professor of Geology; and in 1904 Roxby's appointment at Liverpool was in the Department of Economics, where the new subject was no more than 'an appendix to economics'.[12]

In general, university teaching began with the appointment of a single lecturer. In the oldest of the nineteenth-century institutions (University College London), it took 75 years from the date of foundation for this step to be achieved; in the universities founded in mid-century, the delay was 40 to 50 years; and for those founded in the 1890s, it was only 15 years. The organization of a department, the award of degrees with honours and the creation of a chair commonly had to wait for a further 20 to 30 years after the first lecturer had been appointed (see figure 1). The first such lectureships after Oxford and Cambridge came at Manchester (for clear commercial reasons) in 1892. The first decade of the new century saw geography appointments at Aberystwyth (1906), Reading (1907), Sheffield, Liverpool and Edinburgh (1908), and Leeds and Glasgow (1909). Southampton followed in 1913, the London School of Economics and Political Science in 1918, and Aberdeen in 1919. By 1920, however, there were still only six chairs: at University College London (1903), Reading (1907), Oxford (1910), Aberystwyth and Liverpool (1917), and at Birkbeck

[10] W. H. Parker, *Mackinder: geography as an aid to statecraft* (Oxford: Clarendon Press, 1982); see also chapter 5.

[11] H. J. Mackinder, 'On the scope and methods of geography', *Proceedings of the Royal Geographical Society*, n.s. 9 (1887), 141–160.

[12] J. S. Keltie, 'The position of geography in British Universities', *American Geographical Society Research Series*, 4 (1921), 1–33; reference on p. 19.

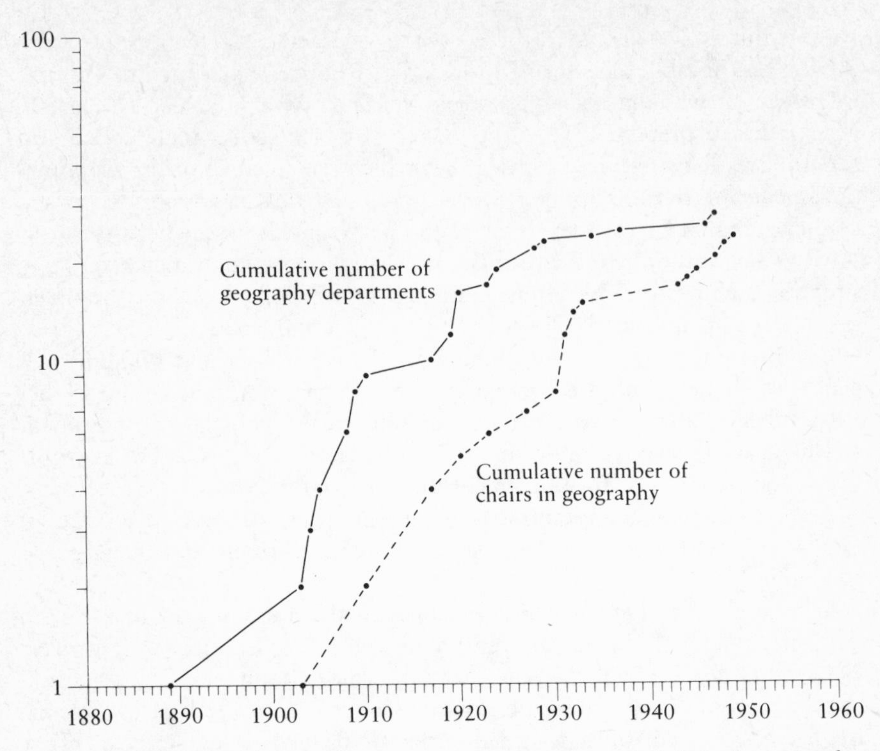

Figure 1 The foundation of departments of geography and of chairs in geography in British universities.

College, London (1920). The first honours courses leading to formal qualifications came at this same time: first at Liverpool in 1917, then at the London School of Economics and at Aberystwyth (1918), at University College London and at Cambridge in 1919, at Manchester in 1923, at Sheffield in 1924, and, belatedly, at Oxford in 1933, 46 years after Mackinder's appointment and 34 years after the organization of the School of Geography. The recency of many of these developments – many in the lifetimes of those still with us – is salutary, and the protracted history of developments at most of these institutions gives some indication of the difficulties which the subject often faced.

The successes in the universities, slow and episodic as they were, followed on the heels of revolutionary changes elsewhere in the educational system. The easy optimism of the Great Exhibition of 1851 had largely given way in

Britain to doubts about the permanence of industrial supremacy by the time of the Paris Exhibition a decade later, and to a growing awareness of the efficiency and strength of the German educational system. The Parliamentary Select Committee on Scientific Instruction under Bernhard Samuelson in 1868 and the Duke of Devonshire's Royal Commission on Scientific Instruction and the Advancement of Science in 1870–1875 took a sobering view of the immediate state and longer-term consequences of education in Britain. The Science and Art Department, set up in 1853, instituted a national system of examinations, but it was the Education Act of 1870 which truly revolutionized elementary education, and in a few years added one and half million new places in the elementary schools. Thomas Henry Huxley's *Physiography* in 1877 both met a need and set a pattern in school instruction, with quite astonishing success: at one time, a quarter of all examination papers being taken at the national level were in physical geography.[13] For Huxley, geography was a matter of direct experience: it was to be learned in the village and the countryside, not read about in books. The field trip and the specimen were the means to knowledge, with the aim an understanding of the world in which we live. Keltie, riding this wave of excitement and enthusiasm, was careful to emphasize the functional linkages between different levels in the educational system: geography in the universities could not succeed unless it were well done in the schools, and it could not be taken seriously in the schools unless it could also be pursued in the universities.

The 1870s and early 1880s were the period, we should recall, of the *Challenger* expedition of 1872–1876, and of John Wesley Powell's epic exploration of the Colorado Canyon in 1875. But it was also the period of Oskar Peschel's *Neue Probleme der vergleichende Erdkunde*,[14] of the first volume of Ratzel's *Anthropo-Geographie*,[15] of Richthofen's *Aufgaben und Methoden der heutigen Geographie*,[16] and of Davis's first essay on the cycle of erosion.[17] And 1877, the year of Huxley's *Physiography*, was both the year of Vidal de la Blache's first appointment in Paris and also –

[13] See chapter 8.

[14] O. Peschel, *Neue Probleme der vergleichenden Erdkunde als versuch einer Morphologie der Erdoberfläche* (Leipzig: Düncker und Humblot, 1870).

[15] F. Ratzel, *Anthropo-Geographie öder Grundzüge der Erdkunde auf die Geschichte* (Stuttgart: J. Engelhorn, 1885).

[16] F. von Richthofen, *Aufgaben und Methoden der heutigen Geographie* (Leipzig: Akademie Antrittsrede, 1883).

[17] W. M. Davis, 'Geographic classification, illustrated by a study of plains, plateaus, and their derivatives', *Proceedings of the American Association for the Advancement of Science*, 33 (1885). 428–432.

symbolizing the relentless advance of a new technological civilization, and symptomatic perhaps of a new phase in man's relationship with the earth – it was the year in which the last Tasmanian died.

FOUNDERS

Who were the men and women who seized these new opportunities, both organizational and intellectual, and who made academic geography in Britain what it is today? I will mention only five, though I could readily add many others of comparable stature who are still alive. The first is, of course, Mackinder, a restless and powerful advocate and entrepreneur. Not only did he establish with great flair the School of Geography at Oxford, he did the same at Reading, and for a time was director of the London School of Economics and Political Science as well as reader in geography there. As a young man he had carried the crusade for geography out into the country, travelling some 30,000 miles (*c.* 48,000 km) and giving 600 extension lectures on physiography and the new geography in the 1880s. The philosophy of geography attached to his name was not so much original as a powerful and coherent reformulation of the views of Freshfield and Keltie, though he also did much to bring to general attention the achievements of German geographers from Humboldt to Richthofen. It is particularly interesting that it was Mackinder, in conjunction with Freshfield and Bentham Dickinson, who was responsible for the foundation of the Geographical Association, for school teachers, in 1893. Yet in spite of his early achievement in the establishment of academic geography, Mackinder remains something of an enigma. Beatrice Webb found him a 'coarse-grained' man, and suggested he had negro blood, and Bertrand Russell (who thought him 'brutal') called him 'the head Beast of the School of Economics'.[18] Mackinder's biographer notes that 'most of the causes he worked for were betrayed or came to nought.'[19] Perhaps his driving ambition and ferocious energy simply overrode his talent, and prevented him from every really cementing the advances of 1887.

Mackinder came to geography with an Oxford background in history, biology and law. H. J. Fleure, my second figure, was for ten years simultaneously professor of zoology and lecturer in geography at Aberyst-

[18] For these comments, see N. and J. MacKenzie, eds, *The diary of Beatrice Webb*, volume 2: *1892–1905* (London: Virago, 1983), p. 252 and *passim*; and B. Russell, *The autobiography of Bertrand Russell 1872–1914* (London: Allen & Unwin, 1967), p. 176.

[19] W. H. Parker, *op. cit.* (note 10), p. 258.

wyth, before translating to a joint chair in geography and anthropology.[20] His catholicity of talent is reflected in over 200 publications; in his fellowship of the Royal Society for his anthropological work; in his concern for teaching reflected by his secretaryship of the Geographical Association and editorship of *Geography* for 30 years; and in his successful struggle to retain recognition of the subject in senior school examinations at the end of the first world war. His major work *The corridors of time* placed him squarely in the great tradition of European humanism, even though he was by training a marine biologist, and it is such bridging of the gap between the arts and the sciences that has always been one of the strengths and attractions of our subject.

S. W. Wooldridge, though a very different personality, had much in common with Fleure.[21] His training was as a geologist, and for his geology and geomorphology he too was elected a fellow of the Royal Society. A long series of research papers, some of them highly technical, did not prevent him from gaining renown as a teacher, not only in the lecture room but also in the field. For him research *was* fieldwork, and through the Field Studies Council and its staff he introduced thousands of schoolchildren to field techniques and field problems.[22] Just as Mackinder and Fleure gave their time to the Geographical Association, so it was largely Wooldridge, in association with R. O. Buchanan, who initiated the moves just 50 years ago which led to the formation of the Institute of British Geographers – partly, it must be admitted, in reaction to the policies and personalities of the R.G.S.[23]

Particular mention must be made of that small but distinguished band of women, lineal descendants of the formidable lady travellers of the 1880s and 1890s, who entered academic geography early in the present century and who had simultaneously to fight for simple equality in the universities:

[20] 'H. J. Fleure, D.Sc., Ll.D., M.A., F.S.A., F.R.S.', *Transactions of the Institute of British Geographers*, 49 (1970), 201–210; A. Garnett, 'Herbert John Fleure 1877–1969', *Biographical Memoirs of Fellows of the Royal Society*, 16 (1970), 253–278; E. G. Bowen, 'Herbert John Fleure and western European geography in the twentieth century', *Geographische Zeitschrift*, 58 (1970), 28–35; J. A. Campbell, 'Some sources of the humanism of H. J. Fleure', *University of Oxford Department of Geography Research Paper*, 2 (1972), 1–45.

[21] J. H. Taylor, 'Sidney William Wooldridge 1900–1963', *Biographical Memoirs of Fellows of the Royal Society*, 10 (1964), 371–388; M. J. Wise, 'Three founder members of the I.B.G.: R. Ogilvie Buchanan, Sir Dudley Stamp, S. W. Wooldridge. A personal tribute', *Transactions of the Institute of British Geographers*, n.s. 8 (1983), 41–54.

[22] M. J. Wise, 'Professor S. W. Wooldridge', *Geography*, 48 (1963), 329–330.

[23] D. R. Stoddart, ed., 'The Institute of British Geographers 1933–1983', *Transactions of the Institute of British Geographers*, n.s. 8 (1983), 1–124; R. W. Steel, *The Institute of British Geographers: the first fifty years* (London: Institute of British Geographers, 1984).

it is startling to recall that at my own University of Cambridge, women were not admitted to degrees on the same terms as men until 1948. Perhaps the most distinguished, independent and productive of all was Professor Eva Taylor, who was born in 1879 and lived through the expansion I have described until her death in 1966. She began as a student and then assistant of Herbertson, and became our leading scholar of the history of geography and of navigation.[24]

And finally there is L. Dudley Stamp, who in large degree and in his time personified geography for a lay public not only here but in many parts of the world.[25] He too was trained as a natural scientist, taking (like Wooldridge) first-class honours in geology and a D.Sc. as a very young man, and then going on to repeat the performance in geography. His first appointment was as a geologist (though he took the opportunity to publish widely in ecology too), and he followed this with the feat of being made reader in economic geography at the London School of Economics while only 28. But this diversity simply demonstrated the breadth of his expertise as a geographer, in which subject he held chairs for the last 20 years of his life. His greatest works were two. The Land Utilization Survey, mapping the land of Britain with 22,000 field assistants in the early 1930s, had the often unremarked outcome that its field directors helped to establish the Ph.D. as a normal and necessary qualification in academic geography; and it also heralded Stamp's lifelong preoccupation with practical matters and planning. The other was his educational work. Beginning with his textbook on *The world*, he produced some 80 texts, generally in multiple editions and often translated into many languages, an achievement which recalls the work of Mackinder's successor, A. J. Herbertson, whose own textbooks had sold by the hundred thousand earlier in the century. It is perhaps fashionable today to decry regional textbooks of the kind Stamp wrote, but I can still recall the pleasure and the stirrings of imagination provoked by reading of Nigeria and New Zealand in Stamp's *Intermediate geography* when I was a boy. Stamp's industry and breadth were reflected in over 400 publications, nearly one-third of them books, and brought him the presidency of the R.G.S. as well as of the International Geographical Union.

[24] 'E. G. R. Taylor: an appreciation', *Journal of the Institute of Navigation*, 20 (1967), pp. 94–101; G. R. Crone, 'Professor E. G. R. Taylor, D.Sc.', *Geographical Journal*, 132 (1966), 594–596.

[25] M. J. Wise, 'Sir Dudley Stamp: his life and times', *Special Publications of the Institute of British Geographers*, 1 (1968), 261–269; R. O. Buchanan, 'The man and his work' [L. Dudley Stamp], *Special Publications of the Institute of British Geographers*, 1 (1968), 1–11.

CHARACTERISTICS

What did these geographers have in common? How far do they reflect the dominant concerns of British geography from 1887 to the middle of the present century? First, they were pragmatic, concerned with practical issues and usually with the physical aspects of the subject, rather than with methods and philosophy. Second, they were pedagogic: all were deeply concerned with education in the schools and with the training of teachers: they wrote textbooks, struggled over what needed to be taught, and examined. Third, they brought to bear on the problems selected for study an almost bewildering range of formal training and interest, obtained before geography itself had become established as a formal discipline, but unified through a shared belief in geographical methods and objectives. Fourth, whether on the physical or the human side, they emphasized fieldwork, especially in the local area and especially in the British Isles – Wooldridge going so far as to grumble that we know more of villages in the Punjab than of villages in Britain and roundly asserting that 'the eyes of a fool are in the ends of the earth.' It is a quotation often made, but elsewhere he lamented the little attention given to the then dependent territories by geographers: we must recall how few in number geographers were until recent years, how little money was available for overseas research, and how difficult world conditions were for study abroad before 1950.) Finally, they were concerned with physical planning and the practical issues of man's relationship to land in increasingly crowded islands. Perhaps it is this very eclecticism and diversity of interest which makes it difficult to define a 'British school of geography' as easily as one might, for example, a French.

ACHIEVEMENT

Where have these achievements led? Let us answer this question first by looking at the simple statistics of educational advance.[26] Since 1950–1951, when data are readily available, the cumulative number of higher degrees awarded in British universities has increased at a remarkably constant rate, doubling every 6.5 years: some 10,000 higher degrees are now awarded

[26] *Index to theses accepted for higher degrees by the universities of Great Britain and Ireland* (London: Aslib, annual since 1950–1951); *Retrospective index to theses of Great Britain and Ireland 1716–1950*, ed. R. R. Bilboul (Oxford: European Bibliographical Centre, 1975–1976); *Statistics of education* (London: H.M.S.O., annual since 1966). See also D. R. Stoddart, 'Growth and structure of geography', *Transactions of the Institute of British Geographers*, 41 (1967), 1–19.

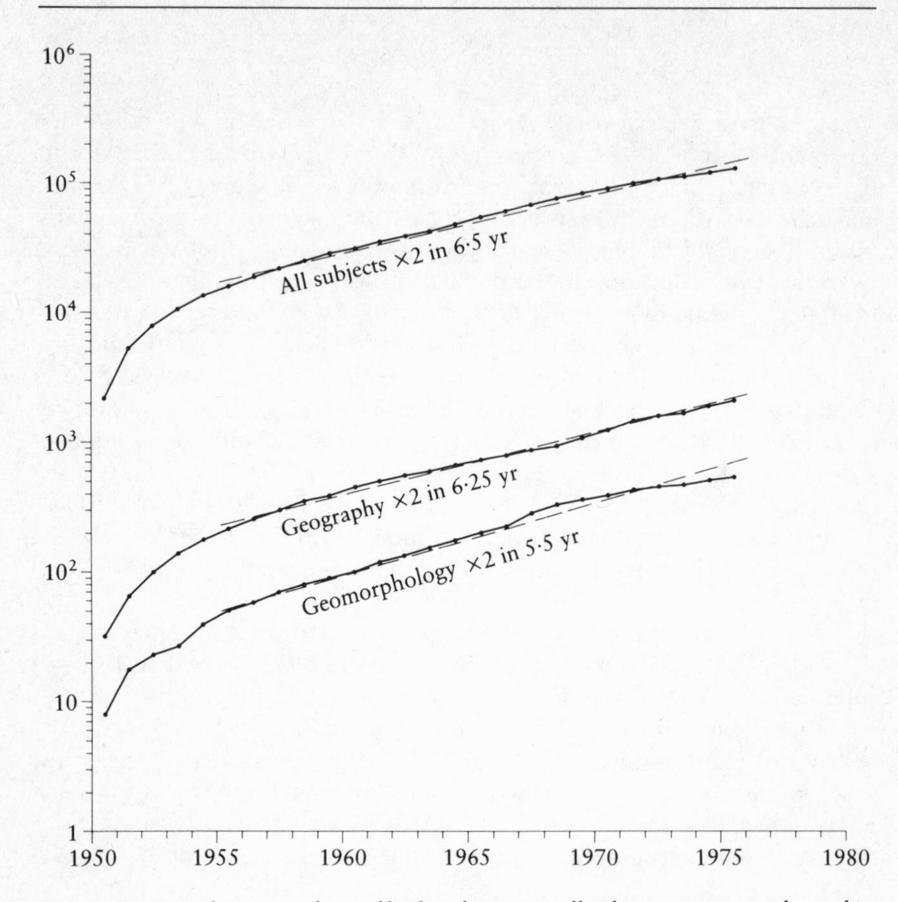

Figure 2 Cumulative numbers of higher degrees in all subjects, in geography and in geomorphology, awarded by British universities since 1950–1951.
Source: *Index to theses accepted for higher degrees by the universities of Great Britain and Ireland* (London: Aslib, annual since 1950-1951).

annually in all subjects (see figure 2). Geography degrees are increasing in cumulative number at the same rate, and are now awarded at the rate of about 200 per annum. Geomorphology (mostly carried out in departments of geography) accounts for 30 to 40 higher degrees each year. Overall, geography accounts for 1.6 per cent and geomorphology for 0.4 per cent of all higher degrees awarded since 1950–1951 (or 2,600 out of a total of 128,500). Very similar trends are apparent in the cumulative totals of

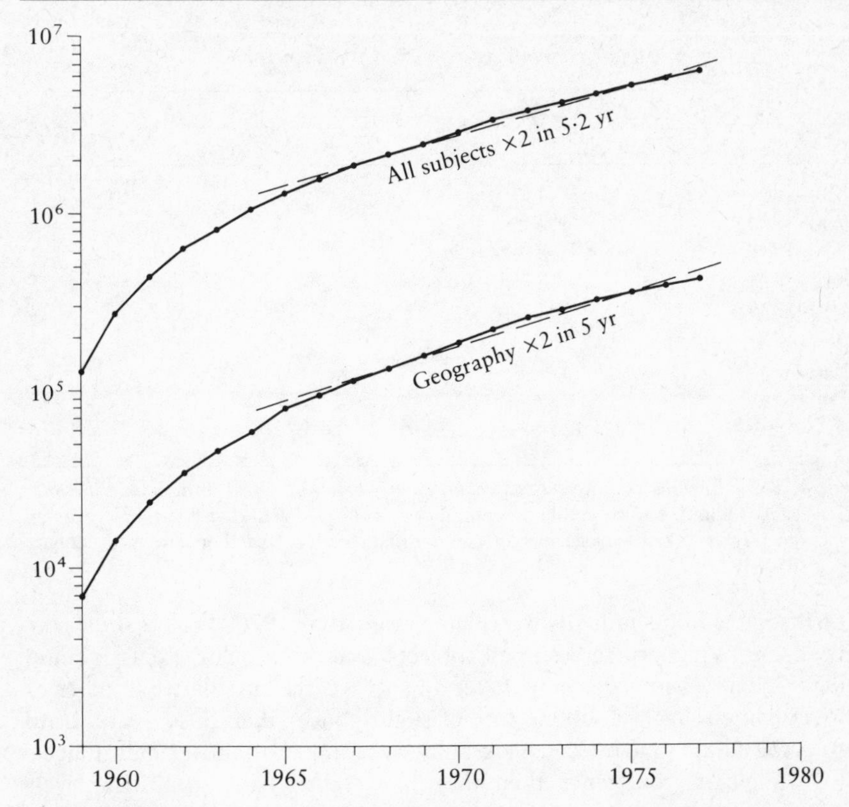

Figure 3 Cumulative numbers of Advanced Level G.C.E. passes in all subjects and in geography since 1964.
Source: *Statistics of Education* (London: H.M.S.O., annual since 1966)

G.C.E. Advanced Level results since 1959 (see figure 3). The cumulative total number of passes has a doubling period of 5.2 years, and of passes in geography of five years. Geography passes now account for 6 to 7 per cent of the total of all subjects, and now number nearly 40,000 per annum in a total approaching half a million.

From the longer set of data summarized in table 1, showing ten-year totals of higher degrees in Britain, it is clear that the first major phase of expansion of the subject in the universities was during the 1930s, spurred on, as we have seen, by the Land Utilization Survey, but with the major development after 1950.

In the United States, where the first doctorates were granted at Yale in

Table 1 Higher degrees in the United Kingdom

Decade	*Geography*	*Geomorphology*	*Total*
1890–1899	1	–	1
1900–1909	1	1	2
1910–1919	8	3	11
1920–1929	18	7	25
1930–1939	105	19	124
1940–1949	66	16	82
1950–1959	391	91	482
1960–1969	679	268	947
1970–1979	1254	222	1476

Source: R. R. Bilboul, ed., *Retrospective index to theses of Great Britain and Ireland, 1716–1950* (Oxford: European Bibliographical Centre, 1975–1976); *Index to theses accepted for higher degrees by the universities of Great Britain and Ireland* (London: Aslib, annual since 1950–1951).

1861, some 35,000 annually were being granted by 1970. Table 2 shows the expansion of higher degrees in all subjects, and of bachelor's, master's and doctor's degrees in geography. Over all subjects the cumulative number of higher degrees has a doubling rate of slightly more than three years. Until 1970, the cumulative rates in geography were similar to those of all subjects (3.0–3.7 years), but since then they have fallen back to British levels (5.5–6.5 years). Nevertheless, of the nearly one and a half million degrees awarded annually in the United States, 4500 are at bachelor's level in geography, 850 at master's level, and 200 at the doctoral level.

Table 2 Higher degrees in the United States

			Geography		
	All degrees	*All doctorates*	*Bachelor's*	*Master's*	*Doctor's*
1947	318,749	3,989	357	157	17
1957	440,304	8,942	849	184	56
1966	871,832	23,091	2,163	463	79
1969	1,072,581	29,872	3,747	637	145
1975	1,344,581	34,076	3,755	655	168

All this represents a quite extraordinary increase, even by the standards of 30 years ago. The expansion is represented in the growth of the R.G.S. itself, which now has over 7000 fellows and members and has never had more. The Institute of British Geographers has nearly 2000 professional members,[27] and the Geographical Association numbers over 8000 teachers. All are flourishing, active and expanding. Geography is now taught in 40 British universities, and employs over 600 faculty members. More is being taught, more is being discovered; and even allowing for the influences of demographic and economic variables on the educational process, geography as a profession is an altogether different creature today from what is was in the time of Mackinder and Fleure, or even of Wooldridge and Stamp. How is all this activity to be embodied in the corpus of knowledge? How are the new discoveries to be diffused through the hierarchy of the education system? Has geography itself changed in its aims and methods during this period of massive expansion and success?

POTENTIAL AND POSSIBILITIES

We have over the past 20 years seen a succession of 'new geographies' follow each other at increasingly short intervals, beginning with the so-called quantitative revolution of the late 1950s. One might be forgiven for thinking that, at the call of 'Charge!', the geographical horsemen thunder past at an increasingly frenetic rate, only shortly to learn that the real onslaught needs to be made elsewhere. Many have certainly felt concerned at both the permanence and the relevance of some of these developments; others have done their best to incorporate what they can at each level of the educational system. It is, I believe, time to look beyond passing fashion and to reaffirm what our founders stood for and what they believed – in other words, to restate our central concerns.

First, we are concerned with earth's diversity; without it, there is no geography. Chimborazo, El Dorado – these are powerful stimuli to the imagination of the young today, and indeed to all of us, just as they were to the mature Alexander von Humboldt. Accounts of the exotic, spectacular and remote have been central to geography since at least the time of Hakluyt and Purchas. I do not see why we should not be proud to acknowledge it and seize advantage from it, and to recognize in the work of explorers in the Royal Geographical Society and elsewhere a parallel achievement to our own more academic pursuits. And it is nonsense to say that exploration is

[27] See the references cited in notes 2 and 22.

over and the record now complete: with little difficulty, one could name areas in every continent where the hand of man has scarcely been felt, and where our knowledge is woefully inadequate.

Second, the use of maps. 'Show me a geographer', said Carl Sauer, 'who does not need them constantly and want them about him, and I shall doubt whether he has made the right choice of life.'[28] I do not understand that attitude to maps which requires them to be locked in drawers like medieval manuscripts. Maps are to be used, cut up, discarded if necessary; above all, they are to be made. Show *me* a geographer who cannot make a map, and I shall doubt *his* credentials. Mapping as an activity goes beyond the simple recording of topography, of course, and may be considered as a broader activity of analysis of patterns and distributions on the earth's surface. As such it includes spatial mathematics and the methods of automated cartography. In restrospect it will seem strange that, in this third decade of the Space Age, we have still made so little use as a profession of satellite imagery and other remote sensing devices: the enormous potential of systems such as Landsat has, on the whole, yet to be recognized by geographers.

Third, fieldwork. Wooldridge and many others in our British tradition have emphasized the need to develop an eye for country, to interpret the landscape in the field, to discern what others cannot see. Fieldwork takes many forms, of course, but John Leighly has suggested to me that one of a geographer's greatest gifts is good and trained eyesight, to apprehend as well as to comprehend. Geography is emphatically not, as Huxley knew so well, a subject to be learned from books. Perhaps we have not entirely avoided the trap which he foresaw in education, where students 'work to pass, not to know; they do pass, and they don't know.' Perhaps our libraries are too well stocked with books simply xeroxed out of other people's books. Geography as a discipline of critical observation ought to be free from this.

I take these concerns with earth's diversity, with maps and with fieldwork as fundamental to our subject, at the elementary as well as at the research level. For, as geographers, we have many characteristics in common. Many of us have, it seems to me, a highly visual imagination: Mackinder, it is said, shaped whole landscapes with his hands while lecturing. We take pleasure in an aesthetic and even emotional appreciation of the landscapes we perceive and project our imaginations on to them. Often we think this to be a modern development, though Humboldt treated it at length in the

[28] C. O. Sauer, 'The education of a geographer', *Annals of the Association of American Geographers*, 46 (1956), 287–299.

Cosmos.[29] And it leads to the sense of wonder which we all must feel, on the coral reef, in the redwood forests, in high mountains or in the desert, and indeed in Los Angeles or Calcutta, Istanbul or Peking, in the realization not only of earth's diversity but of man's ingenuity in coping with it in the process of making a living. Finally, no one can be a geographer without caring profoundly about man's terrestrial inheritance and the ways in which it is used, a theme of conservation which goes back to George Perkins Marsh, and of social concern to Kropotkin and Reclus.[30] We must try to use our skills as best we can, both to alleviate human misery and to work for the conservation and stability of the planet's ecosystems.

We can, of course, add to these concerns the intellectual satisfaction which we derive from our theories and models, from new explanations and understandings. This has always been so, and perhaps we overestimate the novelty of the advances made in these fields during the conceptual and technical developments of the last 20 years: those who think that geographical wisdom began in 1960 should glance at the writings of Cayley, Spottiswoode and Galton in geographical journals a century ago.[31] In retrospect, I wonder greatly at the speed with which some of these technical developments have swept downwards into the schools, with analytical techniques providing puzzles for children to solve and perhaps leaving them little wiser about the world itself. Let us not allow an arid formalism to replace the true attraction of geography, either at the elementary or at the research level. There is, I hope, little chance of this happening: nothing is more heartening than the seriousness and sense with which the role of the new geographies is being examined at the school level within our profession.

I have deliberately not said anything about the philosophy of the subject, and particularly about geography as an integrating discipline. It would need more than this chapter to consider the topic adequately, and, frankly, I am not sure that it would make a great deal of difference. But there is a very practical point to make. Much geographical work is now carried out by national and international organizations – our own Land Resources Development Centre in the Overseas Development Administration, for example, and by such bodies as F.A.O., UNESCO and U.N.E.P. Often these include a diverse collection of natural and social scientists, and I know from

[29] A. von Humboldt, *Cosmos: a sketch of a physical description of the universe*, trans. E. C. Otté (London: H. G. Bohn, 2 volumes, 1848–1849), especially volume 2, '*Incitements to the study of nature*', pp. 370–465. Also E. V. Bunkse, 'Humboldt and an aesthetic tradition in geography', *Geographical Review*, 71, (1981), 127–146, and D. R. Stoddart, 'Humboldt and the emergence of scientific geography', in P. Alter, ed., *Humboldt* (London: Heinemann, 1985).

[30] D. R. Stoddart, 'Kroptkin, Reclus, and "relevant" geography', *Area*, 7 (1975), 188–190.

[31] See chapter 5.

experience that in such teams a geographer can often supply a perspective which few of the others by their training possess. This is a practical kind of integration which is going to increase in scale in the future, and in the longer term it will make a greater impact than much of the argument which goes on about geography at the philosophical level.

The field I have outlined is both enormous and ambitious. 'Geography', said Mackinder, 'is inherently not an elementary but an advanced subject. It postulates both scientific and humane knowledge.'[32] Sauer went even further in his feeling that undergraduate instruction in geography was not a very satisfactory basis for those wishing to do research.[33] It was Mackinder, too, who drew the conclusion that 'the time has come when specialist research and teaching for general education should part company.'[34] We should be clear what he meant. Teaching and research are to a large degree interdependent activities, and most of our great geographers have made this one of the central features of their scholarly lives. But perhaps there is a case for giving the teaching profession some respite from the need to be abreast with every advance on the research frontier, while at the same time relieving the researcher of the constant textbook-writing and examining which took so much of the energies of Herbertson and Dudley Stamp. In saying this I do not wish to diminish the importance of teaching in any way: on the contrary, what is taught in the schools in a sense codifies our accepted knowledge, and, in large degree, the teacher is thus the custodian of truth.

Geographers, I suspect, are born and not made. They come to our universities to learn, not to be taught. Their latent talent needs the stimulus and encouragement of teachers, associates and students, rather than regimentation and contraint. The best research comes from those who strike out on their own and who follow their star into lands of delight. This is what the Royal Geographical Society has always been about, ever since, 150 years ago, its founders drew attention to the universal interest of geography, its practical importance and its potential for what they then called 'rational amusement'. Wooldridge, in his Presidential Address to the Institute of British Geographers 30 years ago, spoke of what he termed 'the high calling of geographer':[35] let us remind ourselves of it, and seek to be worthy in both teaching and research of what he and our precursors have attempted and achieved.

[32] H. J. Mackinder, 'Geography as a pivotal subject in education', *Geographical Journal*, 57 (1921), 376–384.

[33] C. O. Sauer, *op. cit.* (note 28), p. 297.

[34] H. J. Mackinder, 'The music of the spheres', *Proceedings of the Royal Philosophical Society of Glasgow*, 63 (1938), 171–181.

[35] S. W. Wooldridge, 'Reflections on regional geography in teaching and research', *Transactions of the Institute of British Geographers*, 16 (1950), 1–11.

4

The R.G.S. and the 'New Geography'

The foundation of the Royal Geographical Society in 1830 can, in retrospect, be seen as but one among a variety of moves in the early decades of the nineteenth century by people of like interests to recognize and organize their fields of study, and ultimately to delineate a territory within the universe of knowledge within which each alone held sway. The Linnean Society (founded in 1788) was something of a precursor in this movement, though at least initially (in the aftermath of Cook's voyages) its scope was broader than at present. The R.G.S. itself was preceded by societies such as the Geological in 1807, the Civil Engineers (1818), the Royal Astronomical (1920), the Royal Asiatic (1823) and the Zoological (1826). It was soon to be followed by the Entomological, the Statistical and the Microscopical, and in 1841 by the Chemical. Many of these societies have broadly similar histories, and the R.G.S. has as much in common with them as it has with the more often cited sister societies in Paris (1821), Berlin (1828) and New York (1852).[1]

Given the continued role of the R.G.S. in the emergence of professional geography in the nineteenth century, it is pertinent to ask questions about social as well as simply society history. Such questions can be stated much more readily than they can be answered. Who were the members who considered themselves to be geographers? How did the membership change as the century progressed? What, indeed, was meant by geography itself for those who assumed the label? How did the Society accommodate itself to external social pressures as well as to the better-known personal rivalries and intrigues of the early years (so well, if so decorously, documented by Hugh Robert Mill[2])? Above all, how was the territory of geography

[1] D. R. Stoddart, 'Growth and structure of geography', *Transactions of the Institute of British Geographers*, 41 (1967), 1–19.

[2] H. R. Mill, *The record of the Royal Geographical Society 1830–1930* (London: Royal Geographical Society, 1930).

delimited and defended against the predatory claims of rival subjects, especially in the natural sciences, at a time when subject boundaries were in a constant state of flux? In this chapter, I do not pretend to answer these questions. I will, however, show how the Society changed in subtle yet significant ways in the 70 years from 1830 to 1900, and I will show how leading figures, such as Freshfield and Keltie, were instrumental in claiming for geography a subject matter, a method and a role, against the powerful claims of sister sciences.

THE FELLOWSHIP

John Scott Keltie notes that the original 460 members of the Society in 1830 were 'almost entirely . . . men of high social standing', and that 'everybody who was anybody was expected to belong.'[3] Names such as Castlereagh, Peel and Lord John Russell were included among the 3 dukes, 9 earls, 6 viscounts, 3 marquises, 14 miscellaneous lords, and 38 knights and baronets in the original list. It was a pattern maintained, with the addition of some maharajahs and sultans, to the end of the century (see table 3). The Society in origin was thus a social network of a very specific kind, and differed in this respect from the more severely technical societies such as the Geological and Astronomical. To this extent, it long retained the mark of its origins in the African Association and the Raleigh Club.

Table 3 Composition of the fellowship of the R.G.S.

	1830	*1872*	*1900*
Total	460	2387	4031
Duke	3	7	6
Earl	9	20	21
Other peers	24	45	81
Baronets, knights	38	112	169
Naval officers	32	133	142
Army officers	55	289	552
F.R.S.	124	105	44
F.G.S.	19	18	15

[3] J. S. Keltie, 'Thirty years' work of the Royal Geographical Society', *Geographical Journal*, 49 (1917), 350–372, discussion 373–376; reference on p. 350.

Secondly, it had a great military emphasis throughout the nineteenth century. The original list contained 55 army and 32 naval officers – 19 per cent of the total. The proportions remained stable through the century (18 per cent in 1872, 17 per cent in 1900), though the absolute numbers grew as the Society expanded to a total of over 4,000 fellows at the beginning of the new century. However, the early naval emphasis, established by Sir John Barrow and Admiral Smyth, gradually gave way, especially following the abolition of the Bombay Navy in 1862: over the first 70 years the number of army officers increased by a factor of 10, but of naval officers only by a factor of 4.4.

Both of these components of the fellowship suggest a somewhat amateur, if not dilettante, approach to a subject not yet established in professional terms. Indeed, it is a commonplace criticism of the Society in its early decades that it was concerned with exploration and travel to a degree detrimental to other more scientific or educational pursuits. It was an emphasis fostered by successive presidents. Mill[4] records that Murchison returned from Russia in 1845 'strengthened . . . in his confidence in his own powers and in importance as a member of high society . . . with a trace of arrogance in his manner towards humbler folk . . . [and an] amiable weakness for honours and decorations'. Murchison's foibles are well known, but the same point may be illustrated by reference to Mountstuart Elphinstone Grant Duff, president 1889–1893, who after he retired as governor of Madras (C.I.E., G.C.S.I.) devoted himself 'to the cultivation of social amenities. . . . He was in the habit of meeting or corresponding with almost everyone of any eminence in social life in England. . . . He was a member of almost every small social club of the highest class . . . [and] frequented the society of rulers, ambassadors, authors, and other remarkable people.'[5] It is said of him that when he presided over council, he used to imagine himself back in the gubernatorial chair in Madras.[6] He regularly attended the Society's meetings in full evening dress so as not to be delayed in rushing off to royal and similar functions when the meetings ended.

But there was another and ultimately more important component to the R.G.S. than this. The list of fellows in 1830 contains noteworthy names in the sciences, and not only the hydrographers and naval explorers such as Beechey and Belcher, Beaufort and Hall, King and Owen. They included the cartographer John Arrowsmith; the botanists Bentham, Brown and

[4] H. R. Mill, *op. cit.* (note 2), pp. 54, 64.

[5] H. Stephen: 'Sir Mountstuart Elphinstone Grant Duff', *Dictionary of national biography*, 2nd supplement, volume 2, 150–151.

[6] J. S. Keltie, *op. cit.* (note 3), p. 355.

Hooker; the surveyors Everest and Sabine; and especially a distinguished roll of geologists – Broderip and de la Beche, Buckland and Fitton, Greenough, Murchison and Sedgwick. And they were joined both by distinguished academics (such as Whewell) and by amateurs (such as Lubbock) whose talents can hardly now be categorized precisely in an age of specialization. In this first list, there were at least 124 fellows of the Royal Society and 19 of the Geological Society. Of the great figures of nineteenth-century natural science, Darwin was elected in 1838 on his return from the voyage of the *Beagle*, as, later, were both Wallace and Huxley.

Yet if the Society began as a club for travellers and explorers, supported by gentlemen and made intellectually respectable by the scientists, it could not long resist the social pressures of the developing century.[7] The strains to which these led were perhaps most readily apparent in the bitter arguments over the admission of women, and here again the R.G.S. was not unique (though to its credit the Royal Scottish Geographical Society had admitted them since its foundation in 1884, and it was pointed out in *The Times* at the height of the controversy that women were already allowed into the Zoological, Botanical, Statistical, Asiatic, Hellenic, Anthropological and 'every other geographical society in the empire').[8] Women had been allowed into meetings of the Geological Society since 1860, but a formal proposal to elect them as fellows was defeated in 1899 and they were not accepted until 1904. The Linnean would not have them until 1905. The issue was first discussed in the R.G.S. in 1887 – the year of Mackinder's paper on the 'new geography' – but came to a head with raucous commotion in 1892–1893, when the council's action in electing ladies was attacked by the future Lord Curzon and a variety of retired admirals and generals and finally disavowed by a general meeting: Curzon ranked some lady travellers among 'the horrors of the latter end of the nineteenth century'. Both the president and the secretary resigned, the latter (Freshfield) going off to found and preside over the Geographical Association. His waspish lines[9] demonstrate the tensions the issue aroused:

> The question our dissentients bellow
> Is 'Can a lady be a Fellow?'
> That, Sirs, will be no question when
> Our Fellows are all gentlemen.

[7] M. Gowing, 'Science, technology and education: England in 1870', *Notes and Records of the Royal Society of London*, 32 (1977), 71–90.
[8] *The Times*, 29 May 1893.
[9] H. R. Mill, *An autobiography* (London: Longmans, Green, 1951), p. 96.

Punch danced on the sidelines with its song to the Royal Geographical Society:[10]

> A lady an explorer? a traveller in skirts?
> The notion's just a trifle too seraphic:
> Let them stay and mind the babies, or hem our ragged shirts;
> But they mustn't, can't, and shan't be geographic.

And in 'The admiral's room':[11]

> Let 'em darn socks, boil taters, or make tea,
> But out from *us* they go! What can she-creatures know
> Of the science of Ge-o-gra-phee?

Women were, in fact not finally admitted to the fellowship as of right until 1913, ironically under Curzon's presidency, and but 15 years before they were nationally enfranchised.

GEOGRAPHY AND EDUCATION

These controversies were but part of wider social and economic changes, however. The Great Exhibition of 1851, which alerted the nation to European technical achievement, was soon to be followed by *On the origin of species*, which radically transformed man's view of the natural world and of how knowledge about it might best be organized. The voyage of the *Challenger* during 1872–1876 gave considerable impetus to the scientific movement in geography, both in the British Association[12] and in the R.G.S.[13] The R.G.S. went so far as to agree a grant of £500 for the promotion of special scientific branches of geography in 1876, on the proposal of Strachey and supported by scientists of the distinction of Galton, Hooker, Darwin, Wallace, Huxley and Murray. These developments coincided with the Education Act of 1870, which created the machinery of national education and with it a demand for teachers and

[10] *Punch*, 10 June 1893.

[11] *Punch*, 17 June 1893.

[12] R. Strachey, 'Address to the Geographical Section of the British Association, at Bristol, August 26th, 1875', *Proceedings of the Royal Geographical Society*, n.s. 20 (1876), 79–89; F. J. O. Evans, 'Address delivered on the opening of the Geographical Section, at the Glasgow meeting of the British Association, September 7th, 1876', *Proceedings of the Royal Geographical Society*, n.s. 21 (1877), 66–77.

[13] R. Strachey, 'Introductory lecture on scientific geography', *Proceedings of the Royal Geographical Society*, n.s. 21 (1877), 179–203.

textbooks, leading to the upsurge in popularity of 'physiography' following the publication of Huxley's book of that title in 1877[14]; and with the Devonshire Commission of 1870–1875, which focused national anxieties on the lamentable state of scientific and technical education. It was in this context that the last three decades of the century saw the emergence of the 'new geography' in Britain; a growing professionalization of the subject, especially within the R.G.S.; and a growing movement to establish geography as a formal university discipline.

Within the Society the prime movers in the field of geographical education were Francis Galton and Douglas Freshfield.[15] Galton's main initiative – a gold medal in geography for boys at certain selected public schools – did not produced the required results:[16] a large number of the medals went to Liverpool College, of which the headmaster was his brother-in-law, and there was no evidence of any general improvement in standards. The headmaster of Bath College said of two of his own students who had been awarded medals that 'they were boys of singular inaptitude for studies of a nobler sort, and he could not but think . . . that he had been indulging them . . . in a weakness that he ought to have corrected.'[17] After operating for 15 years, the scheme was discussed in council in 1884. Galton, 'habitually in committee more critical than constructive', 'was ever at hand to point out ingenious objections to schemes that were substitutes for his own, which had admittedly failed'.[18] After some acrimony Galton retired from the fray and the medals were abandoned.

The Society's educational initiative was taken up, transformed and brought to success in the universities,[19] and the field of school education was increasingly left to the teachers, especially after the foundation of the Geographical Association. It is perhaps worth remembering, however, that universal education was by no means at that time a generally accepted goal, even among those whose liberal and intellectual credentials were beyond reproach. Hooker exclaimed in exasperation in 1892 that he could not see

[14] See chapter 8.

[15] M. E. Grant Duff, 'Retrospective view of educational efforts. The annual address on the progress of geography: 1891–92', *Proceedings of the Royal Geographical Society*, n.s. 14 (1892), 353–382.

[16] F. Galton, *Memories of my life* (London: Methuen, 1908), pp. 211–212.

[17] T. W. Dunn, 'Discussion', *Proceedings of the Royal Geographical Society*, n.s. 9 (1887), 166–167.

[18] D. W. Freshfield, 'Obituary: Sir John Scott Keltie', *Geographical Journal*, 49 (1927), 285–286; reference on p. 285.

[19] See chapter 5; also D. I. Scargill, 'The R.G.S. and the foundation of geography at Oxford', *Geographical Journal*, 142 (1976), 438–461.

'why on earth I am to be taxed to have every stupid child taught . . . the geography of every place under the sun'.[20] And both Sir Charles Warren[21] and Sir Halford Mackinder[22] foresaw with remarkable prescience the consequences for the old social and economic order of the increasingly rapid spread of knowledge to all sectors of the population.

The driving force behind the university initiative, as behind the admission of women, was Douglas Freshfield. Tom Longstaff, in his remarkable memoir of Freshfield,[23] refers to the establishment of geography at Oxford and Cambridge as 'perhaps the greatest work we have accomplished, for exploration would have gone on somehow even had our Society never existed' – a view Freshfield himself shared.[24] It was largely Freshfield who persuaded the Society in 1884 to commission John Scott Keltie to report on the state of geographical education in Britain. Keltie's powerful and comprehensive document was completed in late 1885.[25] Its publication coincided in December and January 1885–1886 with a series of public lectures by men of national distinction and a public exhibition of maps and appliances. By what was in retrospect the happy coincidence of the death of the Society's librarian from smallpox in 1885 and of its secretary from influenza in 1892, Keltie's services were obtained for the R.G.S. for 30 years.[26] Before considering the content and reception of the Keltie Report, however, we must first note the changing circumstances of geography itself, both within and outside the Society.

CHANGES IN THE R.G.S.

Without question the dominant and dominating figure in the Society at mid-century had been Sir Roderick Murchison, twice president of the

[20] L. Huxley, *Life and letters of Sir Joseph Dalton Hooker, O.M., G.C.S.I.* (London: John Murray, 2 volumes, 1918), reference in volume 2, p. 329.

[21] C. Warren, 'Address', *Report of the 57th Meeting of the British Association for the Advancement of Science 1887* (1888), 785–798, reference on p. 793.

[22] H. J. Mackinder, 'Geography as a pivotal subject in education', *Geographical Journal*, 57 (1921), 376–384; reference on p. 383.

[23] T. G. Longstaff, 'Douglas Freshfield 1845–1934', *Geographical Journal*, 83 (1934), 257–262; reference on p. 259.

[24] D. W. Freshfield to M. Conway, 6 March 1921: Conway Papers, Cambridge University Library Add. 7676/N20.

[25] Royal Geographical Society, *Report of the proceedings of the Society in reference to the improvement of geographical education* (London: John Murray, 1886).

[26] H. R. Mill, 'Obituary: Sir John Scott Keltie', *Geographical Journal*, 49 (1927), 281–287; reference on p. 282.

Geological Society and president of the R.G.S. in 1843–1845, 1851–1853 and finally in 1862–1871.[27] Murchison's impact, especially towards the end, however, was social rather than scientific, and when he died the Society changed rather rapidly in two main ways. First, its direction passed either to much younger colonial administrators (Frere, Strachey, Grant Duff, Rawlinson, Goldie) or to men of affairs (Lords Dufferin, Aberdare, Northbrook, and the Duke of Argyll), and the latter group at least had comparatively small impact on its fortunes and policies. Secondly, geologists disappeared almost entirely not only from the Society's council but even from the Society itself: in 1872 only 18 out of 2,448 fellows were also fellows of the Geological Society. This revolution, hitherto largely unremarked, reflects a sudden self-awareness in both subjects of their discrete roles, formalized in the new structure of education and research emerging in the final decades of the century. Nor were the geologists alone in their withdrawal, for scientific men in the broadest sense almost disappeared after 1870. Francis Galton served continuously on council, but as an inspired if cantankerous amateur rather than as a professional scientist. Working natural scientists in the last 30 years of the century are represented on council only by Sclater, Hooker, Seebohm, Ball, Lubbock, and – the only real geologist – W. T. Blanford, former director of the Geological Survey of India. After Murchison's death in 1871, Blanford was the only geologist concerned with R.G.S. affairs until the first decade of the new century brought in Bonney, Garwood, Watts and Strahan. And to mention these names is itself sufficient to indicate the social and professional gulf which separated their geology from Murchison's.

Thus by 1900 only 44 fellows of the R.G.S. were also fellows of the Royal Society (a fall from 27 per cent in 1830 to 1 per cent), and only 15 were fellows of the Geological Society (a fall from 4 to 0.4 per cent). Partly this reflected a widespread disenchantment with the social flummery and even ill-judgement which the Society too often showed over African explorers, especially in Murchison's day: Hooker, disgusted, came to 'dislike and despise the Geogr. Soc.' for its 'lion-hunting, toadying and tuft-hunting'[28] and both Darwin and Wallace thought similarly.[29] Partly, too, it reflected

[27] A. Geikie, *The life of Sir Roderick I. Murchison, K.C.B., F.R.S.* (London: John Murray, 2 volumes, 1875), see volume 1, pp. 290–306; E. W. Gilbert and A. S. Goudie, 'Sir Roderick Impey Murchison, Bart., K.C.B., 1792–1871', *Geographical Journal*, 137 (1971), 505–511; J. A. Secord, 'King of Siluria: Roderick Murchison and the imperial theme in nineteenth-century British geology', *Victorian Studies*, 25 (1982), 413–442.

[28] L. Huxley, *op. cit.* (note 20), volume 1, pp. 406–407.

[29] H. R. Mill, *op. cit.* (note 2), p. 80.

deep-seated changes in the composition of bodies such as the Royal Society itself, after the mid-century reforms. But above all, it reflected a crystallization of allegiances as a result of the emergence of professionalism in the aftermath of the Darwinian revolution, not simply in geography but across the field of knowledge. Without this realization, it is impossible to understand either the nature or the significance of the debate attending the emergence of the 'new geography'.

Hence we have the paradox that in the 1880s and 1890s, British geography emerged as a scientific discipline without the benefit of aid from strictly scientific men. The paradox is reinforced when we realize that the leading figure in British geology in that period was a geologist who professed himself to be a physical geographer, yet who apparently never played any significant part in organized geography. Archibald Geikie, born in 1835, was himself a protégé of Murchison and occupied the Murchison Chair of Geology at Edinburgh from 1871 to 1882. He was elected a fellow of the Royal Society at the age of 30, and was the last earth scientist to be its president. Both in this position and as director of the Geological Survey from 1882 to 1901, his powers of patronage and influence were enormous. Just as the debate on geography and education was at its height, he published a manual on *The teaching of geography* (1887), and his immensely influential *Physical geography* (1873) sold over half a million copies. In 1879, on the twentieth anniversary of *On the origin of species*, he lectured to the R.G.S. on 'Geographical evolution'.[30] His honours included the K.C.B. in 1907 and the Order of Merit in 1913. Yet in spite of being the most powerful earth scientist of his day, he remained aloof from the Society, was never a fellow, never served on its Council, never held office, never received any of its honours. Like the Sherlock Holmes story of the dog that barked in the night, an understanding of Geikie's position, be it cause or be it effect, is crucial to an understanding of the development of academic geography in its formative years.

GEOGRAPHY AND GEOLOGY

The great expansion in university geology in Britain came after 1860, preceding that in geography by two decades, with chairs established at

[30] A. Geikie, 'Geographical evolution', *Proceedings of the Royal Geographical Society*, n.s. 1 (1879), 422–443. On Geikie as geographer, see W. E. Marsden, 'Sir Archibald Geikie (1835–1924) as geographical educationist', in W. E. Marsden, ed., *Historical perspectives on geographical education* (London: University of London Institute of Education, 1980), 54–65.

Leeds, Birmingham, Manchester, Sheffield, Newcastle and Bristol, and rejuvenated at Oxford and Cambridge. Porter[31] has shown how the heavy teaching loads associated with these new duties, combined with increasing technical specialization within geology itself, transformed the subject from one concerned with fieldwork and mapping to one concerned with laboratory study, lecturing, examining and textbook writing. The period of expansion in geography in the 1880s also coincided with severe retrenchment in the Geological Survey – by far the largest career outlet for geologists – where staff were cut from 73 in 1880 to 34 in 1885.[32]

The structure of the Geological Society also reflects growing professional concerns.[33] After Huxley, their presidents were increasingly research geologists rather than social figures (though the Duke of Argyll makes an appearance here, as in the R.G.S.). They included Prestwich, Duncan, Sorby, Bonney, Judd, Blanford, Geikie, Woodward, Whitaker, Teall, Lapworth and Marr. Honours, too, tended to go strictly to working geologists, in contrast to the catholicity of choice of the R.G.S. Of the medallists after 1870, only Richthofen (who took the Wollaston medal in 1892) might be considered a geographer, though Lake received a moiety of the Murchison Award in 1896.

Hence, as geology became more professional, it also became more parochial and less able to reach out to wider scientific fields. Geologists became known as mineralogists, petrologists and palaeontologists rather than as field surveyors and explorers or, indeed (as they recently had been) as interpreters of the frontiers of scientific advance. It is, therefore, perhaps not surprising that geologists did not view the moves for an autonomous geography in the universities and schools with marked enthusiasm. Their own job opportunities were rapidly shrinking; their own opportunities for research had been dramatically cut by growing teaching loads; and technical developments in their subject were removing it from the comprehension of the general public. And it was precisely that section of geology – the study of landforms – which had previously had widest popular appeal and least technical content that the geographers proposed to usurp. The magnitude of the threat must have been recognized in the astonishing success after 1877 of the new subject of physiography.

[31] R. Porter, 'The natural sciences tripos and the "Cambridge School of Geology", 1850–1914'; R. Porter, 'Geology and the universities in Britain 1850–1914'; R. Porter, 'Gentlemen and geology: the emergence of a scientific career, 1660–1920', *Historical Journal*, 21 (1978), 809–836.

[32] E. B. Bailey, *Geological survey of Great Britain* (London: Thomas Murby, 1952), p. 105.

[33] H. B. Woodward, *The history of the Geological Society of London* (London: Geological Society of London, 1907).

Geologists, in fact, regarded both the subject matter (landforms) as part of geology, and the method (fieldwork and mapping) as peculiarly geological. Hence, as Keltie recognized:[34] 'the strongest opposition to geography as a distinct field of research comes from the side of geology: the pure geologist, especially, is unwilling to admit that geography has any existence apart from geology, which simply shows his ignorance of the great fields that have been worked by Continental geographers.' Freshfield[35] called geology 'the most forward of the would-be "chuckers-out" of geography from the Hall of Education', and he argued with some bitterness against its rival claims:

> Archaeology might as well claim to be equivalent to history, history to political economy, or surgery to medicine, as geology to geography. A geologist is as incapable of giving any complete picture of a country, as an archaeologist of a period. A geological diagram is quite a different thing from a geographical picture, as different as an anatomical diagram from a portrait. His science leaves off where ours begins.

Keltie, in a public lecture in December 1885, reinforced Freshfield's point about professors of geology:

> They seem to think that when they have made a few cross-country sections, mapped the outcrop of rocks, noted the action of water, and rain, and atmosphere, catalogued the animals and plants of the recent period, they have done with geography. That is only the beginning of the subject, the basis on which the varied superstructure has to be reared.[36]

The problem was compounded by a particular difficulty which beset the geographers. Geography itself appeared vague and diffuse, part belonging to history, part to commerce, part to geology. How could some coherence be found which would persuade the universities of its value, answer Sir George Darwin's charge that he could not see 'how geography pure and simple could be made a subject of intellectual training',[37] and overcome the 'contemptuous hostility' as well as indifference which Ravenstein[38] claimed

[34] Royal Geographical Society, *op. cit.* (note 25), p. 31.

[35] D. W. Freshfield, 'The place of geography in education', *Proceedings of the Royal Geographical Society*, n.s. (1886), 698–714; reference on p. 704.

[36] J. S. Keltie, 'On appliances used in teaching geography', in Royal Geographical Society, *op. cit.* (note 182–198, discussion 199–203; reference on p. 182.

[37] G. Darwin, 'Discussion', in Royal Geographical Society, *op. cit.* (note 25), 106.

[38] E. G. Ravenstein, 'The aims and methods of geographical education', Royal Geographical Society, *op. cit.* (note 25), 163–176, discussion 176–181; reference on p. 163.

to detect? The strategy adopted was to abandon historical and commercial geography, at least in the short term, and to concentrate on physical – or 'scientific' – geography. But it was exactly that which made the conflict with the geologists most acute. 'All the best geographers', wrote Keltie,[39] 'have approached the subject from the side of science,' whereas 'the historian pure and simple' – doubtless emphasizing the simpleness rather than the purity – 'has seldom done much effective work.'

Moseley, the naturalist on the *Challenger*, was most explicit. In a public lecture he was clear about the drawbacks of a general geography and the vagueness with which the subject was defined. There was, however, a chance of success with physical geography in the universities. The fact that geologists like Geikie, McKenny Hughes and Boyd Dawkins already lectured on physical geography showed that there was a need and that it could be done; at the same time, geology itself was getting too big for one man to handle. Having held out the olive branch, Moseley then prostrated himself by admitting that if physical geography could be accepted in the universities, 'it is obvious that any professor who could hold such a chair must be a geologist.'[40] He was speaking as a distinguished zoologist who did not, perhaps, have much sympathy with some of the more extreme claims the geographers were making.

His argument was persuasive but, to the sceptical geologists, not compelling. McKenny Hughes, Woodwardian Professor of Geology at Cambridge, who had taken offence at some passages in Keltie's report,[41] argued on the one hand that his own department did all the necessary lecturing and fieldwork in physical geography already, and on the other that if 'physical geography' was proposed, it was a subject for schools rather than for university. The university was a place to specialize, whether in history, zoology or geology, and to use a knowledge of geography already obtained at school. He was willing to concede, with scarcely veiled contempt, that general geography could be 'a useful and interesting subject . . . for the politician, the traveller, and the man of leisure': but that was not what the debate was about.[42] At least Hughes reinforced his claims by actually doing some geography, for he later wrote the Cambridgeshire volume in Guillemard's series of *Cambridge county geographies*. Archibald

[39] J. S. Keltie, in Royal Geographical Society, *op. cit.* (note 25), 32.

[40] H. N. Moseley, 'Scientific aspects of geographical education', in Royal Geographical Society', *op. cit.* (note 25), 225–231, discussion 231–236; reference on pp. 227–229.

[41] F. Galton to T. McK. Hughes, 27 November 1885: McKenny Hughes Papers, Cambridge University Library Add. 5944(22).

[42] T. McK. Hughes, 'Discussion', in Royal Geographical Society, *op. cit.* (note 25), 233–234.

Geikie simply accepted the argument that Moseley put for scientific geography, but without its corollary that new teaching posts were needed: 'In its true sense geography is a branch of geology; and where there is already a Professor of Geology, I would not, for the present at least, advocate the establishment of a separate Chair of Geography.'[43]

It is worth noting that this debate was carried on without any reference to papers of extraordinary prescience in theoretical geography which had appeared in the Society's journals or had been authored by its fellows. Perhaps the first of these was Spottiswoode's paper 'On typical mountain ranges: an application of the calculus of probabilities to physical geography';[44] another was Cayley's 'On the colouring of maps',[45] a classic attempt at the four-colour problem which still diverts mathematicians; and another (among many), Ravenstein's paper on the cartographic depiction of distributions, for example of Roman Catholics in Britain.[46] And this is to say nothing of more purely pragmatic researches, such as Buchanan's[47] on spatial variations in the distribution of oceanic salinity. These pioneer papers were not explicitly physical in derivation or methodology, and all were overlooked for decades. For too long 'scientific geography' continued to mean simply the observation and delineation of topography, in the sense of Sir Richard Strachey.[48]

Not all geologists reacted in the same way to the claims of the geographers. Some, such as Hughes, were actively opposed to an independent university geography. Some, such as Archibald Geikie, while apparently sympathetic, failed to become actively involved in its support. There were those like the petrologists and palaeontogists who were quite uninterested and who played no part in the debate. There were those, like Philip Lake, who started as geologists (Lake began his career in the Geological Survey of India, and was rapidly invalided home) but who failed to secure geological appointments and who ultimately became lecturers in

[43] A. Geikie, 'Discussion', in Royal Geographical Society, *op. cit.* (note 25), 32.

[44] W. Spottiswoode, 'On typical mountain ranges: an application of the calculus of probabilities to physical geography', *Journal of the Royal Geographical Society*, 31 (1861), 149–154.

[45] G. Cayley, 'On the colouring of maps', *Proceedings of the Royal Geographical Society*, n.s. 1 (1879), 259–261.

[46] E. G. Ravenstein, 'Statistics of Roman Catholicism in Great Britain', *Geographical Magazine*, 1 (1874). 102–106.

[47] J. Y. Buchanan, 'On the distribution of salt in the ocean, as indicated by the specific gravity of its waters', *Journal of the Royal Geographical Society*, 47 (1877), 72–86.

[48] R. Strachey, *op. cit.* (note 13); J. S. Keltie, 'Some geographical problems', *Geographical Journal*, 10 (1897), 308–323 (also as 'The function and field of Geography', *Smithsonian Institution Annual Report*, 1897, 381–399).

geography *faute de mieux*. And there were also those who saw a reciprocal benefit to both sides through collaboration. One such was Lapworth. At a British Association for the Advancement of Science discussion in 1893 (at which Geikie somewhat astringently rejected Markham's division of roles of the two subjects), Lapworth looked forward 'to the time when the physical geographer, who knows nothing of stratigraphical geology, will be consigned by the geologist to his museum of extinct monsters, and the geologist who has no knowledge of and gratitude to physical geography, will be looked upon by the geographer as a candidate for a lunatic asylum'.[49] None of the protagonists, of course, considered that geography extended beyond the physical.

But there was also another, more personal component in the confrontation, the extent of which must remain to some degree conjectural. The Survey geologists, especially in the 1870s, were generally fairly rugged outdoor types, typified by Joseph Beete Jukes and James Geikie, bearded, perhaps somewhat uncouth (Jukes managed to kill himself by falling downstairs one night in a public house in the south of Ireland),[50] with their own forthright criteria of competence and efficiency. The first academic geographers were, in general, the antithesis to this. Guillemard, the first lecturer at Cambridge, was a somewhat precious and affected aesthete, a chronic invalid, exquisitely dressed, a connoisseur of wine, old silver and watercolours, and a friend of Housman.[51] His successor, Buchanan, an extremely wealthy man, was cold, difficult and reserved, with no gift for friendship, and with a character that led to his exclusion from contributing to the reports of the *Challenger* expedition on which he had been chemist. The third lecturer (and first reader), Oldham, was a weak and incompetent man who finally had to be replaced. Lake, the fourth, while intellectually distinguished, was shy and retiring. All four were lifelong bachelors, though Guillemard made a brief and disastrous marriage to an invalid cousin who left him to become a nun. On a purely personal level, it is difficult to imagine the geologists having a great deal of sympathy for such people.

Halford Mackinder, the most successful of the 'New Geographers', in a sense also epitomized weaknesses of another kind. It was said of him 'by

[49] C. Lapworth, 'The limits between geology and physical geography (discussion)', *Geographical Journal*, 2 (1893), 518–534.

[50] C. A. Browne, *Letters and extracts from the addresses and occasional writings of J. Beete-Jukes . . . with connecting memorial notes* (London: Chapman & Hall, 1871). Also P. C. Beaton, 'A martyr to science', *Good Words*, 6 (1868), 425–429.

[51] O. J. R. Howarth, 'The centenary of Section E (Geography)', *Advancement of Science*, 8 (1951), 151–165; reference on p. 160.

historians that he knew no history, by geologists that he knew no geology, by climatologists that he knew no climatology':[52] ultimately his career lay not in the intellectual development of the subject or in the extension of knowledge but in political life, both as an academic entrepreneur (as principal at Reading and as director of the London School of Economics and Political Science) and in the House of Commons.[53]

There was, however, one exception to this generalization of the lack of sympathy on personal level between geologists and geographers. Douglas Freshfield, who had insisted while on the council of the R.G.S. on the importance of geography in education as well as on the admission of women as fellows, was an outstanding alpinist and had also explored in the Caucasus and in the Himalaya: he was largely responsible for the development of the cult of mountaineering in Britain towards the end of the century. Freshfield was a wealthy man, with no pretensions to being a scientist, but because he was a mountaineer he was able to meet at least some of the geologists on their own ground.[54] It is perhaps not surprising that the one geologist of the new academic breed who was fully accepted in the R.G.S. was Edmund Garwood, who held the chair at University College London from 1901 to 1931. In addition to being an alpinist, Garwood, who spent much of his time at his villa on Lake Como, was wealthy, sophisticated, played the cello, and was on close terms with Elgar and other composers:[55] it is said that Archibald Geikie kept him out of the Royal Society for years. The power still exerted by the fraternity of explorers and mountaineers, few of whom had the vision of Freshfield, can be recognized when we recall that Mackinder, the first reader in geography at Oxford, deliberately set out to be the first man to climb Mount Kenya in order that his ideas on the nature of geography might command respect: 'At that time',

[52] B. W. Blouet: 'Sir Halford Mackinder 1861–1947: some new perspectives', *Oxford University School of Geography Research Papers*, 13 (1975), 1–49; D. R. Stoddart, 'Mackinder: myth and reality in the establishment of British geography', *Preprints International Geographical Union Commission on the History of Geographical Thought, Leningrad Symposium* (August 1976); W. H. Parker, *Mackinder: geography as an aid to statecraft* (Oxford: Clarendon Press, 1982).

[53] See p. 48.

[54] For the two aspects of Freshfield's work, see E.L.S., G. Yeld, H. G. Willink and E. J. Garwood, 'Douglas Freshfield (1845–1934)', *Alpine Journal*, 46 (1934), 166–176; and L. J. Jay, 'Douglas Freshfield's contribution to geographical education', in W. E. Marsden, ed., *op. cit.* (note 30), 43–53.

[55] T. G. Longstaff, 'Edmund Johnston Garwood 1864–1949', *Alpine Journal*, 57 (1949), 239–240; see p. 240.

he wrote, 'most people would have no use for a geographer who was not an adventurer and explorer.'[56]

CONSEQUENCES

Thus for largely tactical reasons the 'New Geography' came to emphasize physical geography as its main concern, and this led necessarily to an almost axiomatic recognition of a fundamental duality in the subject, with the physical often linked to the human by the most simple chains of cause and effect. The sensitivity of the geologist to what they saw as poaching by the geographers provoked its own reaction, however, in that some of the latter made a virtue of abandoning physical geography altogether. 'The geographer has in consequence damaged his science,' argued Mackinder,[57] at the same time launching a powerful counterattack on what the geologists had managed to achieve with physical geography. 'The rivalry must be well known to all here present,' he suggested in his most famous lecture on 31 January 1887:

> It has been productive of nothing but evil to geography. . . . Perhaps nowhere is the damage done to geography . . . better seen than in the case of physical geography. The subject has been abandoned to geologists, but has in consequence a geological bias. . . . but such a science is not really physical geography . . . [which should aim] at giving us a causal description of the distribution of the features of the earth's surface.

Mackinder then invented a simple yet attractive formula, redolent of Hutton, to allow both the geologists and geographers to share a common subject matter: 'the geologist looks at the present that he may interpret the past; the geographer looks at the past that he may interpret the present.'[58]

Later, when success had been achieved, Mackinder went further, and insisted that, by an exclusive concern with the past, geologists had failed to develop one of the most powerful branches of modern natural science, the 'whole great new science of geomorphology', as Herbertson called it in

[56] E. W. Gilbert, *Sir Halford Mackinder 1861–1947: an appreciation of his life and work* (London: London School of Economics [Mackinder Centenary Lecture], 1947), reference on p. 14.

[57] H. J. Mackinder, 'On the scope and methods of geography', *Proceedings of the Royal Geographical Society*, n.s. 9 (1887), 141–160; reference on p. 145.

[58] *Ibid.*, p. 146.

1901.[59] Hence physical geography had become no more than an empty adjunct of historical geology, and geography had been left as an empty 'manual of names'. Humboldt's vision was compared with Lyell's to the latter's evident disavantage.[60] Much later, he reminisced that 'in this country . . . the geologists had captured Physical Geography, and had laid it out as a garden for themselves, while the remnant known as "General Geography" was a no man's land, encumbered with weeds and dry bones.'[61]

Mackinder's attempt to resolve the dilemma of duality had only limited success, for reasons which cannot be explored here. The geographers indeed counterattacked with such success that the study of landforms rapidly became an almost exclusively geographical preserve, though at the expense of human and economic geography – a development which Mackinder came to regret.[62] When the geologists responded, as at University College London, by formally preventing the geographers from lecturing about landforms at all, the fragmentation of the new subject was both enforced and resented. Lyde, for example, the first professor of geography at University College, came to speak of geomorphology as 'mere morbid futility',[63] devoting himself entirely to regional human geography. Yet it must not be forgotten that, during the last three decades of the century, geologists continued to make important research contributions in physical geography: indeed, the old century ended with James Geikie's *Earth sculpture* in 1898, and the new began with Marr's *Scientific study of scenery* in 1900. The longer-term consequences of the debate with the geologists belong to a later stage in the history of our discipline.

CONCLUSION

Geography in the years between 1870 and 1900 thus encountered many

[59] O. J. R. Howarth, *op. cit.* (note 51), p. 155.

[60] H. J. Mackinder, 'Modern geography German and English', *Geographical Journal*, 6 (1895), 367–379; reference on p. 373.

[61] H. J. Mackinder, 'Geography as a pivotal subject in education', *Geographical Journal*, 57 (1921), 376–384; reference on p. 377; 'The content of philosophical geography', *Report and Proceedings of the International Geographical Congress* (Cambridge 1928), 305–311; see pp. 306–307.

[62] H. J. Mackinder, 'Presidential address to the Geographical Association, 1916', *Geographical Teacher*, 8 (1916), 271–277; reference on p. 277.

[63] S. W. Wooldridge, 'On taking the ge- out of geography', *Geography*, 34 (1949), 9–18; reference on p. 13.

difficulties, before emerging triumphantly as an independent field of knowledge in the early twentieth century. First, the impetus to develop it as a university subject came from outside the universities, almost entirely from the R.G.S., and was met with indifference or opposition within them. Secondly, the case for the subject was largely put by men in the R.G.S. who were talented amateurs like Galton and Freshfield rather than by professional geographers: apart from the universities, and in contrast to geology, there was no other institution at that time which employed geographers as professional people. Thirdly, the strategy for the 'new geography' was largely pinned on Huxley's conception of a liberal education,[64] but tactically the best that could be hoped for in the 1880s was for geographers to be allowed to assist geologists in the technical study of landforms. And all of these difficulties were reinforced by two other considerations; the first, that geology had itself become professionalized, and saw physical geography as but one among many topics with which it was naturally and properly concerned; and the second, that the crisis in the 1880s within the Geological Survey led to substantial fears about the durability of the professional advances that the geologists themselves had already made.

I have sketched the way in which, during its first 70 years, the R.G.S. adapted itself to, and seized advantage from, changes in the intellectual climate and in social and economic conditions, and how with subtlety and determination its representatives secured a position for the 'new geography' in a time of rapid change. In this interpretation there is both a challenge and an enigma. The challenge is the analysis of Freshfield's work as a leader for reform, both in the R.G.S. and the Geographical Association. No one now makes more than incidental reference to him, but his position and his initiatives were clearly crucial. The enigma concerns the role of Archibald Geikie. No one wrote more fluently on physical geography and the science of landforms in the period with which we are concerned, and perhaps no one spanned the educational and scientific range between the elementary school and the presidency of the Royal Society with greater ease than he did. Yet he never joined the ranks of the geographers and he remained aloof from the R.G.S., in striking contrast to his mentor Murchison. Perhaps a deeper appreciation of both the difficulties and the achievements which arose as geography became established will await a fuller understanding of the lives and works of Freshfield and Geikie than has yet been attempted.

[64] M. E. Grant Duff, 'A plea for a rational education', in *Miscellanies political and literary* (London: Macmillan, 1878), 164–213; 'Earth knowledge', in *Some brief comments on passing events, made between February 4th 1858 and October 5th 1881* (Madras: R. Hill, 1884), 241–243.

5

The Foundations of Geography at Cambridge

Cambridge in the 1850s was just beginning the long process of reform which led to the evolution of a modern university. Teaching was still almost entirely concentrated on mathematics and classics: the former had had a tripos since 1747, and though a classical tripos was introduced in 1824 it was open only to those already classed in mathematics. Examinations for the degree of bachelor of arts were confined to the Old and New Testaments, Paley's *Moral philosophy*, church history, Euclid, algebra, trigonometry, the rules of arithmetic, and for honours candidates some elementary physics. There was no teaching at all in such subjects as chemistry, zoology, modern languages or geography, as the chancellor, Prince Albert, himself pointed out.[1] Cambridge had changed little since Darwin's day, and it was clear that major administrative changes had to be made if university teaching was to be improved.[2]

Many members of the university were aware of the need for change, and realized that if it did not come from within it was likely to be imposed from without. A syndicate in 1848 had suggested two new honour examinations, in moral sciences and natural sciences, and proposed a board of mathematical studies to regulate work in that subject; these proposals were accepted. A syndicate was also established in 1849 to revise the statutes. But a memorial calling for a commission of enquiry into the state of the university, signed by Charles Lyell, Charles Darwin and J. F. W. Herschel, among others, had already been presented to the Prime Minister, and

[1] Prince Albert to Lord John Russell, 13 November 1847, in T. Martin, *The life of His Royal Highness the Prince Consort* (London: Smith, Elder, 5 volumes, 1876), volume 2, pp. 121-124. Prince Albert had been installed as chancellor of the university on 25 March 1847. For a critical view of the older universities at this time, see C. Lyell, *Travels in North America: with geological observation on the United States, Canada, and Nova Scotia* (London: John Murray, 2 volumes, 1845), volume 1, pp. 261–316.

[2] D. A. Winstanley, *Early Victorian Cambridge* (Cambridge: Cambridge University Press, 1940), pp. 206–208.

in 1850 a royal commission was appointed for this purpose, the commissioners including Adam Sedgwick and Herschel.

Their enquiry revealed an alarming state of teaching and research in the university. Though, as Southey said, Cambridge had professors of everything, most of them did nothing.[3] Many of the 20 chairs were sinecures, and their teaching duties neglected. Thus the Woodwardian chair of geology, founded in 1731, provided for four lectures a year to be delivered and published. But from 1731 to 1818 'no regular lectures had been either delivered or published, and the anxious provision of the Founder for the defence and maintenance of his philosophical opinions altogether failed to elicit a word, whether written or spoken, in their favour, with the single exception of an Inaugural Lecture by the celebrated Dr Conyers Middleton, who first held the Professorship.'[4] Sedgwick, elected in 1818, lectured regularly, but his courses were of no direct value towards a degree.[5] Lectures were also given by the Rev. E. D. Clarke, professor of mineralogy from 1808 to 1822, who in 1816 took lessons in oil-painting to illustrate his

[3] Anon. [R. Southey], *Letters from England by Don Manuel Alvarez Espriella* (London: Longman, 1807), volume 2, p. 298.

[4] *Report of Her Majesty's Commissioners appointed to inquire into the state, discipline, studies, and revenues of the University and Colleges of Cambridge: together with the evidence, and an appendix* (London: W. Clowes and Sons, 1852), p. 62 [hereafter cited as *Royal Commission Report*]. The record of the occupants of the Woodwardian lectureship was particularly dismal. Conyers Middleton ('fiddling Conyers') 'had no knowledge of any department of science' and resigned after three years. Charles Mason, 'though a very ingenious Man, an excellent Mechanic, and no bad Geographer', was 'unhewn, rough, and unsociable', 'coarse and slovenly to excess', and 'disgusted and disgraced Society'; he held office for 28 years and published one single lecture in that time. John Michell was more competent, but left after less than two years; he was 'a little short Man, of a black Complexion, and fat'. His successor, Samuel Ogden, president of St. John's, 'a bald, swarthy, black Man', 'growling and morose' and 'much broken by Gout', was said to have bribed his way into the position. Nothing is known of his successor Thomas Green. John Hailstone, who held office for 30 years, actually published a 'Plan of a course of lectures' but never gave a single one. 'On Hailstone's resignation there was evidently a feeling in the University that Woodward's bequest had not produced the results which might have been anticipated': see J. W. Clark and T. McK. Hughes, *The life and letters of the Reverend Adam Sedgwick* (Cambridge: Cambridge University Press, 2 volumes, 1890), volume 1, pp. 188–197.

[5] Even Sedgwick, straightforward Yorkshireman that he was, was not immune to the constant (and still present) dangers of Cambridge life. Appointed a proctor in his early forties, he dined regularly on kippered salmon and spiced port. Not surprisingly, within a year he could hardly see, had throbbing in his head, was ordered off strong drink, and had to have 12 leeches attached to his left temple to relieve the pressure. Fifteen years later, tormented beyond measure by rheumatic gout, he had become 'sour-faced and ill-tempered, and abominably cross', and was forced to live only on water and vegetables. See J. W. Clark and T. McK. Hughes, *op. cit.* (note 4), especially volume 1, p. 312, and volume 2, p. 42.

remarks, a commendable aim which can hardly have speeded his rate of delivery.[6] A chair in botany had been established in 1724, but there was no teaching at all in zoology, in spite of the fact that the university possessed 'very fine skeletons of the Cape Buffalo, and of the Great Ant-eater'.[7] The new natural sciences tripos of 1851 covered anatomy, physiology, chemistry, botany, geology and mineralogy, but not meteorology, geography, zoology or physics. The Prime Minister was particularly scandalized that the undergraduates of Cambridge would have to wait a century before being allowed to learn of electricity if such college traditionalists as Whewell had their way.[8]

The Royal Commission proposed that there should be special boards of studies for theology, law, medicine, mathematics, classics, moral sciences and natural sciences, with new chairs in engineering, geometry, anatomy, chemistry, law, international law, Latin, divinity and zoology. The success of these measures would of course depend on attracting capable men for the new 'teaching professoriate' to be established rather on the German pattern, and it was many years before all these proposals were put into effect.[9]

Geography as such was not mentioned specifically in the Royal Commission's report, but geographical studies were by no means overlooked. Indeed, in 1855, the first volume of a new series, *Cambridge essays*, designed to reflect the changes taking place in the university, contained Francis Galton's *Notes on modern geography*,[10] and the concerns with which it dealt were also apparent in the Royal Commission's report. The latter's treatment was somewhat ambivalent, however, and, rather surprisingly, treated geographical matters under the head of moral rather than natural sciences. This was largely because of the provisions of Mr Worts's foundation, which for a century and a half had provided two travelling fellowships, the fellows to report on the 'religion, learning, laws, politics, customs, manners, and rarities, natural and artificial, which they should

[6] D. A. Winstanley, *op. cit.* (note 2), p. 32, quoting W. Otter, *Life and remains of the Rev. Edward Daniel Clarke, LL.D., Professor of Mineralogy in the University of Cambridge* (London: George Cowie, 1824), p. 635.

[7] *Royal Commission Report* (*op. cit.*, note 4), pp. 120–121.

[8] Sir Robert Peel to Prince Albert, 27 October 1847, in T. Martin, *op. cit.* (note 1), volume 2, p. 118.

[9] See W. Champion, 'Commissioners and colleges', in *Cambridge essays, contributed by members of the University* (London: J. W. Parker, 1858), 165–225.

[10] F. Galton, 'Notes on modern geography', in *Cambridge essays*, *op. cit.* (note 9), 79–109. For Galton's geographical work, see K. Pearson, *The life, letters and labours of Francis Galton* (Cambridge University Press, 3 volumes, 1914–1930), volume 2, pp. 1–69; F. Galton, *Memories of my life* (London: Methuen, 1908); and notes 6–8 on p. 44.

find worth observing in the countries through which they passed'. The Commissioners queried whether such fellowships were any longer necessary: they found 'no lack of travellers among our countrymen: no want of information from books on the state of foreign countries'. But at the same time there was a need for education for diplomatic and consular appointments, which might combine languages with 'an acquaintance with the History of Treaties, with Political Geography, with the Principles of International Law, the laws relating to aliens, to commerce and to shipping, and perhaps other kindred subjects' to form a new tripos of honour, and to support which the provisions of the Worts foundation might be varied.[11]

But there was a further argument in favour of providing at least some teaching, if not more formal recognition, of a geographical nature, more from social and practical than from academic and scholarly considerations:

> there is a very important class of Students who resort to the Universities with no professional views, and who, from their expectations in life, have no motive for prosecuting the severer studies beyond the prescribed minimum. . . . This is a class which it is in every way desirable both to induce to the University, and to influence for good when there. . . . If there be mean and frivolous ways in which wealth may be squandered and leisure abused, there are also great and noble objects to which they may be applied. . . . Many of this class of Students become travellers, to whom a knowledge of natural science, and a familiarity with many scientific processes and the handling of a great variety of instruments are likely to be of eminent use, and whom the example of a Humboldt may stimulate to bring home records of more importance than the usual reminiscences of travellers for pleasure. The theory, handling, and systematic use of meteorological and magnetic instruments are an element of the scientific education of a traveller of the last importance; a practical acquaintance with photography not less so.[12]

These aims could, the Commission felt, be met by the teaching professors whose appointment it recommended.

It took time to implement the Royal Commission recommendations. The university set up syndicates to consider the academic, administrative and other implications of the report, and in May 1854 proposed graces for the

[11] *Royal Commission Report*, *op. cit.* (note 4), pp. 101–102.

[12] *Royal Commission Report*, *op. cit.* (note 4), pp. 118–119. These views were reflected in Galton's 1858 essay (note 10) and also in his *The art of travel; or, shifts and contrivances available in wild countries* (London: John Murray, 1855), 196 pp.

establishment of a theology tripos, for boards of studies in theology, natural sciences and moral sciences, and for degrees resulting from work in moral sciences or in natural sciences alone. These the university as a whole voted to reject. Although in the Cambridge University Act of 1856 the government appointed new commissioners to revise the statutes of the university, and established new machinery, including the council of the senate, to facilitate the processes of change, the acceptance of the 1852 recommendations was less than immediate. There were still those in Cambridge in 1860 who objected to having 'about half a dozen little Triposes . . . to suit the convenience of men who cannot brace up the nerves of their mind to pen an examination in anything better than butterflies' wings'.[13] Partly because of this, partly because of the state of teaching in the schools, but especially because the examinations did not lead directly to a degree, few candidates presented themselves in either moral or natural sciences (66 and 43 respectively in the first nine years). Not until 1860 were proposals for boards of moral sciences and of natural sciences accepted, and triposes in these subjects leading to the B.A. degree established in the university. Numbers reading natural sciences remained low,[14] but from 1867 college awards were made in natural science subjects, and the first election to a college fellowship was made in that year on the basis of tripos results in the natural sciences. In 1866 the chair of zoology recommended by the 1852 Royal Commission report was finally created,[15] and the building of the New Museums began in 1863. The gift of the Cavendish Laboratory led to a new chair in experimental physics in 1871, and in 1875 the chair of mechanics and applied science signified the establishment of engineering as a university discipline.

During the 1870s teaching in the universities continued to excite public concern. A royal commission into the property and income of Oxford and Cambridge universities in 1872 drew a memorial from the Royal Geographical Society on the need for chairs in geography, but with no immediate result.[16] More important, the standing Royal Commission on Scientific

[13] Letter in *Cambridge Chronicle*, 7 April 1860, quoted by Mary Hesse, 'History and philosophy of science in the early natural sciences tripos', *Cambridge Review*, 84 (1962), 140–145.

[14] The class-list did not exceed 20 until 1875: D. A. Winstanley, *Later Victorian Cambridge* (Cambridge: Cambridge University Press, 1947), p. 190.

[15] *Ibid.*, p. 192.

[16] J. S. Keltie: 'Report to the Council of the Royal Geographical Society on geographical education', *Supplementary Papers of the Royal Geographical Society*, 1(4) (1886), 439–554. Reprinted in Royal Geographical Society, *Report of the Proceedings of the Society in reference to the improvement of geographical education* (London: John Murray, 1886), to which subsequent citations refer.

Instruction and the Advancement of Science took a broad view of its terms of reference. Its third report in 1873 dealt with science teaching at Oxford and Cambridge. It noted that with the exception of elementary mechanics at Cambridge, 'nothing is done, at any part of the course in either university, to exact from all students alike any knowledge, however small, of the elements of the sciences of experiment and observation',[17] and made the revolutionary suggestion that if classical subjects were compulsory for all, so should scientific subjects be. The Oxford and Cambridge professoriate and lecture lists were compared with those of Berlin, to the latter's evident advantage. 'It is impossible not to be impressed with the evidence . . . of the abundance and variety of the scientific teaching given in the University of Berlin by Professors of great eminence.'[18] Under the head of 'natural sciences', for example, the Berlin lecture list for 1872–1873 included Professor Erman three times a week in winter on 'Terrestrial physics; or, mathematical and physical explanations of geographical phenomena', and in the summer, Professor Erman three times a week on 'Theory of geographical and geographico-physical observations' and Professor Poggendorff on physical geography.[19]

By this time, however, in spite of the comparisons which might be made with continental universities, and despite the fact that undergraduates could still leave Cambridge with no knowledge of the sciences at all, teaching within the natural sciences was much more firmly established. Under the statutes proposed in 1879 the Board of Natural Science Studies was divided into two new boards, of physics and chemistry and of biology and geology, and the number of teaching posts was expanded. Interest in physical geography had been greatly stimulated by the *Challenger* expedition of 1872–1876,[20] and both J. W. Judd and Norman Lockyer lectured and examined on physiography at the new Normal School of Science in South Kensington from 1872.[21] With the natural sciences now well established at Cambridge, and with growing popular and educational interest in scientific subjects such as physiography as well as in exploration and travel, the possibility of formal recognition of geography in the university was very much improved.

[17] *Royal Commission on Scientific Instruction and the Advancement of Science: third report* (London: Eyre & Spottiswoode, 1873), p. xii.

[18] *Ibid.*, p. xxii–xxiii.

[19] *Ibid.*, pp. xxii–xxiv.

[20] C. Wyville-Thomson, *Nature, Lond.*, 14 (1876), 197.

[21] A. J. Meadows, *Science and controversy: a biography of Sir Norman Lockyer* (London: Macmillan, 1972), p. 116.

FIRST MOVES BY THE ROYAL GEOGRAPHICAL SOCIETY

The history of academic geography in Cambridge effectively begins, however, not with internal movements for reform, but with the growing interest of the Royal Geographical Society in education, particularly in the schools but also in the universities, under the influence of Francis Galton and Douglas Freshfield during the early 1860s and later.[22] This interest first resulted in the establishment of a scheme of medals for boys at selected public schools, based on examination papers in physical and political geography set by the Society. This scheme began in 1869 and continued until 1884: it was then judged unsatisfactory because the medals were available to only 16 schools and during the period almost half of them went to only 2 schools.[23] One of the first, however, was awarded to a boy called Donald MacAlister, later to become an advocate for geography at Cambridge and a member of the Board of Geographical Studies.[24]

Shortly after this scheme began, the Society addressed a memorandum to the vice-chancellors of both Oxford and Cambridge, dated 3 July 1871 and signed by the president Sir Henry Rawlinson, on the subject of the treatment of geography in the universities' school examinations. The Society argued for the importance of geography in education and gave 'practical and commonsense Victorian' reasons[25] for giving it greater attention:

> We would point out the special importance of geography to Englishmen in the present age. The possession of great and widely scattered dependencies, the unprecedented extension of our commercial interests, the increased freedom of intercourse and closeness of connection established by means of the steamship and the telegraph, between our country and all parts of the world, the progress of emigration binding us by ties of blood relationship to so may distant communities – all these are circumstances which vastly enhance the value of geographical knowledge. . . .

[22] J. F. Unstead, 'H. J. Mackinder and the new geography', *Geographical Journal*, 113 (1949), 47–57; and also chapter 4.

[23] J. S. Keltie, *op. cit.* (note 16), p. 12. The schools were Dulwich College and Liverpool College.

[24] MacAlister (1854–1934), from Liverpool College, was fellow of St John's College 1877–1934, university lecturer in medicine 1884–1907, principal and vice-chancellor, University of Glasgow, 1907–1929, and chancellor 1929–1934. For obituaries, see *Cambridge Daily News*, 15 January 1934; *The Times*, 16 January 1934; and also E. F. B. MacAlister, *Sir Donald MacAlister of Tarbert* (London: Macmillan, 1935), p. 14.

[25] E. W. Gilbert: 'The R. G. S. and geographical education in 1871', *Geographical Journal* 137 (1971), 200–202; reference on p. 201.

> We speak of geography, not as a barren catalogue of names and facts, but as a science that ought to be taught in a liberal way, with abundant appliances of maps, models, and illustrations. . . . We look to the Universities, not only to rescue geography from being badly taught in the schools of England, but to raise it to an even higher standard than it has yet attained.[26]

Three years later the Society followed this letter with a memorial to the Royal Commissioners enquiring into the universities of Oxford and Cambridge, and to the universities themselves. This urged the need for 'the establishment of Geographical Professorships at both Universities' and developed the argument of the letter of 1871.[27]

> In speaking of geography the Council use the word in its most liberal sense, and not as an equivalent to topography. The word Geography, they desire it to be understood, implies a compendious description of all the prominent conditions of a country, such as its climate, configuration, minerals, plants, and animals, as well as its human inhabitants; the latter in respect not only to their race, but also to their present and past history, so far as it is intimately connected with the peculiarities of the land they inhabit. Each locality has its characteristic features, which it is the province of the geographer to describe with the utmost clearness. He should convey to others the salient ideas that could not otherwise be acquired except by a highly-skilled observer in all branches of knowledge, after a long residence.
>
> Scientific geography does not confine itself to such a description of separate localities as may be found in gazetteers. Having collected similar cases, it proceeds to group them together. It studies antecedent conditions, and concerns itself with the actions of concurrent phenomena upon one another in the same locality, showing why they tend to stability, and to give to each country its characteristic aspect. Thus the geographical distribution of plants and animals, and the light it throws on the early configuration of the surface of the earth, is one of the very many problems with which scientific geographers are accustomed to deal. Another of the problems is concerned with the reciprocal influence of man and his surroundings; showing on the one hand the influence of external nature on race, commercial development and sociology, and on the other, the influence of man on nature, in forest destruction, cultivation of the soil, introduction of new plants

[26] J. S. Keltie, *op. cit.* (note 16), pp. 79–80.
[27] *Ibid.*, p. 81.

> and domestic animals, extirpation of useless vegetation, and the like. This mutual relation of the objects of the different sciences is the subject of a science in itself, so that scientific geography may be defined as the study of local correlations.
>
> Geography, thus defined, does not tend in any degree to supersede the special cultivation of independent sciences, but rather to establish connections which would otherwise be unobserved, and to intensify the interest already felt in each of them, by showing their general value in a liberal education. It is through geography alone that the links can be seen that connect physical, historical, and political conditions; and it is thus that geography claims the position of a science distinct from the rest, and of singular practical importance.

Carl Ritter's influence is apparent in the philosophy so expressed, and the memorial, in demonstrating the success of geography in European universities, drew attention not only to Ritter's at Berlin, but also to the departments at Halle, Marburg, Strasbourg, Bonn, Göttingen, Breslau, Neuchâtel, Geneva, Zurich, Paris, Bordeaux, Caen, Lyons, Clermont-Ferrand, Nancy and Marseilles, a list in sharp contrast to the situation in Britain.

The memorial also stressed the practical advantages of a geographical training, stressing the lack of qualifications for observation of too many travellers and missionaries, and concluded that:

> there is no country that can less afford to dispense with geographical knowledge than England . . . [yet] there are few countries in which a high order of geographical teaching is so little encouraged. The interests of England are as wide as the world. Her colonies, her commerce, her emigrations, her wars, her missionaries, and her scientific explorers bring her into contact with all parts of the globe, and it is therefore a matter of imperial importance that no reasonable means should be neglected of training her youth in sound geographical knowledge.[28]

There was no immediate response to this approach, though Adolphus Ward, later editor of the *Cambridge modern history*, was simultaneously urging the inclusion of historical geography in the proposed new history tripos.[29] No further steps were taken until the Society was compelled to

[28] *Ibid.*, p. 83.

[29] A. W. Ward, *Suggestions towards the establishment of a history tripos* (Cambridge: Cambridge University Press, 1872); J. O. McLachlan, 'The origin and early development of the

re-examine its educational policy following the abolition of the public schools medals scheme in 1884. Several alternatives were then suggested: the appointment of an R.G.S. travelling professor was proposed by Galton, a course of lectures given by 'some carefully selected person' (who turned out to be himself) was suggested by Strachey, and the foundation of a geographical professorship at Oxford was urged by Freshfield.[30] The whole question was examined in detail by J. Scott Keltie, appointed the Society's inspector of geographical education, who reported on 17 May 1885.[31] Keltie's report covered geography in both schools and universities: he argued that it was not well taught in the former because it was not recognized by the latter. Most good school 'geography' was in fact physiography, and little or nothing was done about other branches of the subject. Keltie contrasted the situation in Britain with that in France, Germany and many other countries.

In preparing his report, Keltie corresponded with a number of senior members of the university at Cambridge, particularly Professor Alfred Newton. Keltie took the view that the prospects for geography were more favourable at Cambridge than at Oxford, especially since the Special Board for Geology and Biology had already recommended (6 February 1883) that university teachers should eventually be appointed in this among other subjects.[32] Keltie's approaches met with cautious interest. The Rev. Coutts Trotter,[33] active in university reform, found difficulty in understanding what

---

Cambridge Historical Tripos', *Cambridge Historical Journal*, 9 (1947), 78–105; G. Kitson Clark, 'A hundred years of the teaching of history at Cambridge, 1873–1973', *Historical Journal*, 16 (1973), 535–553.

[30] *Schemes for the promotion of the study of geography, in place of public schools' prizes: memorandum by the president on the future educational policy of the council* (R. G. S. Archives).

[31] Keltie, *op. cit.* (note 16).

[32] *Cambridge University Reporter* [hereafter *CUR*] 1882–1883, p. 885.

[33] Coutts Trotter (1837–1887), vice-master of Trinity, should not be confused with Coutts Trotter (1831–1906), a Bengal civil servant who contributed geographical articles on the islands of the western Pacific to *Blackwood's Magazine* and the *Quarterly Review*. Hugh Robert Mill's reference to 'a gentleman of singular courtesy and generosity' is probably to the latter: 'Recollections of the Society's early years', *Scottish Geographical Magazine*, 50 (1934), 269–280; reference on p. 276. The process of reform at Cambridge was often spoken of as the 'Trotterization' of the university: J. W. Clark, *Old friends at Cambridge and elsewhere* (London: Macmillan, 1900), p. 316. A. E. Shipley tells the story of how Coutts Trotter and J. W. Clark set themselves on fire while testing the inflammability of preserving fluids in the University Museum of Zoology: both had their outer clothing destroyed and Trotter had to be taken back to Trinity wrapped in blankets: *'J' A memoir of John Willis Clark* (London: Smith, Elder, 1913), p. 279.

a geographer would do, and felt that the minute study of regional geography would be 'both uninteresting and unprofitable' to those engaged in it.[34] Professor George Darwin wrote that 'in the hands of a mediocre man it would almost certainly degenerate into a descriptive catalogue,' and he could not 'see how geography pure and simple could be made a subject of intellectual training'.[35] But Darwin felt 'a really accomplished man' could well be employed by the university, and Freshfield's proposal that courses of lectures might be given was favourably received by members of the council of the senate approached by Alfred Newton. It was understood, 'of course', that the geography envisaged was physical and not political.[36] This provoked some annoyance among Cambridge geologists, and McKenny Hughes[37] pointed out that regular lectures on physical geography were given as part of the geology course. The feeling among geologists against a separate provision for geography was indeed so widespread that Freshfield termed them 'the most forward of the would-be "chuckers-out" of geography from the Hall of Education'.[38]

Following Keltie's report, the R.G.S. mounted a large 'Exhibition of educational appliances in geography' in London, Manchester and Edinburgh, and four public lectures were given in London in December 1885 and January 1886. E. G. Ravenstein spoke on 'The aims and methods of geographical education', Keltie himself on 'Appliances used in teaching geography', James Bryce on 'Geography in its relation to history' (proposing new branches of the subject, including ethnological, legal and sanitary geography), and H. N. Moseley, the naturalist of the *Challenger*, on 'Scientific aspects of geographical education'.[39]

[34] J. S. Keltie, *op. cit.* (note 16), pp. 105–106.

[35] *Ibid.*, p. 160.

[36] A. Newton, 21 February 1884, in *ibid.*, pp. 106–107.

[37] T. McK. Hughes, 'Discussion', *Proceedings of the Royal Geographical Society*, n.s. 8 (1886), p. 201. McKenny Hughes was also in correspondence with George Butler and with Galton. The latter in particular was less than supportive of Keltie: 'He speaks throughout entirely for himself. The Council have not even taken his report into consideration and I hope they will not until the Exhibition is over and till he has to shew as well as to tell.' Galton to Hughes, 27 November 1885; also Butler to Hughes, 5 and 25 February 1879 (Cambridge University Library, Add. 5944 (22)). Within the university it was also being argued that a geographical tripos would be premature, since it 'would be only adding complexity to the machinery here without really meeting the main difficulty'. 'A plea for geography', by F., *Cambridge Review*, 7 (1886), 382–383.

[38] D. W. Freshfield, 'The place of geography in education', *Proceedings of the Royal Geographical Society*, n.s. 8 (1886), 698–714, discussion 714–718; reference on p. 704.

[39] The lectures are printed in Royal Geographical Society, *Report*, *op. cit.* (note 16), 163–236.

The Society decided once more to approach the universities. The president, Lord Aberdare, wrote to the vice-chancellor of Oxford on 9 July 1886, and a similar letter from his successor, General Strachey, followed to Cambridge on 9 December. This stressed the need for the 'establishment in our Universities of Chairs or Readerships similar to those held in Germany, viz. by Karl Ritter at Berlin, and Professors Peschel and Richthofen at Leipzig'.[40] In order to accomplish this, the Society made two alternative proposals: a lecturer or reader could be appointed and paid for by the Society and given the status of 'a Reader attached to the university'; or the cost of the readership could be met jointly by the Society and the university and the appointment made by a university committee on which the Society would be represented. If either scheme were acceptable, the Society would also be prepared, in alternate years, to award an exhibition of £100, 'to be spent in the geographical investigation (physical or historical) of some district approved by the Council, to a member of the University of not more than eight years' standing, who shall have attended the Geographical Lecturer's courses during his residence'.[41]

THE R.G.S. LECTURESHIP: F. H. H. GUILLEMARD AND J. Y. BUCHANAN, 1887–1893

The year 1887 was auspicious for British geography. It opened with Mackinder's lecture to the R.G.S. in January on the 'Scope and methods of geography,[42] and before the year was out Mackinder had been appointed reader at Oxford and agreement had been reached for a lectureship at Cambridge. As at Oxford, the first step in response to Strachey's letter was for a meeting to be arranged between a university committee and representatives of the R.G.S. The university nominated the master of Gonville and Caius, Dr MacAlister, Dr Michael Foster and the Rev. G. F. Bourne, and the Society put forward Freshfield, Galton, General Walker and Sir T. F. Wade. They met on 18 February 1887, four days after Mackinder's paper was discussed at the R.G.S. Their *Report on the teaching of geography*[43] was presented to the university on 14 March. It proposed the establishment of a lectureship for five years, with a stipend composed of

[40] *CUR* 1886–1887, pp. 330–332.
[41] *CUR* 1886–1887, pp. 331–332.
[42] H. J. Mackinder', 'On the scope and methods of geography', *Proceedings of the Royal Geographical Society*, n.s. 9 (1887), 141–160, discussion 160–174.
[43] *CUR* 1886–1887, pp. 516–518.

£50 a year from the university and £150 a year from the Society, and the acceptance of the Society's offer of an exhibition of £100 (or prizes of £50 and £25) in alternate years. Matters to do with the lectures, the exhibition and the prizes would be under the control of a joint committee of management.

In the discussion which followed on 3 May a variety of arguments were deployed against the report.[44] It was felt that it was improper for the university to join in such a way with an outside body; it was argued that far from geography being a science it was in fact 'quite an elementary art'; and it was argued that 'the University already had sufficient opportunities for encouraging the study of Geography.' The Rev. Coutts Trotter repeated his doubts about what a professor or reader or lecturer in geography ought to do. But powerful aid came from Professor Alfred Newton, over the years one of the most persistent champions of the cause of geography in Cambridge. In many ways his support was surprising. Guillemard, later the first lecturer in the subject, recalled that 'to put it mildly, Newton was no progressive. In his eyes alteration of any kind was the one unpardonable sin; change little short of a crime. I feel sure that the donning of a new suit caused him actual pain. . . . [He was] a Tory of the Tories.'[45] His lectures were 'desperately dry',[46] physically he was ponderous and slow, and in his college, Magdalene, he was involved in a series of acrimonious disputes over

[44] *CUR* 1886–1887, pp. 669–670.

[45] F. H. H. Guillemard, 'Recollections of Newton', in A. F. R. Wollaston, *Life of Alfred Newton, Professor of Comparative Anatomy, Cambridge University, 1866–1903* (London: John Murray, 1921), 265–274; reference on pp. 270–271.

[46] A. E. Shipley, 'Dr Shipley's Reminiscences', in A. F. R. Wollaston, *op. cit.* (note 45), 93–109; reprinted as 'Professor Newton (1829–1907)' in Shipley's *Cambridge Cameos* (London: Jonathan Cape, 1924), 148–172; reference in Wollaston, p. 104. Newton used to read his lectures, often to an audience of one, from minute script written with a quill pen: from time to time he would come to the outline of a glass drawn in the margin, which was a signal to him to refresh himself. With advancing years he ultimately had to nominate a deputy, and chose William Bateson of St. John's. Although a distinguished geneticist, Bateson himself was somewhat odd. Guillemard recalls that after he had travelled to Turkestan, Bateson lived in a Turki tent erected inside his college rooms, with all appropriate furnishings, and his guests were obliged to accept his hospitality on the floor: F. H. H. Guillemard, 'The years that the locusts have eaten' (typescript, 7 volumes, 1927–1930; copies in the Cambridge University Library and the Bodleian Library, Oxford), see volume 6, p. 136. Wollaston notes that nothing vexed Newton so much as any innovation (*op. cit.*, p. 262), and the strength of his support for the new subject of geography is perhaps surprising. He was, however, one of the first zoologists to accept Darwin's views, and for many years he was a dominant figure in the natural sciences in Cambridge. The subject perhaps needed the support of such 'a fiercely intolerant and prejudiced man, impatient of opposition, and convinced of his own unassailable exactness' (*op. cit.*, p. 248).

the introduction of a harmonium, hymns and ultimately even women into the chapel.[47] He became firmly established as a college reactionary and by the time of his death, as Guillemard tells, he was 'accounted an extinct type, as extinct as the Greak Auk and Dodo of which he loved so much to write'.[48] But in his earlier years especially he was a great ornithologist and carried much scientific weight. He was the first holder of the chair of zoology, founded in 1866 after the reforms, and which he occupied for over 40 years. At the time of the discussion of the report he was also chairman of the Special Board of Studies for Biology and Geology, one of twelve established under the new statutes in 1882, and hence in a position to make his views known.

Each Sunday evening, from the early 1870s, Newton held open house for selected undergraduates in his Magdalene rooms. Sedgwick, Bateson, Frank Darwin, Marr, Shipley, Balfour, Lydekker and others with scientific and especially zoological interests were there, and also the university sub-librarian, Crotch, a 'mighty beetle hunter', notorious for having had cooked and presented for dinner a finger he had accidentally severed while chopping wood.[49] At these evenings Newton's own championing of

[47] A. C. Benson, 'Professor Newton', in *The leaves of the tree: studies in biography* (London: Smith, Elder, 1911), 132–162; originally published in *Cornhill Magazine*, n.s. 30 (1911), 334–349.

[48] F. H. H. Guillemard in A. F. R. Wollaston, *op. cit*, (note 45), p. 273. Guillemard relates that when the time came Newton died 'inch by inch'. He finally insisted on being placed on his chair, so that he died as Bradshaw, the university librarian, had done in such a celebrated fashion almost 20 years before: Wollaston, *op.cit.*, pp. 273, 292; G. W. Prothero, *A memoir of Henry Bradshaw* (London: Kegan Paul, Trench, 1888), pp. 312–322.

[49] F. H. H. Guillemard in A. F. R. Wollaston, *op. cit.* (note 45), p. 268. On Crotch, see G. W. Prothero, *op. cit.* (note 48), pp. 89–90, and J. Smart and B. Wager, 'George Robert Crotch, 1842–1874: a bibliography with a biographical note', *Journal of the Society for the Bibliography of Natural History*, 8 (1977), 244–248. Bradshaw himself was university librarian and a man of such shyness that he is said to have made contact with others in the courts of King's by slipping roses into their hands as he passed them without a word: Prothero, *op. cit.*, p. 412. Monty James tells the story of one of Bradshaw's late-night soirées in his rooms, when 'all sorts of dons and undergraduates might be there. . . . It did sometimes happen, though, that one stayed late and got him almost alone. That was delightful. . . . [There was] a curious habit of his, a liking for having the palms of his hands tickled. . . . I remember sitting doing this for quite a long time one night to one hand, while somebody else attended to the other, and Bradshaw sat and purred.' With a sudden burst of enthusiasm, James adds: 'Would that I had gone oftener to his rooms than I did.' M. R. James, *Eton and King's – recollections, mostly trivial, 1875–1925* (London: Williams & Norgate, 1926), pp. 110–111. This was, of course, the time of the scandalous Oscar Browning, who had a bath installed in his rooms and supervised his students in pairs, so that one could dry him when he emerged and the other could meanwhile play the violin (H. E. Wortham, *Oscar Browning* [London: Constable, 1927], p. 248; for a more recent and less restrained view see I. Anstruther, *Oscar Browning. A*

geography was essentially pragmatic. He was himself, in spite of his lameness, an outdoor man and a traveller in northern Europe; he was also one of the first Darwinians – a zoologist, as Shipley said, not a necrologist.[50] The criticisms of the 1887 report by Trotter and others he abruptly dismissed:

> Whether it was an Art or a Science he did not care. It was a study to be seriously pursued; but until something was done in the way of encouraging it in the University, he believed that the present disgraceful ignorance of Geography among so-called educated people would continue. He had found himself unable to make his lectures on the geographical distribution of Animals intelligible to some members of his class, because of the lamentable ignorance of common Geography in which they had been allowed to grow up.[51]

In spite of the opposition the report was approved on 9 June 1887, and it was arranged that the formal connection of the Society with the university be marked by a course of lectures designed to 'set before the University the claims of the Science of Geography to rank as one among the studies of the University, and . . . [to] show that the subject is capable of treatment in courses of lectures well worth the attention of students of academic distinction or promise'.[52] Four lectures on 'Principles of geography' were accordingly given by the president of the R.G.S., General Richard Strachey,

---

biography [London: John Murray, 1983]). Even the Browning stories pale before that of Lowes Dickinson instructing his students to wear heavy boots, so that during supervisions they could take turns to lie on the floor with booted feet upon each other (D. Proctor, ed., *The autobiography of G. Lowes Dickinson* [London: Duckworth, 1973]). Magdalene, of course, was not King's, even then, but it had its share of eccentrics, beginning with Benson, who became its master in 1915. Benson had repeated bouts of black depression, and his sister had to be locked away permanently after a homicidal attack on their mother with a carving knife. Someone asked Benson late in life whether he had ever kissed a woman. 'Yes,' was his reply, 'once, on the brow.' D. Williams, *Genesis and exodus: a portrait of the Benson family* (London: Hamish Hamilton, 1979) is more frank on these matters than D. Newsome, *On the edge of paradise, A. C. Benson: the diarist* (London: John Murray, 1980). Certainly Benson knew Newton well, and was acquainted with Guillemard. Deeper understanding of the nature of Newton's Sunday evening gatherings may emerge from study of Benson's own diaries. My purpose here is to indicate the curious flavour of the circles in which the founders of Cambridge geography habitually moved: but let it also be recalled that Clerk Maxwell gave his inaugural lecture in 1871, and that the events here recorded coincided with the triumphant development of Cambridge physics, culminating in Cockcroft and Walton's splitting of the atom in 1932.

[50] A. E. Shipley, *op. cit.* (note 46), reference on p. 105 of Wollaston.

[51] *CUR* 1886–1887, p. 670.

[52] *CUR* 1886–1887, p. 517.

in February 1888. Strachey, then aged 71, had made his geographical reputation while serving with the Bombay Engineers. He was elected to the Royal Society in 1854, and later made a second distinguished career for himself in public works in India, retiring in 1871. By his seventies, however, the general 'had grown rather mild in spirit, a little vague and detached from all that was going on even in the closest proximity'.[53] His lectures, subsequently published,[54], were unremarkable in content and apparently made little impact at the time. Joseph Hooker, who went up to Cambridge for the occasion, reported to T. H. Huxley that 'the matter was excellent but very dry – as the Frenchman said of English meat, which he bought from the dog's-meat man.'[55]

The university then proceeded to take the necessary steps for the appointment of a lecturer. Committees of appointment and management were set up following a further report on 14 May 1888,[56] and on 12 June 1888 F. H. H. Guillemard, M.A., was appointed as lecturer for five years from midsummer 1888. Francis Henry Hill Guillemard (1852–1933) was then 36 years old, a fellow of the R.G.S. since 1876. He studied medicine at Gonville and Caius College, attending Newton's lectures on vertebrate zoology, and while an undergraduate he travelled in Lapland to oberve birds. After taking his degree and touring Europe, he spent the years 1877–1878 on an expedition to South Africa, making his way in a wagon drawn by 16 oxen from Bloemfontein to the Limpopo and the Elands River. During this expedition, which ended shortly before the attack at Isandhlwana, he caught a severe infection up the Limpopo, and later made the detailed study of his symptoms the subject of his M.D. thesis in 1882.[57] His

[53] M. Holroyd, *Lytton Strachey: a critical biography*, volume 1: *The unknown years (1880–1910)* (London: William Heinemann, 1967), pp. 14–34; reference on p. 18. The family could reasonably be called eccentric. One of the general's brothers was eaten by a bear; another, who had previously visited India, spent the rest of his life on Calcutta time and regularly donned galoshes and oilskins at the presumptive approach of the summer monsoon; it was the latter who left Lytton his embroidered silk underwear.

[54] R. Strachey, *Lectures on geography delivered before the University of Cambridge during the Lent term 1888* (London: Macmillan, 1888), 211 pp. Previously published as 'Lectures on geography, delivered before the University of Cambridge, 1888', *Proceedings of the Royal Geographical Society*, n.s. 10 (1888), 146–160, 205–220, 220–234, 275–293.

[55] L. Huxley, *Life and letters of Sir Joseph Dalton Hooker, O.M., G.C.S.I.* (London: John Murray, 2 volumes, 1918), volume 2, p. 342.

[56] *CUR* 1887–1888, p. 662.

[57] F. H. H. Guillemard, *On the endemic haematuria of hot climates caused by presence of Bilharzia Haematobia* (Baillière, 1882). The degree was awarded in 1881. 'A.B.', the subject of the dissertation, was Guillemard himself; the thesis includes minute descriptions of the changing condition of his own sexual organ.

examiners obviously feared embarrassment during his examination, since they asked him no questions. Later he spent some time in a private asylum, where he was astonished to find that one of his patients was an earl, but he never practised as a doctor. His geographical reputation was made by his work as a naturalist on the cruise of the *Marchesa* with C. T. Kettlewell during 1882–1884. He visited Kamchatka, Japan, the Ryukyu Islands, New Guinea and Borneo, and on his return spent a year writing an account of the voyage.[58] At the time of his appointment at Cambridge he was working on a biography of Magellan,[59] and had started work on the zoology of Cyprus with Lord Lilford.[60] Guillemard owed his appointment to Alfred Newton,

[58] F. H. H. Guillemard, *The cruise of the Marchesa to Kamschatka and New Guinea* (London: John Murray, 2 volumes, 1886; 2nd edition, 1 volume without coloured plates, 1889). Guillemard's connection with the *Marchesa* began somewhat melodramatically. Two brothers Kettlewell had each been left £800,000 by their grandfather, a Leeds carpet manufacturer. One shortly died, and the other, Charles, became known to Guillemard through a friend named Shuter. Shuter was anxious that Kettlewell and his money should remain in good hands, and decided that in his best interests Kettlewell should take a prolonged cruise. He therefore telegraphed Guillemard at Majuba Hill, not long before the attack, and suggested that he returned to England to accompany Kettlewell. Kettlewell's legal guardian, however, was one Major Bowyer Bowyer-Lane of the 53rd Regiment – in Guillemard's words, 'a scoundrel'. Bowyer-Lane took Kettlewell to Vienna at the age of 17 and devised a plot to obtain his money. In Vienna Bowyer-Lane had a liaison with a woman called Lina Stern, a former mistress of the Emperor Franz Josef's son, Prince Rudolph (later to commit suicide at Mayerling). He proposed that Kettlewell marry Lina's sister Anna, and that all four of them henceforth lived together. It was this prospect that led to Shuter's proposal of a cruise to remove Kettlewell from the scene. But by the time that Guillemard was back in England and had finished his thesis, and a boat had been chosen, the marriage devised by Bowyer–Lane had already been announced in *The Times*. It was decided to press on with the cruise nevertheless. Shortly before they left England, Guillemard took part in a shoot on Skye when Bowyer-Lane was also present. Guillemard was convinced that Bowyer-Lane was determined to murder him in a shooting 'accident' and went to great lengths to avoid being alone with him, finally fleeing to London. But then Bowyer-Lane was arrested, charged with forgery and jailed for three years, and the *Marchesa*, under the command of Captain Richard ffolliott-Powell, R.N., and with Guillemard and Kettlewell on board, sailed for the Far East. The party also included Captain Vesey Bunbury of The Buffs. No hint of this bizarre story appears in Guillemard's book; it is given in his unpublished autobiography, *op. cit.* (note 46), volume 7.

[59] He had been asked to write a biography of Cook for the series 'The World's Great Explorers', but said he preferred to do one of Magellan instead. This was published as *The life of Ferdinand Magellan and the first circumnavigation of the globe* (London: G. Philip and Son, 1892). Guillemard's manuscript notes for this work are in the Department of Geography, Cambridge. He later recalled that he was so overcome by writing of Magellan's death that he was almost unable to eat dinner in Hall that night (Guillemard (1927–1930)), *op. cit.* (note 46), volume 7, p. 147.

[60] *Lord Lilford: Thomas Littleton, fourth baron, F.Z.S., president of the British Ornithologists' Union. A memoir by his sister* (London: Smith, Elder, 1900), 290 pp. Thinly

whose lectures he had attended in 1872. Newton had been impressed with Guillemard's work on birds of paradise in New Guinea, and had got to know him well as an attender at his Sunday evening discussions ever since his early Lapland journey.[61]

Guillemard's tenure of the lectureship[62] lasted only six months, however: his resignation was announced on 11 December 1888. At the age of 80 Guillemard recalled the circumstances in a long, unpublished, autobiography, 'The years that the locusts have eaten':[3]

> The death of my beloved mother shortly after my return from Cyprus materially affected my health, which had been much undermined by the various disorders I had acquired in Africa and elsewhere. I was ordered abroad for the winter and felt the wisdom of the prescription. I was, indeed, quite unfit to inaugurate the teaching of a subject which was not only new but demanded numerous 'properties' of all kinds and sort of which we possessed not even so much as a Bradshaw. Nor had we any domicile. Enquiries as to where I should obtain some place to establish myself brought little more than the advice that I should obtain the part-time loan of a small lecture-room. On my demurring to this arrangement it was tentatively suggested that if I were to build my own lecture-room the University would probably find me a site! All this was not very encouraging, and meanwhile it was getting time for me to leave England. I accordingly proposed that I should give my inaugural lecture in the May Term next ensuing, turning on as soon as possible to the acquisition of the most necessary books, maps, and instruments. At the same time I offered to hand over the whole of my first year's salary towards the defraying of these expenses if the University would only provide a suitable home for its new science. This was agreed to with thanks and I departed happily to Madeira in search of my missing health and spirits.
>
> I was not left long in peace, however. I got a letter asking if I could not reconsider the matter and begin earlier. I was yielding, with

disguised portraits of Guillemard and others appear in W. H. Mallock, *In an enchanted island: or, a writer's retreat in Cyprus* (London: R. Bentley, 1889), 298 pp. Guillemard appears as Dr Guillaume in chapter 19, pp. 262–272. Edward Heawood wrongly identifies 'Mr Adam' in this book as Guillemard in *Geographical Journal*, 83 (1934), 351. For other identifications see pencilled annotations in the Cambridge University Library copy of Mallock.

[61] F. H. H. Guillemard in A. F. R. Wollaston, *op. cit.* (note 45), pp. 266–269.

[62] Not readership, as stated by Guillemard (1927–1930), *op. cit.* (note 46), chapter 31, and *Geographical Journal*, 83 (1934), 351. The same error was made by his successor (note 78).

[63] F. H. H. Guillemard (1927–1930), *op. cit.* (note 46).

> considerable reluctance I admit – when I got another communication asking me if I could not see my way to commencing a regular course of lectures in the Lent Term. Had I been left to carry out my original plan, I daresay I could have managed, but this was too much. Worried and out of health, I felt I could face neither the climate nor the work, and I sent in my resignation forthwith.
>
> I do not know that geographical teaching at the University lost much by my secession.[64]

Guillemard spent the rest of his life in Cambridge, editing a number of volumes of travel, writing the section on 'Malaysia and the Pacific archipelagoes' for Stanford's Compendium of Geography and Travel, and acting as General Editor for the Cambridge Geographical Series and the *Cambridge County Geographies*, both published by the University Press. He later served on the Cambridge Board of Geographical Studies[65] and on the council of the Royal Geographical Society, and his resignation of his lectureship does not seem to have affected his relationship with either the university or the Society. His last major foreign excursion was with Sir Charles Euan Smith's mission to the court of Morocco in 1892, and it is said that 'one of his best stories was of his washing and ironing a dress shirt on this occasion.'[66] This must have been an unusual experience for Guillemard, 'one of the fortunate few who could indulge their natural bent untrammelled by the need to earn a living'.[67]

After 1898 he lived at the Old Mill House, Trumpington, and became a notable Cambridge figure. He lived elegantly, with exquisite food, wine, glass and silver, in a house lacking gas and electric light, and filled with objects from all over the world, particularly watercolours. He usually went to bed at 2 a.m. and breakfasted at 10. The house was run by three maids in long black dresses and caps, whose combined service at the end of his life totalled more than 100 years. Guillemard was of Huguenot origin,[68] and one of his chief items of furniture was an old wardrobe in which an ancestor had been smuggled across the Channel from France. In addition to his

[64] *Ibid.*, volume 7, pp. 106–108.

[65] Guillemard was one of the most regular attenders at meetings of the board from its first meeting on 27 May 1904 to 5 March 1920, and occasionally thereafter until the end of 1922.

[66] *The Times*, 27 December 1933, p. 12.

[67] *Geographical Journal*, 83 (1934), p. 351. 'At the outbreak of the War [1914], however, he was hard put to it, for his money was in the Antwerp waterworks; but the death of his cousin Walter Guillemard, and of his neighbour, Dr T. B. Bumstead, placed him in affluent circumstances' (*The Times*, 27 December 1933). He left estate of £47,131 gross (*The Times*, 15 February 1934).

[68] F. H. H. Guillemard, *Pedigree of the Huguenot family of Guillemard* (privately published, 1924).

paintings he made other collections: one, of 357 photographs of Famous Waterfalls of the World, he donated to the Map Room of the Royal Geographical Society in 1910.[69]

Guillemard inevitably became out of touch with geography and geographers. An assistant at the Press helping with the *Cambridge County Geographies* later recalled a characteristic incident:

> Himself one of Nature's aristocrats, he was frequently puzzled by the background of some of the contributors to the series. Many of them used post-cards, a form of communication which he deplored. 'The fella's written to me on a post-card,' he would say to me in his plaintive purring voice, 'and do you see where he lives? He lives at *Ealing*! Now fancy a fella living at Ealing. . . .' It was something outside Guillemard's sociological experience.[70]

He became somewhat eccentric, affecting an elaborate mock-Arabic style in personal correspondence.[71] It is said that 'he made a point of attending every sale of a Great Auk egg or skin held in Britain.'[72]

In retrospect it is clear that Guillemard was temperamentally unsuited to the task of establishing geography in the university. Edward Heawood in his *Geographical Journal* obituary notice refers to him as 'a charming companion to those admitted to intimacy' but 'of a somewhat retiring nature.'[73] The master of Pembroke, writing in *The Times* after Guillemard's death, found him:

> a man of refined and delicate taste, keenly alive to the beauties of nature and of art . . . [who] looked with mild and tolerant disapproval on many of the developments of modern life, but it was impossible for his gentle and sensitive nature to do or say anything which might give pain to others. Invariabily courteous, he was a thoughtful and charming host, and the most loyal of friends. . . . His death leaves a gap in the lives of many friends, and a fragrant and precious memory of a very perfect gentleman.[74]

[69] *Geographical Journal*, 36 (1910), 511–512.

[70] S. C. Roberts, *Adventures with authors* (Cambridge: Cambridge University Press, 1966), reference on p. 27. The story about postcards may be apocryphal: there are postcards sent by Guillemard himself in the archives of the Cambridge University Press.

[71] For example, *ibid*., p. 28.

[72] G. E. Hutchinson, 'Cambridge remembered', in *The enchanted voyage and other studies* (New Haven: Yale University Press, 1962), 149–156; reference on p. 155. Professor Hutchinson tells me that he himself took instruction from Guillemard in taxidermy.

[73] *Geographical Journal*, 83 (1934), 351.

[74] *The Times*, 29 December 1933.

Guillemard had married a cousin – also a Guillemard – in 1890, but they soon separated (she becoming a Catholic and spending the rest of her life in a wheelchair), and he lived alone until his death on 27 December 1933. In addition to the masters of several colleges, his funeral was attended by the geographers Oldham, Debenham and Wordie, the geologist Cowper Reed, the zoologist Forster-Cooper, and A. E. Housman. He bequeathed 'His collection of the photographs of the cruise of the yacht *Marchesa* and all other of his photographs of foreign places, together with any works on Geography and Travel that the Reader in Geography may wish to choose from his library to the Museum of the Geographical Department of the University of Cambridge' and his medallion bust to the Royal Geographical Society. His pictures he left to the Fitzwilliam Museum.[75]

On his resignation, the lectureship was advertised with tenure for five years.[76] There is some evidence that Hugh Robert Mill was among those considered,[77] but on 23 May 1889 the appointment was announced of John Young Buchanan, F.R.S.[78] 'J.Y.', as he was known, was a scientist of distinction, but an unfortunate choice to rescue the subject from the Guillemard débâcle. Hugh Robert Mill described Buchanan's 'reserved and retiring disposition', though he added that 'his friendship was greatly esteemed by the few congenial souls whom he admitted to his confidence.'[79] In his *Autobiography* Mill was more forthright: he referred to Buchanan as 'an acquaintance of many years' standing, but always cold and distant, with no tendency towards friendship'.[80] The fault may have been partly Mill's, however, for *The Times* on his death called him 'a charming and stimulating companion, having a humour all his own'.[81]

Buchanan was a Scot, educated at Glasgow University under Crum Brown; he had a great facility with foreign languages, and had also studied at Marburg, Leipzig, Bonn and Paris. He made his reputation by his work as chemist and physicist on the voyage of the *Challenger* in 1872–1876.[82] During the expedition he demonstrated that Huxley's famous *Bathybius*

[75] *The Times*, 15 February 1934.

[76] *CUR* 1888–1889, p. 255.

[77] H. R. Mill to J. S. Keltie, 11 June 1903 (R.G.S. Archives).

[78] *CUR* 1888–1889, p. 754; not reader as he himself often stated.

[79] *Geographical Journal*, 66 (1925), 575.

[80] H. R. Mill, *Autobiography* (London: Longmans, Green, 1951), p. 50.

[81] *The Times*, 17 October 1925, and for Guillemard ([1927–1930], *op. cit.* [note 46], volume 7, p. 156 [verso]), 'he was not a very handsome and attractive man, but a very nice one.'

[82] D. R. Stoddart, 'Buchanan – the forgotten apostle', *Geographical Magazine*, 44 (1972), 858–862.

was merely a gelatinous precipitate of alcohol in seawater. He also published the first map of oceanic salinity ever produced[83] and was the first to demonstrate the nature of manganese nodules on the ocean floor. Later, in 1883 and 1886, he carried out oceanographic work round the Canaries and along the West African coast, and discovered the Congo submarine canyon.[84] He first visited the Alps in 1867, and during his later life went there each year, working on the physics and chemistry of ice.[85] He published well over 100 papers, some of which appeared in collections in 1913, 1917 and 1919.[86]

Buchanan was a man of very substantial means,[87] able to maintain private laboratories in Edinburgh and London, a large private yacht which he used for oceanographic work, and a house in Park Lane, London. He was on good terms with Prince Albert of Monaco, to whom he had been introduced by Henry Guillemard, and they often cruised together.[88] For the last ten years of his life, after the outbreak of war in 1914, he lived in Cuba and the United States, but for twenty years before that he lived in Christ's College. He never married, though Guillemard refers to an unsuccessful love affair in the 1880s.[89]

It fell to Buchanan to initiate geographical teaching in Cambridge. His inaugural lecture, 'Geography: in its physical and economical relations', was given in October 1889: it was concerned with the effects of the development of railways on the trade and population of new countries, particularly Argentina, which Buchanan visited in 1885 and 1888 and where he had substantial commercial interests.[90] During his lectureship, he gave courses on general geography; the distribution of land and water on the globe; physical geography and climatology; physical and chemical geography, with special reference to land surfaces and their development under climatic and other agencies; and oceanography.[91] In 1889–1890 his

[83] 'On the distribution of salt in the ocean, as indicated by the specific gravity of its waters', *Journal of the Royal Geographical Society*, 47 (1877), 72–86.

[84] 'The exploration of the Gulf of Guinea', *Scottish Geographical Magazine*, 3 (1887), 217–238.

[85] 'In and around the Morteratsch Glacier: a study in the natural history of ice', *Scottish Geographical Magazine*, 28 (1912), 169–189.

[86] *Scientific papers*, volume 1 (all published) (1913); *Comptes rendus of observation and reasoning* (1917); *Accounts rendered of work done and things seen* (1919); all published in Cambridge by the Cambridge University Press.

[87] Estate of £166,854: *The Times*, 26 January 1926, p. 17.

[88] Guillemard (1927–1930), *op. cit.* (note 46), volume 7, verso of p. 108.

[89] *Ibid.*, volume 7, verso of p. 156.

[90] Buchanan (1919), *op. cit.* (note 86), pp. 1–20.

[91] *CUR* 1889–1890, pp. 88, 638; 1890–1891, pp. 289, 420, 737, 764; 1891–1892, pp. 23, 56, 405, 431; 1892–1893, pp. 45, 376, 714.

general geography lectures in the Michaelmas term attracted 8–12 students, and oceanography in the Easter term 5–6. In 1890 he also lectured on physical geography and climatology in both the Lent and Easter terms.[92] In his last report to the R.G.S., on 7 May 1893, Buchanan reported that 'in the Michaelmas Term the attendance at the lectures was very satisfactory, and one student attended the laboratory for practical work. The attendance in the present term is not so favourable. . . . The room in the New Museums, which as been fitted up as a geographical laboratory, has already been taken advantage of by one or two students, and continuous work is being done in it by my assistant, Mr Heawood.'[93] Edward Heawood later became librarian at the R.G.S.

From his published work and from these reports it is clear that, distinguished as he was in oceanography and glaciology, Buchanan was not arousing the enthusiasm for geography, broadly defined, in Cambridge that Mackinder was stimulating in Oxford. 'He was, perhaps', wrote Guillemard,[94] 'too good for his post: too much of a specialist and inclined to be over the heads of his audience, but being a rich man I think he helped the School and his name carried considerable weight. He remained in office for some years, but later became much out of health and spirit . . . [and] gave up the Readership.' To Mill he was 'an eager investigator, and full of ideas for new researches which he pursued until he had accumulated enough data to be worth discussing, but then he either lost interest, or could not overcome some innate inertia, and often went no further. I never knew a man who did so much work and wrote so little about it.'[95] Buchanan's tenure of the lectureship came to an end in 1893: in the advertisement for a replacement it was stated that he was himself eligible for reappointment,[96] and he was invited to continue by Freshfield. But after consideration, he decided to withdraw.

> I have now had five years' experience of it, and I have not been able to observe any growth in it and the result, however it is looked at, is disappointing. In these circumstances I have with great regret decided not to apply for re-appointment. . . . If I may express an opinion I think that the new lecturer should, under all circumstances, be a Cambridge man, and preferably a young man who has recently got a

[92] *Notes on the lectures in geography in the University of Cambridge in the year 1889–90 by J. Y. Buchanan* (R.G.S. Archives).
[93] *Geographical Journal*, 2 (1893), 27–28.
[94] Guillemard (1927–1930), *op. cit.* (note 46), volume 7, p. 109.
[95] Mill (1951), *op. cit.* (note 80), p. 50.
[96] *CUR* 1892–1893, p. 787.

fellowship. Such a man is necessarily in touch with the undergraduates and with all the younger members of the University and would start with considerable advantages in the ways of creating a class.[97]

Buchanan died in London on 17 October 1925.[98] He had been elected a Fellow of the Royal Society in 1887, received the Keith medal of the Royal Society of Edinburgh, and in 1912 the gold medal of the Royal Scottish Geographical Society.[99] He left extensive manuscript notes on his work, including diaries illustrated with his own watercolours, but of these only his letters from the *Challenger* and some personal mementoes still survive.[100]

H. YULE OLDHAM AS LECTURER, 1893–1898

Towards the end of Buchanan's tenure of the lectureship, the R.G.S. approached the university to renew its offer of £150 a year towards the stipend of a lecturer for a further five years. The council of the senate recommended this to the university for acceptance in a report on 6 March 1893,[101] together with the offer of an exhibition or prizes which had never been taken up at Cambridge. The council felt, however, that a 'more precise definition of duties' for the lecturer was desirable, and also that the lecturer should be connected with one of the special boards of studies. The report aroused no discussion, and was approved on 27 April 1893.[102]

Nevertheless the vacancy came at a difficult time, for the R.G.S. was beginning to consider what its long-term attitude to higher education should be. A memorandum by the president, Sir Clements Markham, in May 1895 stated that the university subsidies formed 'a much larger sum than our finances will bear. It cripples our legitimate work, deranges our system of

[97] J. Y. Buchanan to D. W. Freshfield, 20 May 1893 (R.G.S. Archives).

[98] For biographical details: J. Peile, *Biographical register of Christ's College 1505–1905 and of the earlier foundation God's House 1448–1505* (Cambridge: Cambridge University Press, 1913), volume 2, p. 889; J. A. Venn, *Alumni Cantabrigienses* (Cambridge: Cambridge University Press), part II, volume 1 (1940), p. 429; *The Times*, 17 and 20 October 1925; *Geographical Journal*, 66, (1925), p. 575; *Proceedings of the Royal Society of London*, A, 110 (1926), pp. xii–xiii; *Nature, Lond.*, 116 (1925), pp. 719–720; G. Kutzbach, 'John Young Buchanan', in C. C. Gillispie, ed., *Dictionary of scientific biography* (New York: Scribners), volume 2 (1970), 557–558.

[99] *Scottish Geographical Magazine*, 28 (1912), p. 669.

[100] I am indebted to Mr W. Pollok-Morris for the loan of Buchanan's surviving diaries and papers; these latter include such items as invitations to take tea with Queen Mary.

[101] *CUR* 1892–1893, pp. 596–598.

[102] *CUR* 1892–1893, pp. 678, 783.

accounts, and it cannot be continued without serious detriment to the Society's interests.' He argued that 'our object was to obtain the same recognition for geography as a science, and as a subject for which honours may be conferred, as it holds in foreign universities. After ten years, and an expenditure of nearly £2000, we shall have entirely failed in this object.'[103] The assistant secretary, now J. S. Keltie, formerly the Society's inspector of geographical education, suggested that 'Council may conclude they have now done enough for the universities, and that these should now stand on their own legs.'[104] Markham reported that Mackinder was 'convinced that Oxford is not the place for planting our educational tree',[105] and there seemed 'slight prospect' of success at Cambridge. Feeling in the Society grew that funds should be devoted to a London geographical institute or school under the Society's direct control,[106] and although this ultimately came to nothing it seemed for a time that it would usurp the support hitherto given to the universities.

The question of a successor for Buchanan was thus a pressing one. Buchanan had been a fellow of Christ's College, where one of his colleagues was the professor of Arabic and editor of the *Encyclopaedia Britannica*, Robertson Smith. Smith was undoubtedly aware of the problems over the lectureship and he was also a close friend of the Russian *émigré* geographer Prince Kropotkin. It seems that Smith tried to persuade Kropotkin to apply for the position, but Kroptkin felt his freedom of action as an anarchist would be compromised and he declined.[107] The lectureship was accordingly advertised in April 1893, and on 8 June 1893 H. Yule Oldham was appointed.[108]

Oldham taught in Cambridge for 28 years, formally retiring in 1921. He lived for another 30 years, dying in his ninetieth year in 1951, but in spite of this he remains an elusive figure.[109] He wrote no books and published few

[103] C. R. Markham, *Memorandum . . . on the future educational policy of the Council, May 1895* (R.G.S. Add. Pap. 95).

[104] *Notes on the educational policy of the Society, October 1895* (R.G.S. Add. Pap. 95).

[105] C. R. Markham, *op. cit.* (note 103).

[106] L. M. Cantor; 'The Royal Geographical Society and the projected London Institute of Geography 1892–1899', *Geographical Journal*, 128 (1962), 30–35.

[107] J. Mavor, *My windows on the street of the world* (London: Dent, 2 volumes, 1923), volume 1, p. 23; G. Woodcock and I. Avakumović, *The anarchist prince: a biographical study of Peter Kropotkin* (London: Boardman, 1950), pp. 228–229. There is unfortunately no confirmation of this anecdotal evidence in the most recent full biography by M. A. Miller, *Kropotkin* (Chicago: University of Chicago Press, 1976). The appointment would of course have been bizarre.

[108] *CUR* 1892–1893, p. 1024.

[109] For biographical details: J. J. Withers, *A register of admissions to King's College,*

papers, and his death went unnoticed by the *Geographical Journal* and other periodicals. He took his B.A. at Oxford in 1886, and immediately spent a year as private tutor to the Duc d'Orléans. He was then for two years an assistant master at Hulme Grammar School, Manchester, and for one year at Harrow (when Winston Churchill was there), before spending the year 1891–1892 at Berlin University under Richthofen. Returning to England he was appointed to the R.G.S. lectureship at Owens College, Manchester, organized in conjunction with the Manchester Geographical Society.[110] In addition to his courses in Manchester, Oldham gave a series of 36 extension lectures for Victoria University, courses of lectures in Lancaster and Liverpool, and single lectures at other places.[111]

His inaugural lecture at Cambridge, on 'The progress of geographical discovery', was delivered on 24 October 1892;[112] he gave a second formal lecture the following year, with the vice-chancellor present, on 'A new discovery of America'.[113] Oldham offered courses on physical geography each year; on the geography of Europe in 1896–1897; and on central Europe in 1897–1898. These lectures, given at first in the chemical laboratories, attracted an initial audience of 15, rising after a few years to more than 30, and in the Lent term 1897 to more than 40.[114] In addition he continued to lecture at Owens College, Manchester, until the middle of 1894, and he travelled throughout England, almost rivalling Mackinder,[115] giving extension lectures at Bishop Auckland, Burton on Trent, Scarbor-

---

Cambridge, 1797–1925 (Cambridge: Cambridge University Press, 2nd edition, 1929), p. 227; J. A. Venn, *op. cit.* (note 98), part II, volume 4 (1951), p. 586; *Annual Report of Council*, King's College, Cambridge, November 1951, pp. 23–24; J. N. L. Baker, *Jesus College, Oxford, 1851–1971* (Oxford: Principal and Fellows of Jesus College, 1971). Oldham is unmentioned by L. P. Wilkinson, *Kingsmen of a century 1873–1972* (Cambridge: King's College, 1980). He was, however, commodore of the Cambridge University Cruising Club, where his name is still remembered (R. H. Tizard, personal communication).

[110] T. N. L. Brown, *The history of the Manchester Geographical Society 1884–1950* (Manchester: Manchester University Press, 1971).

[111] *Geographical Journal*, 2 (1893), 28.

[112] *CUR* 1893–1894, p. 54.

[113] *CUR* 1894–1895, p. 55; published in *Geographical Journal*, 5 (1895), 221–240.

[114] *Geographical Journal*, 4 (1894), p. 30; 6 (1895), p. 27; 8 (1896), pp. 62–63; 9 (1897), pp. 654–655; 12 (1898), p. 8. Details of lectures are given in *CUR* 1893–1894, p. 54; 1894–1895, p. 55; 1895–1896, pp. 88, 389; 1896–1897, p. 427; 1897–1898, pp. 57, 396. See also H. Y. Oldham to J. S. Keltie, 2 February 1897 (R.G.S. Archives). After 1904 the lectures were mainly given in the new Sedgwick Museum.

[115] E. W. Gilbert, 'Seven lamps of geography: an appreciation of Sir Halford J. Mackinder', *Geography*, 36 (1951), 21–43.

ough, York, Leamington, Chester, Leicester, Hull, Stepney, Market Drayton and Wellington. For the first time Cambridge and the Society had a lecturer interested primarily in education rather than travel or research. Over half a century later a member of his college recalled:

> Oldham in his prime, the pioneer whose enthusiasm drew large audiences, not so much in Cambridge as in the great manufacturing towns of the north of England, where his discourse – racy, anecdotal, scarcely perhaps by modern standards systematic enough to pass as academic – kindled an interest in such mysteries as the measurement of the curvature of the earth, and still keener interest in the story of the sailors and explorers who opened up the New World by their curiosity and courage.[116]

In the university Oldham nevertheless realized that 'the attendance at continuous courses will be limited as long as geography is not recognized in examinations',[117] and at least at first he felt hindered by being in a strange university and also by 'the prejudice of a preceding attempt . . . which lacked success'.[118] Moves were made towards the end of his first tenure to place the subject on a more secure basis. Oldham sent Keltie a list of names of influential men in the university who might be approached by the R.G.S. for their help. They included Lord Acton, Professor McKenny Hughes, the provost of King's (the Rev. Augustus Austen Leigh), Dr Cunningham, Dr Keynes, and – 'my chief help and adviser' – Dr MacAlister.[119]

THE READERSHIP ESTABLISHED

On 20 March 1897 Sir Clements Markham, as president of the R.G.S., wrote to the university to suggest that, since the Society had supported the lectureship for ten years, the subject was well established and Oldham's lectures successful, the university might be prepared to continue and support the lectureship itself. During the whole of this period the Society had contributed £150 a year and the university £50 a year to the lecturer's stipend. The university did not feel able to increase its financial commitment, but it proposed to the Society that it would convert the lectureship to a readership if the Society would continue its grant of £150 a year for

[116] *Annual Report of Council*, King's College, Cambridge, November 1951, pp. 23–24.
[117] *Geographical Journal*, 6 (1895), 27.
[118] H. Y. Oldham to J. S. Keltie, 30 March [1897?] (R.G.S. Archives).
[119] H. Y. Oldham to J. S. Keltie, 17 February 1897 (R.G.S. Archives).

another five years, and this the Society accepted. The readership would be connected with the Special Board for Biology and Geology, and the reader would also be a member of the Special Board for History and Archaeology. These proposals were put to the Senate on 9 February 1898 in the *Report of the General Board of Studies on a readership in geography.*[120] The report was discussed on 24 February 1898, when Dr MacAlister referred to the 'three able men' who had held the lectureship.[121] It was approved on 10 March,[122] and the readership was advertised the following day.[123] Oldham was elected for five years on 14 May 1898.[124] The university, without making any absolute commitment, agreed to re-examine the provision it had made for the teaching of geography, if funds were sufficient, and the vice-chancellor, in sending to the president of the R.G.S. the general board's minute on the subject, paid glowing compliments to Oldham and his work:

> may I add at no time since the establishment of the Lectureship . . . has the study of Geography been so active or so fruitful as it is at present. From all sides we receive evidence of the success of Mr Oldham's teaching and I personally am of opinion that it would be a grave misfortune for the University if so flourishing a department of work were to be closed.[125]

From 1898–1899 to 1901–1902 Oldham continued to give as reader the courses he had given as lecturer – on physical geography, the history of geographical discovery, the geography of Europe, and the geography of central Europe. In addition he offered a course in 1900–1901 on the geography of Europe illustrating the influence of physical features on historical development, which undoubtedly reflected a tour he had taken earlier in 1900 through Brittany with W. M. Davis, when he took over 100 photographs;[126] and in 1901–1902 he gave a practical class, fee one guinea, in map-making and map-reading.

As the end of his tenure as reader and the time when discussion between the R.G.S. and the university on any possible renewal of their agreement drew closer, there was a remarkable increase in activity in Cambridge geography. Special courses were organized in the Lent term 1903, 'to serve

[120] *CUR* 1897–1898, pp. 479–481.
[121] *CUR* 1897–1898, p. 563.
[122] *CUR* 1897–1898, p. 612.
[123] *CUR* 1897–1898, p. 595.
[124] *CUR* 1897–1898, p. 782.
[125] Vice-chancellor to president, R.G.S., 10 November 1897 (R.G.S. Archives).
[126] H. Y. Oldham to J. S. Keltie, 9 November 1900 (R.G.S. Archives).

as a training for those who wish to undertake exploration, or who while travelling or being stationed in foreign countries desire to contribute to our knowledge about them'.[127] under the title of courses for 'Explorers, and for Military Officers, Civil Servants, Missionaries and others'.[128] For a fee of three guineas, beginning on 19 January 1903, A. C. Haddon offered anthropogeography and practical ethnology, each twice a week; J. E. Marr, geomorphology and geology, twice a week; E. J. Garwood, plane-table and photographic surveying, weekly; and A. R. Hinks, elementary astronomical surveying, weekly.[129] Meanwhile Oldham continued to offer his usual courses.

These special lectures revealed the interest of members of the Special Boards both of Biology and Geology and of History and Archaeology, and they foreshadowed what were to be for many years the major interests of the Cambridge department. They also provided a skilful and successful means of bringing geographical studies to the attention of the university at a critical time, and of involving such influential men as Haddon.[130] For in

[127] *CUR* 1902–1903, p. 273.

[128] *CUR* 1902–1903, p. 337.

[129] *CUR* 1902–1903, pp. 273, 337.

[130] Haddon was professor of zoology in Dublin from 1880 to 1901; he also held a lectureship in physical anthropology at Cambridge in 1895–1898, and in ethnology in 1900–1909, becoming reader in ethnology in 1909. He was elected to the Royal Society in 1899. His lectures on anthropogeography for geographers lasted from 1903 to 1917. Frank Debenham recalled how Haddon's 'final lecture of the year he called his "Hurrah Lecture". . . . In it Dr Haddon gave a brilliant exposé of all his lectures rolled into one, and he won his audience by his intensity and his manner rather than by his elocution, which was jerky and hesitant, or by his appearance, which was that of a white-haired Papuan . . . or by his mode of speech, which was that of a man with a large pebble in his mouth': Diamond Jubilee Address, *Geography* 38 (1953), 212–220; reference on p. 216. It is said that on occasion his speech impediment was such that he had to abandon his lecture and would simply write on the blackboard an invitation to students to follow him to his room to read his lecture notes. In 1914–1915 he asked his secretary Miss Whitehouse to give the lectures for him, though they were still advertised in his name in the *Reporter*. When in 1920 they were being delivered by another former secretary, Mrs Quiggin, Haddon formalized the arrangement by changing the announcement in the University Lecture List. It was the first time a woman's name had been inserted in the List, 'a profanation which led to intemperate abuse of women in general and Haddon in particular', and it had to be withdrawn. It is of Haddon that the story is first told (it reappears in the history of Cambridge geography as recently as 1958) that while describing the amatory capabilities of the Torres Straits Islanders, and observing some Girton girls leaving by the back door, he called out, 'No hurry, there won't be a boat for some weeks.' See A. H. Quiggin, *Haddon the head hunter: a short sketch of the life of A. C. Haddon* (Cambridge: Cambridge University Press, 1942), see pp. 126–127. Times of course have changed: I am told that it is no longer necessary for lady undergraduates to travel to New Guinea for such experiences.

spite of Oldham's efforts, developments at Cambridge had lagged behind those at Oxford. Oxford had had a department, with a diploma, since 1899; Mackinder's initial appointment had been as reader; and he had been joined as assistant and then as lecturer in regional geography by A. J. Herbertson, soon himself to be reader.[131] Clearly more formal recognition was required at Cambridge if the subject were to succeed.

ESTABLISHMENT OF THE BOARD OF GEOGRAPHICAL STUDIES

In a preliminary *Report of the General Board on Geographical Studies in Cambridge*, dated 11 June 1903, the university was informed of discussion with the R.G.S. whereby the Society would increase its contribution to £200 a year for five years, and the university would match this with an equal sum, continue the readership, and establish a board of geographical studies and a special examination leading to an ordinary B.A. degree.[132] A fuller report was presented on 21 October 1903.[133] This included detailed regulations for the board, the special examination and the diploma. Much of the subsequent discussion[134] centred on whether the university had entered into any commitment or obligation to the R.G.S. by its previous support, to take the steps now envisaged: few speakers were either actively in favour of the subject or against it. The former included Dr MacAlister, nearly 40 years after his public school medal;[135] the latter Dr Mayo, who argued that historical geography must mean ancient geography, and would necessarily involve the study of Strabo and Pausanias in the original language. 'Would they do this ? Of course not. Well, then, where was the sincerity of those who proposed this?' The strongest support for the proposals came from Mr A. E. Shipley – 'a rotund figure who enjoyed his wine'[136] – speaking for the explorers:

> He did not think any university in the world had sent out so many expeditions during the last twenty years as had gone out from Cambridge. Not to mention the expedition of Dr Hickson, of

[131] J. F. Unstead, 'H. J. Mackinder and the new geography', *Geographical Journal*, 113 (1949), 47–57; E. W. Gilbert, 'Andrew John Herbertson 1865–1915: an appreciation of his life and work', *Geography* 50 (1965), 313–331.

[132] *CUR* 1902–1903, pp. 1016–1017.

[133] *CUR* 1903–1904, pp. 90–93.

[134] *CUR* 1903–1904, pp. 178–182.

[135] See note 24.

[136] G. Manley to the author.

> Downing, to the Celebes, and of Mr Caldwell, of Caius, to the interior of Australia, within the last six or seven years the number of Cambridge men who had gone to distant parts of the world and had really added to knowledge was quite astonishing. Only on Monday Mr Budgett returned from his fourth expedition to Africa, and returned with the material which for the last six years he had been diligently seeking on the Gambia, in Uganda, and lately in the delta of the Niger; he was the only man who had successfully obtained the embryos and the eggs of a rare fish of great importance to zoologists, and which an American expedition – so highly organized that it had even had its note-paper heading 'The American Polypterus Expedition' – failed to secure.
>
> Mr Grogan of Jesus was the only man who had ever walked from the Cape to Cairo, and he thought that those who had read his book must have felt how infinitely better it would have been if he had had some training in the sciences which it was now proposed to promote.[137]

Following the discussion, a slightly amended report was put to the university on 18 November 1903,[138] but specific proposals brought forward on 5 December all aroused some opposition. The first, on the establishment of the board and on the increase of the university's grant to £200 a year, to be matched by an equal amount from the R.G.S., was carried by 68 votes to 28. The second, on the stipend and length of tenure of the reader, and the third, on the regulations for the board, for a geographical education fund, the special examination and the diploma, were carried by a majority.[139]

The first Board of Geographical Studies was constituted in February 1904.[140] It comprised Dr Guillemard and Professor Darwin, nominated by the senate; Dr Marr, nominated by the Special Board for Biology and Geology; Dr Bury, Special Board for History and Archaeology; Dr A. W. Ward, Special Board for Economics and Politics; and Sir Clements Markham and Dr J. S. Keltie, for the R.G.S. The board also included the vice-chancellor and the reader of geography. It met for the first time on 27 May 1904. Meanwhile, the readership, with tenure to March 1908, had been advertised at a salary of £200 a year,[141] and Oldham was forthwith

[137] *CUR* 1903–1904, p. 474.
[138] *CUR* 1903–1904, pp. 204–207.
[139] *CUR* 1903–1904, pp. 271–272.
[140] *CUR* 1903–1904, pp. 503–536.
[141] *CUR* 1903–1904, p. 474.

appointed,[142] also becoming secretary of the board. The new board differed from the special boards of studies already in existence largely in that it had no right of representation on the General Board of Studies, but it otherwise had similar powers to regulate teaching and appoint staff. Boards, as opposed to special boards, were appointed for several new subjects at this time, largely to limit the number of bodies with whom the General Board had to deal.

The first task of the new board was to draw up regulations and a syllabus for the special examination and the diploma. In the board's report of 26 November 1904,[143] it was recommended that the examination for the diploma should consist of two parts, of which the first, or Part I, should be identical with the special examination. The special examination would consist of six papers: physical geography, historical and political geography; economic and commercial geography; cartography; history of geographical discovery; and elements of ethnology. Of these the first four would be compulsory and the last two optional. Part II of the diploma also comprised six papers: regional geography, surveying and mapping; geomorphology; oceanography and climatology; history of geography and anthropogeography. Of these the first two were compulsory and candidates had to take at least two of the others. When the detailed schedules for these examinations were finally published, they were signed by Donald MacAlister, himself by this time deputy vice-chancellor.[144] When the report was discussed on 26 November, Dr Mayo observed on the one hand that 'the bare fact of having travelled ought not to be accepted as a claim to any honour or title bestowed by the University,' on the other that the proposed examinations were too hard.[145] But the board's proposals were accepted.

The Michaelmas term 1905 began with an inaugural address on 19 October by Sir Clements Markham,[146] and special lectures were given during the year by Sir Archibald Geikie, Major E. H. Hills and D. G. Hogarth.[147] These lectures attracted audiences of more than 100, while Mr Oldham's ordinary courses attracted more than 80, over twice their previous number.[148] Other standard courses were given by Dr Marr, Dr Haddon and Mr Hinks, clearly following the pattern of the special lectures given in 1903; all the courses were given in the Sedgwick Museum. The first

[142] *CUR* 1903–1904, p. 614. Oldham acted as secretary to the board until December 1908.
[143] *CUR* 1904–1905, pp. 301–303.
[144] *CUR* 1904–1905, pp. 620, 718.
[145] *CUR* 1904–1905, pp. 530–531.
[146] *CUR* 1905–1906, p. 95.
[147] See Appendix for list of special lectures.
[148] *CUR* 1906–1907, p. 343.

formal lecture list was issued by the new board in the Michaelmas term 1905,[149] one year after that issued by the new Board of Anthropology.

THE WORK OF THE BOARD AND THE READERSHIP CRISIS OF 1908–1909

The work of the new Board of Geographical Studies was reviewed in annual reports, the first of which was published in December 1905.[150] The lecture courses by Oldham, Marr, Haddon and Hinks continued, and arrangements were made for the first examinations. The first examiners were appointed in 1906: Oldham and Hinks for Part I, and Oldham, Hinks, W. W. Watts and G. G. Chisholm of Edinburgh for Part II.[151] The examiner's fee was fixed at £5, 'to be increased if there are any candidates':[152] in this first year, there were none. The same examiners were again appointed in 1907,[153] and the first examination took place in the Corn Exchange at the end of that academical year – Part I on 29–31 May and Part II on 6–8 June.[154] C. H. Moore was examined and approved on both parts,[155] and was the first person to be granted the diploma in geography at Cambridge.[156] Efforts were made to organize a small library and equipment store, and a subscription to *The Geographical Journal* was taken out in 1905.[157]

The composition of the board changed from year to year, with members nominated by the senate, the special boards of biology and geology, of history and archaeology, and of economics and politics, and by the R.G.S. The original R.G.S. members were Keltie, whose original report had done so much to draw attention to the condition of geographical education in 1885, and Markham, the latter briefly replaced in 1905 by Sir George Taubman Goldie, his successor as president of the R.G.S. and a colonial administrator with little experience of academic geography. Goldie's appointment was not welcomed in Cambridge, and Markham soon returned.[158] The geologist

[149] *CUR* 1905–1906, p. 68.

[150] *CUR* 1905–1906, p. 331.

[151] *CUR* 1905–1906, p. 603.

[152] *Minute Book*, Board of Geographical Studies [hereafter *MB*].

[153] *CUR* 1906–1907, p. 694.

[154] *CUR* 1906–1907, pp. 959, 997.

[155] *CUR* 1906–1907, pp. 999, 1070.

[156] For a list of diplomas awarded between 1907 and 1912, see *CUR* 1912–1913, p. 159; and thereafter annual lists.

[157] *MB* 31 May 1905.

[158] *CUR* 1907–1908, pp. 919-920. These changes were discussed by the board on 4 February 1908, with Oldham absent (present: the vice-chancellor, Sir George Darwin, Guillemard, Marr, Hinks, Head), and the council of the senate was asked to discuss the changes directly with the R.G.S. (*MB*).

J. E. Marr served continuously for the first few years, as did Guillemard. At the beginning of 1907, two new members appeared, A. R. Hinks and J. Stanley Gardiner, both of whom were giving courses of lectures. Gardiner was then 35 years old and had made himself an impressive reputation as a scientific explorer with expeditions to the Pacific and Indian Oceans; he was also a member of council of the R.G.S.[159] Oldham served as secretary of the board from its inception in May 1904 until December 1908. With the changes in membership of the board in 1907, difficulties which had long been developing became more pronounced, and ultimately led to the most severe crisis in the history of the subject at Cambridge.

There were several strands in this discontent, but they mainly revolved round Oldham's capacity as a teacher and organizer and the kind of geography he wished to teach. Sir George Darwin wrote to Markham in 1903 that 'although we may admit in *private* that he [Oldham] is not a man of first rate force, yet I hear that his lectures on these subjects (literary, historical and political geography) are exceedingly good, and he certainly does his best.'[160] Darwin had in fact interviewed Oldham in November 1902, clearly to stimulate him to greater things: 'he is', he reported to Keltie, 'bucking up tremendously and has clearly a holy terror of me.'[161] Nevertheless Oldham's interests continued to be of a somewhat antiquarian and historical nature, and had little in common with the more technical geography of Marr (now lecturing on the new subject of geomorphology), of Hinks (lecturing on survey), and of Haddon (lecturing on biogeography and ethnology).

Problems arose over the place of survey in the examinations,[162] Darwin being in favour of compulsory survey and Oldham against. But this was broadened into a division between physical and human geography, and

[159] C. Forster-Cooper, 'John Stanley Gardiner 1872–1946', *Obituary Notices of Fellows of the Royal Society*, 5 (1945–1948), 541–553. Stanley Gardiner, who joined the board in 1907, had taken part in the Royal Society's boring expedition to Funafuti and had visited Rotuma; in 1899–1900 he led an expedition to Minikoi and the Maldive Islands, making large collections; and he had just returned from the successful Percy Sladen Trust expedition in H.M.S. *Sealark* to the coral reefs of the central and western Indian Ocean. He lectured several times to the R.G.S. and was elected to its council. He was also active in university affairs, being dean of his college in 1903–1909 and senior proctor in 1907–1908, though at that time his only formal appointment was as demonstrator in zoology. He was not a commanding teacher, though Forster-Cooper (who was with him in the Maldives and subsequently became director of the British Museum (Natural History)) noted that 'there was never any riot' in his lectures.

[160] G. H. Darwin to J. S. Keltie, 21 November 1902 (R.G.S. Archives).

[161] G. H. Darwin to C. R. Markham, 2 June 1903 (R.G.S. Archives).

[162] G. H. Darwin to J. S. Keltie, 21 November 1902 (R.G.S. Archives).

further reinforced by a growing antipathy between Gardiner and Markham, and hence between the university and the R.G.S. members of the board.[163] Markham in 1905 had just come to the end of a 12-year tenure of the R.G.S. presidency; previously he had been secretary for 25 years, and a fellow since 1854.[164] His interests, like Oldham's, were literary and historical, nor was he perhaps in sympathy with the new scientific geography of Gardiner and his colleagues. Mill wrote:

> He was by nature strongly drawn to the traditions of the past, and required time and tactful persuasion to turn his mind to new ideas. Although ready to accept scientific facts he was never able to adjust his mind to scientific modes of thought. Consequently he was disturbed and alarmed by the reforming tendencies of Galton and Strachey, nor was he in sympathy with the views of Freshfield on geographical education. . . . at times he may have failed in the exercise of critical faculty, and his judgement was often swayed by his strong likes and dislikes.[165]

At the age of 75 he was clearly a somewhat autocratic figure, though with his reputation damaged by the embroilment of himself, the Society and the government over the support of polar exploration, during which he had been criticized by the Prime Minister in the House of Commons and had fallen out with some of the country's most distinguished physical scientists, such as Sir John Murray and J. W. Gregory, over the conduct of scientific expeditions.[166] Markham was the chief link between the university and the

[163] H. Y. Oldham to C. R. Markham, 24 November 1904 (R.G.S. Archives).

[164] For biographical details, see A. H. Markham, *The life of Sir Clements R. Markham, K.C.B., F.R.S.* (London: John Murray, 1917), 384 pp., and J. S. Keltie, 'Biographical sketch', *Geographical Journal*, 47 (1916), 165–172.

[165] H. R. Mill, *The record of the Royal Geographical Society 1830–1930* (London: Royal Geographical Society, 1930), pp. 135–136, 137.

[166] Markham's championship of Scott in Antarctic exploration, to the exclusion of others, led to deep-seated antipathies, as well as to the public censure of the Royal Society and the Royal Geographical Society by the Prime Minister, Mr Balfour, in the House of Commons: see D. R. Stoddart, 'Social amenities and scientific explorations: geography and geographers in late Victorian London' in R. McLeod and S. Forgan, ed., *Scientific London* (forthcoming). After Markham's accidental death in 1916 (he was burnt to death while reading by candlelight in the hammock in which he always slept), Guillemard undertook to prepare for the press his last book, a history of polar exploration entitled *The lands of silence*. Sir Ernest Shackleton took offence at many passages in the book, allegedly either derogatory to himself or his expeditions or attributing to Scott discoveries which the latter did not make. A series of passages in the book were alleged by Shackleton to be libellous misstatements, indicating prejudice amounting to malice; withdrawal was called for and action threatened. Guillemard was taken aback, and wrote

council of the R.G.S., and not surprisingly it was to him that Oldham turned over problems such as the role of surveying in the examinations.

It is not difficult to see how Oldham's relative incompetence, Markham's arrogance and Gardiner's ambition led to serious dissension over the kind of geography to be taught. After only a few months on the board, Gardiner was complaining that 'intentionally or unintentionally Sir Clements Markham thros [*sic*] his weight on the Geog. Board here on the side of Historical and Economic Geography as opposed to what we may call Scientific, or Physical and Geological.'[167]

The difficulties came to a head with the prospective end both of Oldham's tenure of the readership in 1908 and of the agreement between the university and the R.G.S. each to provide £200 per annum. Throughout 1907 dissatisfaction with Oldham's teaching grew, though he himself was apparently largely unaware of it, and unperceptive reports by Markham had done little to warn the council of the R.G.S. of the impending trouble.[168] Only Keltie, living up to his reputation as a universal confidante and adviser, seems to have been aware of developments, but also in accordance with his reputation of 'when in doubt, do nothing,'[169] he seems to have let matters take their course.

The issue came to a head at the meeting of the board on 28 November 1907. Markham most injudiciously, as R.G.S. nominee, proposed that a sub-committee of the board be appointed to discuss the renewal of the readership with the R.G.S., and that the members of the sub-committee should be the vice-chancellor, the master of Peterhouse, and Guillemard. 'He has', wrote Gardiner to Keltie, 'finished my patience today' because of this. Concerning the membership of the sub-committee proposed by Markham,

> the V.C. alone is at all in touch with under-graduates and general teaching etc. in Camb. – and he of course knows little or nothing of Geogr. as a 'Science'. I don't suppose that Sir Clements understood

---

at once to Shackleton to express his amazement that any such construction could have been placed on passages which were, in any case, written by Markham and not by Guillemard himself. Messrs Hutchinson to F. H. H. Guillemard, 24 March 1921, and F. H. H. Guillemard to Sir E. Shackleton, undated (Cambridge University Press Archives).

[167] J. S. Gardiner to J. S. Keltie, 28 November 1907 (R.G.S. Archives).

[168] *Report to the council on the meetings of the Cambridge Geographical Studies Board on May 31st 1905; Report on the geographical studies at Oxford and Cambridge, 26 October 1905* (R.G.S. Archives).

[169] H. R. Mill, *op. cit.* (note 165), p. 140.

> that he of all members of the Board should not have proposed a Committee to represent the University. As it was we had to let it pass, while we wanted a Committee of the V.C., Marr and Haddon (or Sir Geo. Darwin).[170]

Gardiner informally asked Keltie to find some way of replacing Markham as R.G.S. representative on the board, and told him that Sir Thomas Holdich was willing to serve in his place. Relations between Gardiner and Oldham were no better than those with Markham: Gardiner told Keltie that Oldham was 'simply not *persona grata*' with other members of the board.[171]

Markham himself had by now realized the nature of the problem, and could 'quite see that a change is desirable' in the readership; nevertheless he found the matter a delicate one, for he had 'been on friendly terms with Oldham for 14 years and in various ways'.[172] Just before Christmas Marr wrote a long confidential letter to Keltie, which must have made it impossible for Oldham's tenure to be renewed:

> First with regard to the Reader's attainments, etc. I have never considered him strong on the physical side of the subject. . . . I have had ample means of judging of this as I still give courses with him and I have heard adverse comments from those who have attended his classes. It seems to be generally agreed by those who are in a position to judge that he is not strong on this side. . . . I further do not think that the present Reader has the energy and capacity necessary for the successful development of a school of geography here: it is a hard task. Consequently, were he now being reappointed for the first time I would not be surprised if those who were concerned with reappointment were to suggest that another Reader should be nominated.[173]

The problem was, as Marr saw it, that Oldham had already several times had his tenure renewed without any deep enquiry into his qualifications. His advancement to reader was 'done I believe after strong testimonials from the Head of a House and a University Professor. Surely after such marks of approbation, the present Reader might consider his tenure secure, and he will understandably find it difficult to start a new career at his time of life.'[174]

170 J. S. Gardiner to J. S. Keltie, 28 November 1907 (R.G.S. Archives).
171 J. S. Gardiner to J. S. Keltie, undated [1907?] (R.G.S. Archives).
172 C. R. Markham to G. T. Goldie, 28 November 1907 (R.G.S. Archives).
173 J. E. Marr to J. S. Keltie, 21 December 1907 (R.G.S. Archives).
174 *Ibid.*

Marr was, however, scrupulously fair to Oldham. He reported that the reader had done his best to familiarize himself with physical geography by travel in America, Africa and Australia, and he reported a historian's opinion that the lectures he gave were useful to history candidates. Finally, in spite of Oldham's limitations, could a better man be found? – 'it will be an exceedingly difficult task to find a really competent successor, who knows both sides of the subject, can teach, and understands Cambridge ways.'[175]

Oldham still knew little of these developments, though he cannot have been unaware of the feelings about him. Keltie asked him for a report on the work in geography at Cambridge, and Oldham supplied a detailed comparison with developments at Oxford, where the funds available were twice as large and the school had been established earlier. He reviewed developments up to the formation of the Board of Geographical Studies and stressed 'the difficulties presented by lack of success on the part of a previous holder' of his post. When the board was finally set up, 'so many conflicting interests had to be met, that the Board had great trouble in drawing up and getting accepted its scheme for the Examinations. I think you will see that our difficulties have been much greater than those at Oxford. Under the circumstances it is surprising that we have been able to do as well.'[176] It was a defensive and rather apologetic paper which Oldham later held had been intended as a purely private letter to Keltie.[177] Nevertheless Keltie circulated it to the R.G.S. council, where it can have done Oldham's cause no good.

Meanwhile, Markham's proposed sub-committee, to which Gardiner had taken such exception, had been replaced by a committee appointed by the General Board, comprising the master of Peterhouse, the vice-chancellor, Guillemard, and – doubtless at Gardiner's instigation – Sir George Darwin.[178] This committee met with the president of the R.G.S., Sir George Goldie, on 19 December 1907, and its discussions were reported to the R.G.S. council on 27 January 1908. It was then resolved:

> That the President be requested – in case it should prove necessary – to inform the Vice-Chancellor that the Council feel unable to alter their decision already intimated to him that the renewal of the Society's contribution of £200 p.a. towards the School of Geography at Cambridge is dependent on that School being placed on a satisfactory footing under a new Reader.[179]

175 *Ibid.*
176 H. Y. Oldham to J. S. Keltie, 25 January 1908 (R.G.S. Archives).
177 H. Y. Oldham to J. S. Keltie, 5 February 1908 (R.G.S. Archives).
178 Vice-chancellor to G. T. Goldie, 2 December 1907 (R.G.S. Archives).
179 Royal Geographical Society, *Minutes of Council*, 27 January 1908.

This remarkable resolution represented complete victory for the 'scientific' geographers led by Gardiner and Marr; the university was officially informed of it the following day.[180]

In retrospect two events probably proved fatal to Oldham's cause. One was the departure of Dr MacAlister to become principal of the University of Glasgow in 1907. MacAlister had supported the subject since the beginning and his earlier high opinion of Oldham's work has already been mentioned. The second was the election of Sir George Goldie as president of the R.G.S. The incisive and uncomprising nature of council's resolution certainly reflects Goldie's own temperament. Dorothy Wellesley's frank memoir of him, while perhaps exaggerated in some respects, certainly gives one no reason to suppose that he would seek a diplomatic solution to the difficulty. She tells of Goldie's 'uncontrollable passions, ruthlessness, indifference to individuals, contempt for sentimentality in any form. . . . He was, moreover, a violent and uncompromising man . . . [who] had a good deal of uncontrolled temper and was lashed into frenzies of impatience by stupidity or incompetence. Never did man suffer fools less gladly.'[181] With MacAlister still in Cambridge and with Markham supporting Oldham in the R.G.S. the outcome might well have been different, but it is difficult to resist the conclusion that the decision taken was ultimately of greater benefit to the university.

Oldham learned of these developments early in February, and he behaved with dignity in an impossible situation. He wrote to Keltie to say that 'the only cause of complaint against our work that you have told me of' was that progress lagged behind that at Oxford. 'I have never been asked to furnish any statement for the R.G.S. Council, and am in entire ignorance of the reasons for the attitude which the V.-C. tells me the Council have adopted. I have no doubt they are sound, but think you ought to have let me know. I am not likely to stand in the way of a solution to the present difficulty.'[182] To the vice-chancellor he wrote: 'I suppose the best thing I can do is to take steps to remove my person,' a remark which the vice-chancellor did not find entirely clear.[183] In February Oldham asked for an interview with Keltie at the R.G.S. Keltie saw him a fortnight later. It was 'a painful interview'. Only

[180] G. T. Goldie to vice-chancellor, 28 January 1908 (R.G.S. Archives).

[181] D. Wellesley, *Sir George Goldie, founder of Nigeria, a memoir* (London: Macmillan, 1934), pp. 112–115.

[182] H. Y. Oldham to J. S. Keltie, 5 February 1908 (R.G.S. Archives). Keltie's reply is an embarrassed one: the decision resulted from 'strong representations from another quarter' and indicates 'no depreciation of your merits as a lecturer and teacher'. J. S. Keltie to H. Y. Oldham, 6 February 1908 (R.G.S. Archives).

[183] Vice-chancellor to J. S. Keltie, 12 February 1908 (R.G.S. Archives).

now did Keltie openly express any opinion: 'personally', he told the vice-chancellor, 'I am sorry that his services could not be retained.'[184]

As the situation developed, the possibility arose of getting rid of Oldham altogether. A new post was to be established at Sheffield, largely at Goldie's initiative, and Keltie wrote to Oldham to tell him of it, at the same time informing the vice-chancellor that 'I believe he would be much more suited for a post of that kind. . . . He would have the thing in his own hands entirely, and would not have, as at Cambridge, to carry out a common scheme along with others. I believe', he continued, 'he is a really good lecturer and I am assured by those who know him that he is a clear and efficient teacher.'[185] Perhaps Keltie was embarrassed by the forthrightness of council's resolution, but his compliments came too late. The vice-chancellor also personally urged Oldham to apply for the Sheffield position. If he would, then the publication in the *Reporter* of the changes in the organization of geography could be delayed until after Oldham's resignation, and public embarrassment hence avoided.[186] The Sheffield closing date for applications was 14 March.[187] But Oldham declined to be pushed out. He stayed where he was while the university and the R.G.S. tried to reach agreement on what would happen next.

At first it was rather assumed that another reader would be appointed. Keltie wrote to W. M. Davis and F. W. Taussig at Harvard to enquire whether Ellsworth Huntington might be considered a candidate, but after consulting H. E. Gregory, Huntington's head of department at Yale, they advised against it.[188] One obvious internal choice, agreed by both the vice-chancellor and, surprisingly, Markham, was Stanley Gardiner. Markham pointed out, however, that 'he has enemies, and as he has rather taken a part against Oldham he might feel a delicacy in taking his place' – a delicacy which Markham thought could be overcome.[189] This was in November 1907.

By mid-February the vice-chancellor was exploring with Keltie whether the university and the R.G.S. could find more money, so that the number of

[184] J. S. Keltie to vice-chancellor, 22 February 1908 (R.G.S. Archives).

[185] J. S. Keltie to vice-chancellor, 13 February 1908 (R.G.S. Archives).

[186] Vice-chancellor to J. S. Keltie, 12 February 1908 (R.G.S. Archives).

[187] Vice-chancellor to J. S. Keltie, 12 February 1908 (R.G.S. Archives). The Sheffield post went to R. N. Rudmose-Brown, who was lecturer there from 1908 to 1931, and professor from 1931 to 1945; he died in 1957. See *Geography*, 42 (1957), 123 (obituary by Alice Garnett), and *Geographical Journal*, 123 (1957), 576–577 (obituary by D. L. Linton).

[188] G. J. Martin, *Ellsworth Huntington, his life and thought* (Hamden: Archon Press, 1973), p. 82.

[189] C. R. Markham to G. T. Goldie, 29 November 1907 (R.G.S. Archives).

teaching posts might be increased and the disparity between Oxford and Cambridge reduced.[190] In considering what was to be done, the General Board informed the university that 'the steady expansion of Geographical Science and the increase of specialization render advisable a reorganization of the Department of Geography and that the number of lectures provided from the Geographical Education Fund should be correspondingly increased'. On being formally approached, the R.G.S. agreed to continue its contribution of £200 a year for another three years, to be used specifically for the stipends of R.G.S. lecturers in regional or physical geography (£150 a year) and surveying and cartography (£50 a year). The board proposed that when Oldham's present tenure came to an end the readership should be suppressed and be replaced by a third lectureship, in historical and economic geography, at a stipend of £150 a year. The university would continue to match the financial contribution to the geographical education fund made by the R.G.S. for another three years. These proposals were recommended to the senate in the *Report of the General Board of Studies on the Department of Geography*, published on 16 May 1908.[191] In contrast to the public opposition when the Board of Geographical Studies was established in 1903, and surprisingly in view of the private dissension which had led to these changes, there was no opposition to the report, which was approved on 11 June 1908.[192]

The three new lectureships were rapidly filled. Oldham's appointment to the university lectureship in political (*sic*) and economic geography was announced on 6 August, at the same time as that of Philip Lake, M.A., to the R.G.S. university lectureship in physical and regional geography. Lake, already 43, was over two years younger than Oldham. He had a distinguished academic career, entering St John's College as a foundation scholar in 1884 and taking first classes in both parts of the natural sciences tripos. He then went to India with the Geological Survey, but returned after three years because of ill-health. He was a shy and retiring but kindly and assiduous man, and his steadying influence was needed after the problems of recent years. During the early years of his lectureship he published his *Textbook of geology* (with R. H. Rastall) and his *Physical geography*, both of which became standard texts and remained in print for over half a century. Perhaps because of his shyness he was never elected to a college fellowship, but his work led to the consolidation and ultimate success of

[190] Vice-chancellor to J. S. Keltie, 12 February 1908 (R.G.S. Archives).
[191] *CUR* 1907–1908, pp. 919-920.
[192] *CUR* 1907–1908, pp. 1051, 1154.

geography in Cambridge.[193]

At the beginning of the Michaelmas term A. R. Hinks, M.A., then chief assistant at the University Observatory, was appointed R.G.S. university lecturer in surveying and cartography.[194] Hinks's appointment, like those of Oldham and Lake, was for three years, but all were extended for a further five years in 1911,[195] and, in the case of Oldham and Lake, for a further five years in 1916.[196] In these extensions Oldham's title was changed from lecturer in political geography to lecturer in historical and economic geography, to correspond with the terms of the 1908 report. Hinks resigned on 18 July 1913 to become assistant secretary of the R.G.S., taking over when Keltie resigned as secretary on his 75th birthday, 5 February 1915: Hinks held the post for 30 years. Though his time in Cambridge was brief, it was nevertheless distinguished: he wrote his two influential books, *Map projections* (1912) and *Maps and survey* (1913) during his tenure, and was elected a fellow of the Royal Society for his astronomical work in the year of his move to London.[197] Hinks was replaced by C. S. Wright, the polar explorer, appointed from 1 January 1914 to 30 September 1916, but in 1915 Wright left for war service and Lake had to deputize for him.[198] Wright was reappointed in his absence for five years from 1 October 1916, but he resigned at the end of the war, and Frank Debenham was appointed, from 1 October 1919 to 30 September 1924, in his place.[199]

The reorganization of 1908 was completed by other minor changes. The *Report of the Council of the Senate on the constitution of the Board of Geographical Studies* of 19 October 1908[200] proposed a widening of the scope of the board to include a member nominated by the Special Board for Mathematics, in addition to those nominated by other special boards, and

[193] For biographical details, see *Geographical Journal*, 114 (1949), 115–116 (by J. A. Steers); *Nature, Lond.*, 164 (1949), 134 (by H. J. Woods); *Proceedings of the Geologists' Association*, 61 (1950), 117; *Quarterly Journal of the Geological Society of London*, 105 (1950), lxxxv-lxxxvi; *The Times*, 16 June 1949.

[194] *CUR* 1908–1909, p. 178.

[195] *CUR* 1910–1911, p. 1355.

[196] *CUR* 1915–1916, p. 807.

[197] For biographical details, see *Geographical Journal*, 105 (1945), 146–151 (by C. F. Arden-Close, G. R. Crone and K. Mason); *The Times*, 19 and 24 April 1945; *Dictionary of national biography*, 1941–1950, 393–394 (by R. N. Rudmose-Brown); and *Obituary Notices of Fellows of the Royal Society*, 5 (1945–1948), 717–732 (by H. Spencer Jones and H. J. Fleure). Hinks never recovered from being blown off the roof of the Royal Geographical Society by a bomb towards the end of the second world war, while fire-watching.

[198] *CUR* 1913–1914, p. 370; grace 4 of 4 June 1915.

[199] *CUR* 1915–1916, p. 807; 1918–1919, pp. 732, 954.

[200] *CUR* 1908–1909, pp. 100–101.

an extension of such appointments from one to four years. These proposals were accepted.[201] In June 1909 the board proposed changes in the examination regulations for Part II of the diploma (abolishing the requirement for two compulsory papers and replacing it by the need to pass in at least four of the papers). There were also accepted, and a detailed revision of the scope of papers was published.[202]

The board's discussions were now free from the unpleasantness of earlier years. Gardiner himself was elected to the chair of zoology in 1909 and became a fellow of the Royal Society in the same year.[203] He remained secretary to the board and worked closely with Lake in the development of the department. In December 1909 Gardiner reported to Keltie that 'the Board doesn't pull at cross purposes now; and they are really united to assist forward the subject.'[204] Though Gardiner ought perhaps to have resigned the secretaryship on taking his chair, he enjoyed himself and felt himself useful. 'I have', he told Keltie, 'perhaps more power in the post than anyone else would be likely to have. In the first place I am absolutely independent of the subject, and in the second, that my three expeditions to the Indian Ocean and your generous recognition of them, by placing me on the Council, have given me a certain indefinable prestige.'[205]

It is intriguing to speculate on the development of the subject if Gardiner had become reader in geography rather than professor of zoology. 'The only way to work the subject in Cambridge is, I believe,' he told Keltie before the crisis began, 'thro' the Natural Science School. . . . I believe it to be the only way to run Geography *here* into anything worth having.'[206] But Gardiner did not become reader, no close links were ever developed with the natural sciences tripos, and ultimately geography had a tripos of its own.[207]

A notable feature of this period of the department's development was the series of lectures given by visiting speakers. These effectively began with Douglas Freshfield's lecture on Ruwenzori in February 1907. Many were given by explorers, including Sven Hedin, Ernest Shackleton, Aurel Stein,

[201] *CUR* 1908–1909, pp. 192, 223.

[202] *CUR* 1908–1909, pp. 1145–1146; 1909–1910, p. 247. Gardiner replaced Oldham as secretary of the Board of Geographical Studies at the beginning of 1909, and remained in the post until mid-1912.

[203] It is said that one of the subject's strongest supporters, the zoologist A. E. Shipley, effectively abandoned the Department of Zoology on Gardiner's election to the chair, and thereafter devoted himself to Christ's College, where he was elected master in 1910.

[204] J. S. Gardiner to J. S. Keltie, 4 December (R.G.S. Archives).

[205] J. S. Gardiner to J. S. Keltie, 8 December 1909 (R.G.S. Archives).

[206] J. S. Gardiner to J. S. Keltie, undated [1907?] (R.G.S. Archives).

[207] See also note 158.

T. G. Longstaff, Sir Francis Younghusband, and Commander Evans of Scott's last expedition, but a few, by George Darwin, Bonney and Hinks himself, were of a more academic nature.[208] The board had long hoped to attract W. M. Davis to come to Cambridge,[209] but when he did speak, in the Senate House on 21 November 1913, on 'The lessons of the Colorado Canyon', it was at the invitation of the Woodwardian Professor of Geology.[210]

The department was also extending its holdings of books and equipment, and feeling the need for accommodation of its own. The board considered in 1909 that the department required a lecture room, a geographical laboratory and three private rooms for its own use,[211] and presumably this need became more acute when it received J. W. Clark's large collection of eighteenth- and nineteenth-century books of travel, bequeathed to the university in 1911.[212] Other miscellaneous possessions also accumulated: thus in the same year the board agreed to pay £12 to the International Congress of Hygiene for a model of Korea, provided it was delivered free of charge to the Sedgwick Museum.[213] The needs of the department were pressed in a *Report of the Special Board for Geographical Studies on the Department of Geography* in 1910,[214] and in the fifth, sixth and especially seventh *Annual reports* (1910, 1911, 1912).[215] The expansion of the department's work, including the arrangement of long-vacation courses for teachers in 1911, attended by 108 students, was also leading to financial difficulties: it was reported in 1910 that of an expenditure of approximately £840 in the year almost £280 was not met from regular income: at least a further £200 a year was thus required, in addition to the sums already being received from the R.G.S. and the university, in order to run the department effectively.[216]

The agreement between the university and the R.G.S. expired in 1911, but the Society yet again offered to contribute £200 a year for a further five years, together with a further additional sum of £100 a year for the same period if the university would likewise increase its contribution. The university was willing to provide £200 a year, but it could not agree to meet

[208] See the appendix.
[209] *MB* 7 November 1911, 18 July 1913.
[210] *CUR* 1913–1914, p. 245.
[211] *MB* 2 February 1909.
[212] *MB* 12 December 1910, 27 February 1911, 26 April 1911.
[213] *MB* 27 February 1911.
[214] *CUR* 1909–1910, pp. 958–960.
[215] *CUR* 1909–1910, pp. 432–433; 1910–1911, pp. 241–243; 1911–1912, pp. 368–370.
[216] *MB* 26 April 1911.

any additional commitment. The R.G.S. then offered to pay the additional £100 for one year only if the university could raise a similar sum. Seventeen senior members of the university, including the three current and two former lecturers, contributed sums of £3–£10 each to provide the money. The R.G.S. made it clear that their offer of an additional £100 a year until 1916 still stood if the university could provide equal sums.[217] It failed to do so again in 1912, when the whole £100 was subscribed by one of the university's M.P.s, Mr J. F. P. Rawlinson,[218] but in spite of this private generosity the financial position continued to deteriorate.

In November 1912 the Board of Geographical Studies presented its case in detail to the General Board. It pointed out that the department was now dealing with 30–40 students a year, that the number taking the special examination and Part I of the diploma had increased from 8 in 1909 to 24 in 1912, and those taking Part II of the diploma from 4 to 10 in the same period. The lecturers, none of whom was a fellow of a college, had inadequate stipends, and they could only hope to cover the syllabus through voluntary assistance with extra courses, for example by Professor Gardiner on oceanography and Dr Marr on geomorphology. The board pointed out what was surely self-evident, that 'the long-continued generosity of the Royal Geographical Society has alone made it possible to establish in the University an efficient Department of Geography,' and it asked the university to increase its contribution by £100 a year to match the increase proposed by the R.G.S. but only so far assured by private generosity.[219] In the *Report of the General Board of Studies on an increase in the grant to the Geographical Education Fund* in May 1913[220] this was recommended to the university and subsequently approved. When the university was again approached by the R.G.S. in December 1915 (this time in a letter written by its new secretary, A. R. Hinks) with a proposal to continue the grant of £300 a year for a further five years, the approach was for the first time immediately welcomed.[221] In the discussion on the proposal, only Dr Mayo, professing an 'entirely neutral and thoroughly independent position', raised objection, suggesting that if the lecturers' stipends were reduced by half the university could avoid making any contribution at all: in spite of this the report was accepted.[222]

[217] *CUR* 1909–1910, pp. 958–960.
[218] *CUR* 1911–1912, pp. 241–243.
[219] *CUR* 1912–1913, pp. 368–370.
[220] *CUR* 1912–1913, pp. 1075–1078.
[221] *CUR* 1915–1916, pp. 524–525.
[222] *CUR* 1915–1916, p. 550. The Rev. James Mayo (1840–1920), 'extremely eccentric',

Geography was now well established in the university. Until the outbreak of war the numbers passing the examinations continued to increase: 32 students passed the special examination or Part I of the diploma in 1913-1914, and 18 students Part II, with 98 attending courses. Though numbers declined during the war years when problems of teaching and fieldwork arose, the scope of the diploma was extended by adding a paper on advanced commercial geography in Part II, to match the economic and commercial geography of Part I, in 1917.[223] At this time also the Board of Geographical Studies began to discuss the need for a geographical tripos, leading to the B.A. degree with honours, not only because of the development of the subject itself in the university but also because of the implications of such a qualification for those taking the civil service examinations.[224] With the end of the war these proposals were formulated in detail.

THE ESTABLISHMENT OF THE GEOGRAPHICAL TRIPOS

The *Report of the Board of Geographical Studies on the establishment of a tripos examination in geography*, dated 24 October 1918,[225] was presented to the university in November. It was argued that while the diploma had been useful in preparing teachers, a further qualification was required for 'those who intend to devote themselves either wholly or in part, to geographical research'. The need was justified in terms very similar to those first used in 1871: it would be a suitable qualification for 'the future statesman, administrator, merchant or missionary', and it would be valuable for the 'Home, Colonial and Indian Civil Services and for the Diplomatic and Consular Services'. A two-part tripos was proposed, with six papers in Part I, all compulsory, and five papers in Part II, of which a candidate would offer not less than two and not more than three. In addition, candidates for both parts would sit a practical examination. The

---

was well known as 'a continual attendant at discussions in the Senate House, always contributing his remarks after stating that he "knew nothing of the subject in question". One curious failing was to sit year after year for Special Examinations, in which he invariably gained first classes. . . . In the early days of wireless he took legal action against his next-door neighbours, alleging they were injuring him by sending "rays" through his person.' J. A. Venn, *op. cit.* (note 98), part II, volume 4 (1951), p. 378. Also *The Times*, 17 April 1920.

[223] *CUR* 1916–1917, p. 607.

[224] *MB* 23 October and 23 November 1917, 28 February, 23 May and 24 October 1918.

[225] *CUR* 1918–1919, pp. 180–183.

form of the papers resembled those of the diploma. In Part I they comprised physical geography; political and economic geography; cartography; history of geography; anthropogeography; and regional geography; and in Part II geodetic and trigonometrical surveying; geomorphology; oceanography and climatology; historical and political geography; and economic and commercial geography. Although there was considerable private discussion over the role of surveying in Part I, the report was approved by the university on 31 January 1919, in contrast to the earlier dissension over proposals for the diploma.[226]

With the new requirements of the tripos and the prospect of a greatly increased number of students after the war (113 students attended in 1919–1920),[227] the whole scale of teaching in the subject had changed. The Board of Geographical Studies considered early in 1919 that the department needed a head, who should be both a reader and a physical geographer.[228] Aware of these discussion, the R.G.S. once more approached the university and proposed the immediate creation of a readership in geography, and on 11 June 1919 the General Board recommended this to the university for an initial period until 30 September 1921, when the existing agreement under which the R.G.S. contributed £300 a year expired. The General Board further proposed that the first reader be Philip Lake, and that his lectureship in regional and physical geography be suspended for the duration of his tenure of the readership.[229] Lake was appointed in November 1919;[230] his tenure as what was called Royal Geographical Society Reader was extended for five years from 1 October 1921,[231] and for another five years from 1 October 1926.[232] Though he resigned prematurely in April 1927 (having attempted to resign earlier in 1922), to be replaced by Debenham,[233] Lake was responsible for the beginnings of tripos teaching and for the enlargement of the department which this entailed.[234]

The first examination for Part I of the tripos was held in June 1920, and for Part II in June 1921. The class-list for Part II, issued on 11 June, included two names in Class I: J. A. Steers (who had taken the diploma in 1917) and

[226] *CUR* 1918–1919, pp. 274, 432.
[227] *CUR* 1919–1920, p. 367.
[228] *MB* 28 February and 9 May 1919.
[229] *CUR* 1919–1920, p. 52.
[230] *CUR* 1919–1920, p. 285.
[231] *CUR* 1920–1921, p. 1198; Lake attempted to resign but was dissuaded in February–May 1922 (*MB*).
[232] *CUR* 1926–1927, p. 288.
[233] *CUR* 1926–1927, p. 1140.
[234] For biographical details see *Geographical Journal*, 114 (1949), 115–116.

J. H. Wellington; A. Bond in Class II; and N. Senepaty in Class III. With this, the Royal Geographical Society's aims were realized. In 1921 it reduced its contribution to £200 a year, and stopped it altogether in 1923.[235] Oldham, who had taught in Cambridge for nearly 30 years, half of the time single-handed, retired at the end of 1920–1921, and though not a fellow, lived in King's for many years afterwards.[236] His replacement, and the other lecturers and demonstrators appointed at that time (R. W. Stanners, J. A. Steers and Mrs. M. Anderson) under Debenham, who became the first professor in 1931, dominated Cambridge geography for the next two or three decades.

CONCLUSION

The establishment of the geographical tripos and the first Cambridge degrees in geography in 1921 completed 50 years of effort by the Royal Geographical Society, beginning with Sir Henry Rawlinson's letter to the vice-chancellor exactly 50 years before. It had taken 10 years for the first lecturer to be appointed, and for another 20 years teaching was in the hands of one man – first Guillemard, then Buchanan, and then Oldham, the latter as reader from 1898 to 1908. The effective organization of the subject began with the formation of the Board of Geographical Studies in 1903, and really effective teaching with the appointment of three lecturers, Oldham, Lake and Hinks, in 1908. The approval of the tripos regulations and Lake's appointment as reader and head of department in 1919 brought the period of the establishment of academic geography in Cambridge to a close. A chair had still to be founded, and a building to house the department obtained, but the form of teaching and indeed many of the teachers were to characterize the Cambridge school for the next 20 to 30 years. Though the R.G.S. took little further active part in Cambridge geography, it must have been well satisfied with its investment, which had totalled some £7,250 over 35 years.

At the same time, it must be admitted that the reputation of the Cambridge department stems from the period after the institution of the tripos, rather than before it, and the contrast between the early history of geography in Cambridge and that in Oxford is marked.[237] Both universities

[235] *CUR* 1922–1923, pp. 976–977.

[236] Oldham, until replaced by R. W. Stanners, continued to give lectures until 1930.

[237] E. W. Gilbert, 'Geography at Oxford and Cambridge', *Oxford Magazine*, 75 (1957), 274–278; J. N. L. Baker, 'The history of geography in Oxford', in *The history of geography*

were approached by the R.G.S. in 1886; a readership was established at Oxford in 1887 and a lectureship at Cambridge in 1888. But there were two major differences between the two universities. The first was simply in the nature and calibre of the men appointed. At Cambridge Guillemard was an aesthete who did not take his appointment seriously, though he later did a great deal for the subject; Buchanan was a rather reserved physical and chemical oceanographer, uninterested in geography in the wider sense and with no gift for teaching undergraduates; and the indefatigable Oldham, while constantly lecturing, seems to have aroused little enthusiasm and made little impact outside educational circles. In contrast, the work of Mackinder and Herbertson at Oxford was establishing geography not only at that university but on a national level, and it was extended by lectures, textbooks and by Herbertson's connection with the Geographical Association.[238] The second reason was that during this early period most of the better students went to Oxford rather than to Cambridge. Partly this resulted from the differences in the reputations of the teachers at each university, but it was also an effect of the establishment of a formal qualification in geography at Oxford as early as 1900 (first awarded 1901) compared with 1903 (and 1907) in Cambridge. Thus of all those granted the Cambridge diploma, only two (Leonard Brooks in 1908 and J. A. Steers in 1917) made reputations in academic geography. Diplomas at Oxford, meanwhile, were being awarded to such distinguished students as O. J. R. Howarth (1902), E. G. R. Taylor (1908), C. B. Fawcett (1912), O. G. S. Crawford (1910) and W. G. Kendrew (1911). As Gilbert pointed out, by 1920 the Oxford school had 'provided the whole of the teaching staff to six departments in universities and university colleges, and one or more member of staff to the staffs of six more'.[239] It was not until the first years of the Cambridge tripos, with the names of J. A. Steers (1921), J. H. Wellington (1921) and G. Manley (1923), that the balance began to be redressed.[240]

(Oxford: Basil Blackwell, 1963), pp. 119–129; D. I. Scargill, 'The RGS and the foundations of geography at Oxford', *Geographical Journal*, 142 (1976), 438–461. The only published accounts of the development of Cambridge geography are notes by P. Lake, 'The geographical school at Cambridge', *Geographical Teacher*, 10 (1919), 80–81; by W. M. Davis, 'Geography at Cambridge University, England', *Journal of Geography*, 19 (1920), 207–210; and by J. S. Keltie, 'The position of geography in British universities', *American Geographical Society Research Series*, 4 (1921), 1–33, see pp. 12–14.

[238] E. W. Gilbert, *Geographical Journal*, 110 (1947), 94–99, and *Geography*, 50 (1965), 313–331.

[239] E. W. Gilbert, *Oxford Magazine*, 75 (1957), p. 274.

[240] For personal accounts of the early post-war period by Alfred Steers, see 'St. Catharine's

There was a macabre ending to the involvement of the Royal Geographical Society with Cambridge. A. F. R. Wollaston, who was well known to Guillemard and who wrote Alfred Newton's biography,[241] became honorary secretary of the R.G.S. and also tutor of King's. Like Guillemard he had collected birds of paradise in New Guinea, and had also explored in the Ruwenzori as well as being a member of the first Mount Everest expedition in 1921. On Monday 2 June 1930 he attended a meeting of the council at the R.G.S. He returned to Cambridge, and on the following day, Tuesday 3 June, he was murdered in his college rooms by an undergraduate named Potts, armed with a stolen .32 Webley automatic, who immediately afterwards used it to kill first a policeman and then himself.[242] It was almost sixty years since the first approach by the R.G.S. to the university had initiated the story of the development of Cambridge geography.

College in, and immediately after, the first world war', *Magazine of St. Catharine's College Society*, 1982, 32–37, and 'Geography at St. Catharine's College', *Magazine of St. Catharine's College Society*, 1983, 20–23.

[241] A. F. R. Wollaston, *op. cit.* (note 45).

[242] See *Evening Standard*, 3–7 June 1930; *Cambridge Daily News*, 3–7 June 1930; and N. Wollaston, 'Potts', *Telegraph Sunday Magazine*, 57 (16 October 1977), 67–72. Guillemard and Oldham both attended the funeral. John Maynard Keynes described Wollaston as a man who could 'break down everyone's reserves, except his own': 'The death of Mr A. F. R. Wollaston', *The collected writings of J. M. Keynes*, volume X: *Essays in biography* (London: Macmillan, 1972), 347–348. For biographical details, see *Geographical Journal*, 76 (1930), 64–66. There is an unsettling anecdote about Wollaston, recounted in Miriam Rothschild, *Dear Lord Rothschild, birds, butterflies and history* (London: Hutchinson, 1983), chapter 19, 'Walter's collectors: A. F. R. Wollaston and N. C. Rothschild', 170–176; reference on p. 175. He once described how, during his New Guinea expedition, 'a mysterious man walked ahead of him, steadily and helpfully during a particularly arduous and dangerous return trip through the jungle. Wollaston never caught up with this man who always kept some distance ahead, but he retained a very clear cut mental picture of him. On his return to England, when he was buying a coat in a clothes shop in London, he glimpsed in a long mirror the familiar, unmistakable, figure of his unknown friend. Very startled, he turned round quickly, and realized it was his own reflection seen from behind.' It is also curious that the weekend before he was murdered, Wollaston told of meeting Lord Rothschild's younger brother Charles while walking on Newmarket Heath; Charles had however committed suicide seven years before. Wollaston wrote to tell Charles's widow of the conversation, but when, greatly distressed, she tried to find out more, Wollaston himself was already dead.

APPENDIX

SPECIAL GEOGRAPHICAL LECTURES GIVEN AT CAMBRIDGE 1888–1914

1888 18 and 25 February, 3 and 10 March. General R. Strachey: Principles of geography.
1889 October. J. Y. Buchanan: Geography, in its physical and economical relations (inaugural lecture).
1893 24 October. H. Yule Oldham: The progress of geographical discovery (inaugural lecture).
1894 24 October. H. Yule Oldham: A new discovery of America.
1895 3 March. Surgeon-Major Sir G. Scott Robertson: The north-west frontier of India.
1905 19 October. Sir Clements Markham: Inaugural address (Board of Geographical Studies).
1905 21 November. Sir A. Geikie: The evolution of a landscape.
1906 5 May. Major E. H. Mills: Geography of international frontiers.
1906 20 February. D. G. Hogarth: Geographical conditions affecting population in the east Mediterranean lands.
1907 14 February. D. W. Freshfield: Ruwenzori (Duke of the Abruzzi's expedition).
1909 12 February. T. G. Longstaff: Explorations in the Himalayas.
4 March. Sven Hedin: Explorations in Central Asia.
10 May. G. Darwin: Tides in the solid earth.
21 October. T. H. Holdich: Some aspects of physical geography.
29 October. Lieut. Shackleton: The last Antartic expedition.
25 November. T. G. Bonney: A desert phase in the development of Britain.
1910 20 January. Aurel Stein: Explorations in Asia.
19 May. T. G. Longstaff: Glacier exploration in the eastern Karakoram Himalaya.
11 November. R. T. Gunther: Earth movements of the Italian coast.
1911 23 February. Sir F. E. Younghusband: Practical geography.
8 May. A. R. Hinks: Recent progress in the measurement of the earth.
24 May. H. G. Lyons: The hydrography of the Nile and the Congo.
1912 1 February. A. Rose: Chinese frontiers of India.
15 November. W. Larden: Agriculture and industry in Argentina.
1913 10 November. Cdr Evans: Captain Scott's expedition.
1914 13 February. R. E. Priestley: The northern party of Captain Scott's Antarctic expedition.
2 March. E. M. Jack: The Mufumbiro Mountains.

6

Humanizing the New Geography

It has become fashionable to interpret the rise of the 'New Geography' at the end of the last century as a response to the demands created by the spread of imperialist policies among the European powers. Hudson[1] has indeed argued that 'the study and teaching of the new geography at an advanced level was vigorously promoted at that time largely, if not mainly, to serve the interests of imperialism in its various aspects including territorial acquisition, economic exploitation, militarism and the practice of class and race domination.' Certainly the political tensions generated by the Franco–Prussian war and the unification of Germany focused attention on the need for accurate geographical knowledge, as did the expansion of colonial activities in Africa and Asia.[2] To the many soldiers, sailors and

[1] B. Hudson, 'The New Geography and the new imperialism: 1870–1918', *Antipode*, 9 (2) (1977), 12–19; reference on p. 12.

[2] For France, see D. V. McKay, 'Colonialism in the French geographical movement, 1871–1881', *Geographical Review*, 33 (1943), 214–232; and N. Broc, 'Nationalisme, colonialisme et géographie: Marcel Dubois (1856–1916)', *Annales de Géographie*, 87 (1978), 326–333. In the English literature the case is made most explicitly by T. H. Holdich, 'The use of practical geography illustrated by recent frontier operations', *Geographical Journal*, 13 (1899), 465–480; and G. T. Goldie, 'Progress of exploration and the spread and consolidation of the Empire in America, Australia and Africa', *Geographical Journal*, 17 (1901), 231–240. The geographical societies of Hamburg and Berlin had a particular emphasis on colonial territories: E. von Drygalski, 'Die Entwicklung der Geographie seit Gründung des Reiches', *Mitteilungen der Geographischen Gesellschaft in Hamburg*, 43 (1933), 1–11; R. Barmm, 'Krieg und Erdkunde unter besonderer Berücksichtigung des erdkundlichen Unterrichtes', *Mitteilungen der Geographischen Gesellschaft in Hamburg*, 30 (1917), 245–277; S. Passarge, 'Das Geographischen Seminar des Kolonial-Instituts und der Hansischen Universität 1908–1935, Erinnerungen und Erfahrungen', *Mitteilungen der Geographischen Gesellschaft in Hamburg*, 46 (1939), 1–104; and for commentary, F. J. W. Bader, 'Die Gesellschaft für Erdkunde zu Berlin und die koloniale Erschliessung Afrikas in der zweiten Hälfte des 19. Jahrhunderts bis zur Gründung der ersten deutschen Kolonien', *Die Erde*, 109 (1978), 36–48. Similar interpretations have been argued for other field sciences, e.g. by L. H. Brockway in *Science and colonial expansion: the role of the British Royal Botanic Gardens* (London: Academic Press, 1979).

colonial administrators who served on the council of the Royal Geographical Society in the eighties and nineties the practical advantages of geographical knowledge must have been self-evident. 'It is a moot question,' said Sir George Goldie (in a paper rather ironically titled 'Geographical ideals'), 'whether war is more useful to geography or geography to war. . . . War has been one of the greatest geographers.'[3]

But there was, simultaneously, a quite different tradition, which can be traced back to the broad and liberal sympathies of the Forsters and of Humboldt: one which stressed the equality of men and the need for compassion and collaboration in the solution of world problems. At the end of the century two men symbolized this view, both in their lives and in their writings: Elisée Reclus and Peter Kropotkin. It is timely to reassess their work and to enquire why their influence on the formal organization of the New Geography was not greater than it was and why in the longer term their views command increasing respect. This I shall do through brief examination of their lives and thought.[4]

Elisée Reclus was born near Bordeaux in 1830 and lived through the century.[5] His father was a strict and unbending Moravian minister, who once censured Elisée for his wickedness in exploring the local countryside: Elisée's childhood was one of 'sadness and dread', according to his sister, 'of which he never spoke without bitterness'.[6] The Moravian observance not surprisingly came to appear to the growing boy a 'disgusting ritual of childish practices and conventional lies'.[7] When he was nineteen, Elisée went to Strasbourg to study, and later walked to Berlin where he came under the influence of Ritter and began to develop a somewhat similar philosophy. It was at this stage, too, that Elisée's deep romanticism became apparent. Many years later his older brother recalled that Elisée become so excited at his first sight of the sea, in 1849, that he bit him on the shoulder

[3] G. T. Goldie, 'Geographical ideals', *Geographical Journal*, 29 (1907), 1–14; reference on p. 8.

[4] This chapter has been expanded from 'Kropotkin, Reclus, and "relevant" geography', *Area* (1975), 188–190, and 'Humane geographer: the enigma of Elisée Reclus', *Progress in Human Geography*, 5 (1981), 119–124.

[5] The main biographical source is G. S. Dunbar, *Elisée Reclus, historian of nature* (Hamden: Archon Books, 1978). See also P. Girardin and J. Brunhes, 'Elisée Reclus Leben und Wirken', *Geographische Zeitschrift*, 12 (1906), 65–79; P. Geddes, 'A great geographer: Elisée Reclus, 1830–1905', *Scottish Geographical Magazine*, 21 (1905), 490–496, 548–555; B. Giblin, 'Elisée Reclus: pour une géographie' (Paris: University of Paris-Vincennes, thesis, 1978), and her 'Elisée Reclus: géographie, anarchisme', *Hérodote*, 2 (1976), 30–49.

[6] M. Fleming, *The anarchist way to socialism: Elisée Reclus and nineteenth-century European anarchism* (London: Croom Helm, 1979), reference on p. 31.

[7] *Ibid*., p. 32.

until blood flowed.[8] Both brothers were involved in protest after the re-establishment of the Empire by Napoleon III in 1851, and were forced to leave France. Elisée went to England and thence to Louisiana, where he was profoundly shocked by the institution of slavery. He travelled elsewhere in North America, before an unsuccessful attempt to settle down in Colombia. He did not return to France until 1857.

This time in South America resulted in Reclus's first important book, *Voyage à la Sierra Nevada de Sainte-Marthe: paysages de la nature tropicale*, published in 1861.[9] It is perhaps curious that Reclus did not make more, as a geographer, of his attempts at pioneer farming in northern Colombia. Partly this must have resulted from his almost total lack of training, either in scholarly observation or in farming practice, when he set out. But partly too it resulted from the utter wretchedness of his last months in Colombia – wracked by malaria, the inhabitants living in ordure, filth and stench, surrounded by lepers and vermin. He wrote to his brother Elie in June 1859 to say that the whole episode had become 'le plus inepte de ma vie'.[10] Thereafter – and perhaps in consequence of this fiasco – Elisée retreated into a purely literary geography.[11]

He then began a prolific writing career, producing guide books for travellers and journalism for the intellectual periodical press, notably for the *Revue des Deux Mondes*. His *La Terre* appeared in two volumes in 1868–1869; his major work, *Nouvelle Géographie universelle: la terre et les hommes*, in nineteen volumes between 1876 and 1894 (originally appearing in weekly parts throughout this entire period); and his culminating synthesis, *L'Homme et la terre*, in six volumes, mostly posthumously, between 1905 and 1908.[12] The latter in particular is very inadequately

[8] *Ibid.*, p. 47.

[9] Paris: Hachette.

[10] E. Reclus, *Correspondance* volume 1 (Paris: Libraire Schleicher Frères, 1911), p. 151.

[11] Dunbar, *op. cit.*, (note 5) makes the barbed comment that whereas Reclus 'always retained the bucolic ideal', he 'never seemed to dirty his hands with anything other than ink'; reference on p. 90.

[12] For a brief commentary, see M. W. Mikesell, 'Observations on the writings of Elisée Reclus', *Geography*, 34 (1959), 221–226. These works were widely disseminated in translation, but their bibliography is very confused because the publisher systematically omitted dates from the title pages of the translations. *La Terre* was translated and published in London in 1871–1873, and the same version, edited with minor changes by the anthropologist A. H. Keane, appeared in two large volumes in 1886–1887. It was also published in New York in 1871–1873. The *Nouvelle Géographie universelle* was translated in weekly parts as it was written. It appeared in London in a total of 29 volumes during 1876–1894: volumes 1–8 edited by E. G. Ravenstein, 9–10 by Ravenstein and Keane, and 11–29 by Keane alone. The same

known and was never translated into English (being rejected by the London firm of Edward Arnold). Dunbar, however, has called the *Nouvelle Géographie universelle* 'the greatest individual writing feat in the history of geography',[13] and Kropotkin called it 'the geographical work which is most representative of our times'.[14]

During this lifetime of intense literary activity, Reclus built himself an equal reputation in radical politics.[15] He was only 21 when his protest at Napoleon's coup d'état led to his voluntary exile, and his attitudes were further sharpened by his revulsion over slavery in the American south. The events leading to his emergence as an anarchist, however, began with a visit to Mikhail Bakunin in Florence in 1865, and his ultimate involvement with the International Workingmen's Association. His crisis came in 1870–1871 with the collapse of the Empire, the siege of Paris, and the establishment on 18 March 1871 of the Paris Commune. Reclus enlisted, though there appears to be no record of his reputed activities as a military balloonist. He surrendered with Communard troops early in April, and on 15 November was sentenced to deportation to New Caledonia. English friends organized a petition on his behalf, and on 3 February 1872 the sentence was commuted to banishment. Five weeks later he was released, and went to live in Switzerland.[16] He could not return to France until the general amnesty of 1880. It was in Switzerland in 1877 that he first met Kropotkin, a few months after Bakunin's death; and it was to Bakunin, Kropotkin and Reclus that the anarchist movement largely owed its intellectual respectability.

In his later life Reclus travelled a great deal – to Turkey in 1883, to the U.S.A. in 1889 and 1891, and to Brazil in 1893. In 1894 he finally left France to take up an appointment at the Free University of Brussels, and it was in Brussels that he died in 1905.

So much for the bare chronology; what of the man? Reclus was without doubt a person of great natural dignity and distinction, smaller than one might have expected and in later life rather frail, and characterized by all

---

translation was published in New York during 1881–1898. The African volumes were separately issued, edited by Keane, in 1899.

[13] G. S. Dunbar, *op. cit.* (note 5), reference on p. 95.

[14] P. Krapotkin [*sic*], 'On the teaching of physiography', *Geographical Journal*, 2 (1893), 350–359; reference on p. 355.

[15] G. S. Dunbar, 'Elisée Reclus, an anarchist in geography', in D. R. Stoddart, ed., *Geography, ideology, and social concern* (Oxford: Basil Blackwell, 1981), 154–164; and M. Fleming, *op. cit.* (note 6).

[16] E. Candaux, 'Un brillant météore a passé dans notre ciel', *Le Globe*, 114 (1974), 115–124.

who knew him as a scholar of humility, compassion and selflessness.[17] He was also oblivious of conventional bourgeois standards of behaviour. His first marriage to Clarisse Brian, who was half-Fulani from Senegal, lasted for 11 years, to her death, and produced two daughters. His second and third marriages were by a process of 'free union' without the assistance of either ecclesiastical or civil authorities; and it was Elisée himself who officiated at the 'marriages' of his two daughters in 1882, a procedure which led to calls for his expulsion from Switzerland. In spite of his Moravian education and the fact that his father had been a priest, Reclus lost all religious belief during his early twenties, and he saw no necessary connection between organized religion and what he termed 'intellectual and moral hygiene'. He was a vegetarian at 25, and for the rest of his life lived only on fruit, nuts, lentils, bread and water.[18]

It is clear that not only was Elisée's life untrammelled by social convention, but that he carried his beliefs past idiosyncracy to the point of eccentricity. He seems to have been, all his life, a profoundly unpractical man. During the Commune, as a soldier, his ineptitude probably made him as a great a danger to his comrades as to anyone else. Like Patrick Geddes, who was his disciple, he lapsed increasingly into unpractical, bizarre and financially impossible proposals, though fortunately he lacked Geddes's frenetic insistence on (or perhaps capability for) getting them underway.[19]

Reclus himself had no obvious formal qualifications for academic life; he was self-exiled from 1851 to 1857; banished between 1872 and 1880; and after 1894 he lectured at the Free University of Brussels, an institution which paid no stipends and whose degrees were unrecognized. It was, of course, in 1877 (when Reclus was first meeting Kropotkin in Switzerland) that Vidal de la Blache took up his first appointment at the École Normale

[17] The portrait (plate 13) of Reclus appears in an unsigned review, 'Elisée Reclus and the Géographie Universelle', *Scottish Geographical Magazine*, 11 (1895), 248–251.

[18] E. Reclus, 'On vegetarianism', *Humane Review*, 1 (1901), 316–324.

[19] In his Brussels years he became obsessed with constructing enormous public globes on scales of 1:100,000 or even 1:50,000. The former would have been nearly 130 m in diameter and the latter nearly 260 m (the dome of St Paul's in London is 34 m across and that of St Peter's in Rome 43 m; the bigger globe would have rivalled the 300 m Eiffel Tower), and the costs were estimated at up to 50 million francs. Not surprisingly these globes never materialized, in spite of interesting discussion as to whether the topographic detail should be on the outside, viewed from external spiral staircases, or on the inside, viewed from the centre, the latter minimizing distortion if maps were to be made from photographs. See G. S. Dunbar, 'Elisée Reclus and the Great Globe', *Scottish Geographical Magazine*, 90 (1974), 57–66. Patrick Geddes was also much involved in the monster globe proposals. See P. Kitchen, *A most unsettling person: an introduction to the ideas and life of Patrick Geddes* (London: Gollancz, 1975), pp. 176–177.

Supérieure, in appointment which led to thirty years' domination of French academic geography, both there and at the Sorbonne. By the relentless and inevitable exercise of the patronage associated with such a position, Vidal provided the manpower through his students and colleagues for the establishment of professional academic geography in France; and in this process Elisée Reclus could play no part at all. Not only did Reclus lack students, his books lacked the rigour of textbooks for courses, and journalism in the *Revue des Deux Mondes* was no substitute for scholarly research in Ludovic Drapeyron's *Revue de Géographie* (after 1877) or Louis Raveneau's *Annales de Géographie* (after 1891). By the standards of the new intellectualism in geography, Reclus was seen as no more than a 'vulgarisateur'.[20]

Second, and perhaps more important – for it isolates him from ourselves as well as from his contemporaries – Reclus was no scientist in either a conceptual or a technical sense. His philosophy throughout his life was that of the *romantische Naturphilosophie* which Ritter had taught him in Berlin, though without Ritter's teleology.[21] Indeed, Kropotkin, in his most sensitive obituary, compares Reclus with Goethe,[22] and Geddes in his with Bernardin de St Pierre, Rousseau, Ruskin, Michelet and even Georges Sand.[23] Yet these were emphatically the years, as Kropotkin knew so well, when 'the foundations were laid of the mechanical theory of heat, the kinetic theory of gases, modern atomistic chemistry, the variability of species and modern biology altogether, anthropology, physiological psychology, and so on.'[24] Reclus was 29 when *The origin of species* was published: perhaps he was too old for its impact to be felt. It is in any case a commonplace that of all countries the Darwinian revolution passed France by,[25] and in retrospect it

[20] N. Broc, 'L'établissement de la géographie en France: diffusion, institutions, projets (1870–1890)', *Annales de Géographie*, 83 (1974), 545–568; see p. 550.

[21] Geddes, *op. cit.* (note 5) said that Reclus's 'essential distinction' was 'to have attained to the universal consciousness of the material world, and most to have promoted this'; reference on p. 492. Whatever this meant it was scarcely a component of the new scientific geography, and it tells as much about Geddes as about Reclus.

[22] P. Kropotkin, 'Elisée Reclus', *Geographical Journal*, 26 (1905), 337–343; reference on p. 338.

[23] P. Geddes, *op. cit.* (note 5).

[24] P. Kropotkin, *op. cit.* (note 22), reference on p. 338.

[25] Y. Conry, *L'Introduction du Darwinisme en France au XXe siècle* (Paris: Libraire Philosophique J. Vrin, 1974); J. R. Moore, 'Could Darwinism be introduced in France?', *British Journal for the History of Science*, 10 (1977), 246–251; R. E. Stebbin, 'France', in T. F. Glick, ed., *The comparative reception of Darwinism* (Austin: University of Texas Press, 1974), 117–163. Darwin himself was not elected to the Académie des Sciences until nearly twenty years after the publication of *The origin*.

is the absence of the coherence and integration supplied by evolutionary ideas that makes *La Terre* seem so anachronistic in its treatment of landforms today.[26]

Reclus was undoubtedly profoundly isolated from such influences. As Dunbar notes,[27] his roots were both Gallic and Mediterranean, and his appreciation of nature as much aesthetic and emotional as scientific, in contrast to the narrower but more influential vision of George Perkins Marsh. Though he enjoyed his period in Berlin and described the German countryside with sensitivity and skill, his lack of sympathy for German thought and the new German science was doubtless reinforced when the Prussians shelled Paris in 1870 and his family had to abandon their home and take refuge all over the city. Reclus's politics also derived from Bakunin rather than from Marx, a distinction the significance of which was perhaps more apparent then than now.[28] And nothing could have been more antithetic to his warm compassion and deep human sympathy than the rigours of a deterministic and materialistic Social Darwinism.[29]

This alone would not have been sufficient to make Reclus's work irrelevant, even in France, had it not been that it was to Germany and to German science that the French turned for intellectual revival after 1870.[30] At its most basic, the German influence was seen in the structure of academic life (in universities, societies and journals) and in the imposition of minimum scholarly conventions in publication, conventions eagerly

[26] Perhaps because it was written in weekly parts, *The earth: a descriptive history of the phenomena of the life of the globe*, ed. A. H. Keane (London: J. S. Virtue, 1886), reads like a series of disconnected essays on such topics as mountains, rivers, earthquakes and lakes. Yet in spite of its lack of any general theoretical structure, its first chapter, on the place of the earth within the universe, represents Reclus's writing at its most lyrical and persuasive, and it is not difficult to see why the book was so popular.

[27] G. S. Dunbar, *op. cit.* (note 5), pp. 15–16.

[28] See especially M. Fleming, *op. cit.* (note 6), chapter 3, and also E. Reclus, 'Anarchy: by an anarchist', *Contemporary Review*, 45 (1884), 627–641. It is interesting that Reclus was Marx's first choice to translate *Das Kapital* into French in 1867. The two men met in 1869 in London, but Marx could not accept Reclus's support for Bakunin and the proposal lapsed: Fleming, *op. cit.*, pp. 62–63.

[29] Reclus was, of course, perfectly well aware of the idea of evolution. His most famous pamphlet, repeatedly reprinted and translated, was entitled *Évolution et révolution* (Geneva, 1880). It was ultimately expanded into *L'Évolution, la révolution et l'idéal anarchique* (Paris: P.-V. Stock, 1898), but this in no way deals with the technical or conceptual implications of Darwinism, and indeed Darwin's name does not appear in it.

[30] N. Broc, 'La géographie française face à la science allemande (1870–1914)', *Annales de Géographie*, 86 (1977), 71–94; C. Digeon, *La Crise allemande de la pensée française (1870–1914)* (Paris: Presses Universitaires de France, 1959; H. Paul, *The sorcerer's apprentice: the French scientist's image of German science* (Gainesville: University of Florida Press, 1972).

adopted by the classicists and historians who became the first French 'new geographers' (Reclus's work is notorious for its almost total lack of either original observation or citation of sources). In addition, Ratzel – himself profoundly influenced by *Darwinismus*[31] – imposed a new analytical rigour on human geography that Reclus's work profoundly lacked. In sum, Reclus's method, like Ritter's, was comparative and synthetic but not evolutionary or analytical, and his mode of explanation, especially in *L'Homme et la terre*, was mainly through historical narrative rather than through functional association: Mackinder remarks on the 'insinuating delicacy' of such causal explanation as was to be found in the *Nouvelle Géographie universelle*.[32] After 1859 this would no longer do: science simply passed him by; and if, as Dunbar terms him, he was a 'historian of nature', it was a type of natural history which did not long survive *The origin of species*. At the same time, Reclus believed profoundly that geography derived not from the library or the study, but from the earth itself – his tragedy was that he never learned the new rules of measurement and observation. He wrote to his mother on 13 November 1855 with what amounted to a geographer's statement of faith:

> D'ailleurs voir la terre; c'est pour moi l'étudier; la seule étude véritablement sérieuse que je fasse est celle de la géographie, et je crois qu'il vaut beaucoup mieux observer la nature chez elle que de se l'imaginer du fond de son cabinet. Aucune déscription, aussi belle qu'elle soit, ne peut être vraie, car elle ne peut réproduire la vie du paysage, la fuite de l'eau, le frémissement des feuilles, le chant des oiseaux, le parfum des fleurs, les formes changeantes des nuages; pour connaître, il faut voir.[33]

It gave Reclus himself deep satisfaction, but had no appeal at all to the scientism of the New Geography.

Yet Reclus emphatically cannot be dismissed. He believed, perhaps naively, in the perfectability of man in society, and his deep humanitarianism was fundamentally optimistic. At one level his writings lift themselves out of their late-nineteenth-century nationalistic frame, as Kropotkin recognized: 'Not only is his work free from absurd national conceit, or of national or racial prejudice; he has succeeded, besides, in indicating in every

[31] See G. Buttmann, *Friedrich Ratzel: Leben und Werk eines deutschen Geographen 1844–1904* (Stuttgart: Wissenschaftliche Verlagsgesellschaft M.B.H., 1977).

[32] H. J. Mackinder, 'Reclus' "Universal Geography"' *Geographical Journal*, 4 (1894), 158–160; reference on p. 159.

[33] E. Reclus, *op. cit.* (note 10), volume 1, p. 109.

branch, stem, or tribe of the human race those features which make one feel what all men have in common – what unites, not what divides them.'[34] At a deeper level, Reclus knew that the important problems in the relationship of man and land were not to do with the enumeration of people or resources, but with the social structures by which the resources were utilized and distributed, not simply in space but through society; in this at least he anticipated Febvre.[35] And it is the passion and conviction with which he argued for his beliefs[36] that command respect but also inspire regret that such issues were to be disregarded for a century by an increasingly positivist 'New Geography'. Without question, if one is to seek the first French *géographe engagé*, it is not Brunhes, as Buttimer[37] curiously claims, but Elisée Reclus. 'He knew how to die poor,' wrote Kropotkin, 'after having written wonderful books.'[38]

Kropotkin, a decade younger, was born to a life of privilege as a page of the Tsar. As a boy he read widely, and developed his own romantic vision of the unity of nature. Like Darwin and many others before him, he was profoundly impressed by Humboldt, 'above all the first volume of . . . [the] *Cosmos*, his *Tableaux de la Nature*', as well as Ritter's monographs on the tea-tree and the camel.[39] He realized 'the greatness of nature, its poetry, its ever throbbing life', and developed 'that awakening love of mankind and

[34] P. Kropotkin, *op. cit.* (note 22), p. 341.

[35] Note the solitary quotation by A. Buttimer, *Society and milieu in the French geographic tradition* (Washington: Association of American Geographers, 1971), p. 81. Reclus wrote in *L'Homme et la terre* (Paris: Libraire Universelle, volume 2, p. 42): 'It is no longer the political, juridical, constitutional armour of past times [that concerns us]. It is their whole life, their whole material and moral civilization, the whole evolution of their sciences, arts, religions, trade, divisions, and social groupings.' There is some discussion of Reclus's implicit structuralism by K. R. Olwig, 'Historical geography and the society/nature "problematic": the perspective of J. F. Schouw, G. P. Marsh and E. Reclus', *Journal of Historical Geography*, 6 (1960), 29–45.

[36] Consider the savagery of Reclus's comment in 1884 (quoted by G. S. Dunbar, *op. cit.* [note 5], p. 88): 'May mothers refuse to conceive so long as the political and social milieu condemns then to give birth only to victims or executioners.'

[37] A. Buttimer, *op. cit.* (note 35), pp. 61, 127.

[38] P. Kropotkin, *op. cit.* (note 22), p. 343.

[39] P. Kropotkin, 'What geography ought to be', *Nineteenth Century*, 18 (1885), 940–956; see p. 951. For biographical accounts of Kropotkin, see F. Planche and J. Delphy, *Kropotkine, descendant des Grandes Princes de Smolensk, page de l'Empereur, savant illustre, révolutionnaire international, vulgarisateur de la pensée anarchiste* (Paris, 1948); G. Woodcock and I. Avakumović, *The anarchist prince: a biographical study of Peter Kropotkin* (London: T. V. Boardman, 1950); M. A. Miller, *Kropotkin* (Chicago: University of Chicago Press, 1976).

faith in its progress which makes the best part of youth and impress man for a life'.[40] 'Gradually the sense of man's oneness with Nature, both animate and inanimate – the poetry of Nature – became the philosophy of my life.'[41]

In May 1862 he joined a regiment of mounted Cossacks on the Amur and began his studies of easternmost Russia. Serfdom had been formally abolished in the Empire the previous year, but its consequences remained. Kropotkin could not tolerate the injustices and oppression which he saw both on his own father's estate and on his expeditions in Siberia, and he became increasingly critical of the social order in Russia. In 1872 he became converted to libertarian socialism during a visit to a community of anarchist watchmakers in Switzerland, having become involved in politics at about the time of the Paris Commune. In 1874, when serving as a secretary of the Imperial Russian Geographical Society in St Petersburg, and shortly after giving a paper before it, he was denounced and arrested for his membership of a clandestine anarchist group. He was imprisoned in the Fortress of St Peter and St Paul, but managed to escape and fled to Hull, one of the few places in the world where the secret police might not think to search for him. From Hull he went to London and met John Scott Keltie, then with *Nature*, and sought journalistic work.[42] Keltie became his closest friend, and in 1921 wrote the only obituary of him to appear in any of the major geographical journals.[43]

Kropotkin in his exile, which continued until the Russian revolution, was seen in a variety of roles. To some he was the Red Flag personified, to others the Sermon on the Mount; a poet and dreamer; a scientist pure and simple; a revolutionist of the most dangerous type; a mixture of St Francis of Assisi, Danton and Don Quixote.[44] Oscar Wilde said that he had the soul of a 'beautiful white Christ'[45] and regarded him as one of the two really happy men he had ever met.[46] James Mavor described him at the age of 42:

[40] P. Kropotkin, *Memoirs of a revolutionist* (London: Smith, Elder, 2 volumes, 1899), volume 2, p. 90.

[41] *Ibid.*, p. 109; also *op. cit.* (note 14), p. 355. Compare Elisée Reclus's essay, 'Du sentiment de la nature dans les sociétés modernes', *Revue des Deux Mondes*, 63 (1866), 353–381.

[42] Kropotkin introduced himself to Keltie under the assumed name of Levashov. He was thrown into a moral dilemma – and finally forced to confess his true identity – when Keltie gave him some of his own books to review for *Nature*. For Kropotkin's unsigned review of his own 'Researches on the glacial period', see *Nature, Lond.*, 16 (1877), 161.

[43] J. S. Keltie, 'Prince Kropotkin', *Geographical Journal*, 57 (1921), 316–319.

[44] These somewhat extreme assessments come from E. Sellers, 'Our most distinguished refugee', *Contemporary Review*, 66 (1894) , 537–549; see p. 537.

[45] Quoted by M. A. Miller, *op. cit.* (note 39), p. 169.

[46] According to W. H. G. Armytage, *Heavens below: Utopian experiments in England 1560–1960* (London: Routledge & Kegan Paul, 1961), p. 306.

> He was short, not more than five and half feet, slight in build, very erect, with unusually small feet, a slender waist and broad shoulders. He had a short neck and a large head. He wore a full brown beard, seldom trimmed and never lacking its distinctive character. The top of his head was destitute of hair, but on the sides and back of it his dark brown hair was ample. His eyes sparkled with genius, and when he was roused became almost incandescent.[47]

He suffered from scurvy while in prison, and as a result lost all his teeth.[48] He had, said Keltie, 'a warm and perhaps too tender heart'.[49]

He was not happy in England. 'Such gloom, this London; I passionately detest this English exile.' He had little sympathy too with what passed for anarchism in England – 'anarchie de salon', he called it, 'epicurean, a little Nietzschean, very snobbish, very proper, a little too Christian'.[50] He scandalized the Royal Geographical Society by declining to drink the health of the King, and ostentatiously kept his seat while others stood to do so.[51] It says much for the breadth of sympathy of the Society's fellows at that time, and especially of Keltie, that Kropotkin could continue to write and publish his geographical work through his anarchist years.

Most of Kropotkin's strictly geographical writing, especially his work on the physical geography of Russia, is now only of historical interest, and the only paper of his ever cited in this field is that 'On the teaching of physiography' in 1893.[52] In 1883, however, while in France, he was arrested at a time of bomb outrages and accused of complicity in an attempt to establish the International in France. In January 1884 he was imprisoned for five years.[53] At this time he was on close terms with Reclus, and his wife lived with Reclus's brother during part of Kropotkin's incarceration. It was while Kropotkin was in prison at Clairvaux that Keltie wrote and published his report on geographical education which led directly to the establishment of lectureships in the subjects at Oxford and Cambridge and to the recognition of geography as an academic discipline in Britain.[54] The report

[47] J. Mavor, 'Prince Peter Alexander Kropotkin, 1886–1921', in *My Windows on the Street of the World* (London: J. M. Dent, 1923), volume 1, pp. 91–106; reference on p. 92.
[48] J. Slatter, 'The Kropotkin papers', *Geographical Magazine*, 53 (1981), 917–921.
[49] J. S. Keltie, *op. cit.* (note 43), p. 319.
[50] M. A. Miller, *op. cit.* (note 39), p. 169.
[51] N. Walter, 'Kropotkin and his memoirs', *Anarchy*, 10 (1970), 84–94; reference on pp. 87–88.
[52] *Op. cit.* (note 14).
[53] In England, however, his only brush with the police was over his failure to buy a dog licence: N. Walter, *op. cit.* (note 51), p. 86.
[54] See chapter 5.

provoked much discussion, mostly of a technical character on the nature of geography and how it should be taught. One paper stands out in this controversy, though it has since been almost completely overlooked. Written in prison, Kropotkin's 'What geography ought to be' appeared in the *Nineteenth Century* for December 1885. It is a passionate and humane statement of the social significance of a geographical education. Yes, it was to do with knowing and explaining the great phenomena of nature. But more than that, it:

> must render . . . another far more important service. It must teach us, from our earliest childhood, that we are all brethren, whatever our nationality. In our time of wars, of national self-conceit, of national jealousies and hatreds ably nourished by people who pursue their own egotistic, personal or class interests, geography must be . . . a means of dissipating these prejudices and of creating other feelings more worthy of humanity. It must show that each nationality brings its own precious building-stone for the general development of the commonwealth. . . . It is the task of the geographer to bring this truth, in its full light, into the midst of the lies accumulated by ignorance, presumption, and egotism.[55]

And he speaks with bitter contempt of the way in which 'civilization' has been spread by bayonets and massacre, whisky and kidnapping. This was a justification for geography, both liberal and scientific, quite different from any other made at this time. It is a most powerful document which speaks directly to us across the decades, and which supplies one of the central traditions of our subject. After his release, Kropotkin lectured on the same theme to the Manchester Geographical Society,[56] but then his contribution was forgotten.

The expansion of geographical education in the later years of the last century was, in itself, part of the social revolution for which Reclus and Kropotkin worked throughout their lives. Sir Charles Warren, speaking as president of Section E at the British Association in Manchester in 1887, argued that the children of wealthy parents would be at a disadvantage as geographical teaching spread, because of their remoteness from the practical business of making a living, and he warned that they would 'in a few years

[55] *Op. cit.* (note 39), pp. 942–943. See also A. J. Fielding, 'What geographers ought to do: the relationship between thought and action in the life and works of P. A. Kropotkin', *University of Sussex, Department of Geography Research Paper in Geography* (1980), 1–15.

[56] P. Krapotkin [*sic*], 'What geography ought to be', *Journal of the Manchester Geographical Society*, 5 (1889), 356–357, discussion 357–358.

be distanced by the sons of the labourers, artisans and shopkeepers'.[57] Mackinder himself agreed that a knowledge of present-day geography was essential if the 'educated classes' were not 'to lose their grip and their influence over the half-educated proletariat'.[58] The 'New Geography' and the social revolution at the end of the nineteenth century spread together.

To what extent did the anarchism of Kropotkin and Reclus influence their geography? Why did their professional studies not have greater significance, either then or now? The conventional view has always been that one could make a clear distinction between their geographical work and their political views. Keltie, in his obituary of Kropotkin, expressed 'regret that his absorption in these [political activities] seriously diminished the services which otherwise he might have rendered to geography'.[59] Yet such an interpretation fails to comprehend the extent to which the geographical work of both men is enlivened and given meaning by the passion and conviction of their beliefs in a manner quite lacking in the work of contemporaries such as G. G. Chisholm.

Conversely, it remains difficult to specify more precisely the positive characteristics of their contributions.[60] Partly this results from the broadly unspecific and naively idealistic nature of their philosophies. Both, in their emphasis on the organic, holistic and historically determined nature of the relationship between man and the earth, drew profoundly on the writings of Humboldt and Ritter. Neither was impressed by the apparent determinism of Darwinian evolution or by its mechanism, and their romantic optimism about the social condition was hard to reconcile either with events in the world around them or with the advances of the positivistic human sciences. Both men, too, stood by as the subject became increasingly professionalized throughout Europe and America, and neither benefited from the new academic bureaucracy of careers, doctorates, textbooks and examinations. Both relied on the interpretation of secondary materials rather than on new data revealed by scholarly research. Even in their politics they remained visionary rather than activist. In Dunbar's words, they served as 'badges of

[57] H. J. Mackinder, 'Geography as a pivotal subject in education', *Geographical Journal*, 57 (1921), 376–384; reference on p. 383.

[58] C. Warren, 'Address', *Report of the 57th Meeting of the British Association for the Advancement of Science 1887* (1888), 785–798; reference on p. 793.

[59] J. S. Keltie, *op. cit.* (note 43), reference on p. 316.

[60] See M. M. Breitbart, 'Peter Kropotkin, the anarchist geographer', in D. R. Stoddart, ed., *Geography, ideology, and social concern* (Oxford: Basil Blackwell, 1981), 134–153; B. Galois, 'Ideology and the idea of nature: the case of Peter Kropotkin', *Antipode*, 8 (3) (1976), 1–16, reprinted in R. Peet, ed., *Radical geography: alternative viewpoints on contemporary social issues* (London: Methuen, 1977), 66–93; A. J. Fielding, *op. cit.* (note 55).

respectability for the anarchists', rather like 'bemedalled old soldiers who were hired as doormen by bankers to disguise the nefarious activities within'.[61]

Perhaps it is not surprising, therefore, that our histories of geography relegate both men to footnotes and concentrate on the academic achievers and textbook writers of the past. Yet merely to read the books and the papers of these two gentle anarchists is to be vividly reminded not only of the aridity of much that passed for research as the New Geography became established, but of the permanence of the human problems which Kropotkin and Reclus attacked so passionately and so optimistically a century ago.

[61] G. S. Dunbar, *op. cit.* (note 5), p. 93.

7

Geography, Exploration and Discovery

'Of all the sciences, geography finds its origin in action.'[1] To many, Joseph Conrad must speak of a different world, before the geographer of action and the geographer of the armchair had parted their ways.

Conrad himself was no stranger to the world of action. He speaks with direct simplicity of his passage under sail through Torres Straits, to Mauritius out of Sydney, and of taking a paddlesteamer along the Upper Congo to within sound of Stanley Falls.[2] As a child he had been impressed by Leopold McClintock's account[3] of the search for Franklin and the lost *Erebus* and *Terror*. McClintock's book was translated into every European language, and in 1859 Mudie's Circulating Library found it three times more popular than Tennyson's *Idylls of the king*.[4] The shock of the Franklin disaster, drawn out as it was, combined with the perceived majesty,

[1] Joseph Conrad, 'Geography and some explorers', *Last essays* (London: J. M. Dent, 1926), pp. 1–31.

[2] For a reconstruction of Conrad's sea-going years, see N. Sherry, *Conrad's eastern world* (Cambridge: Cambridge University Press, 1966). Born just before *The origin of species* was published, Conrad was not only greatly influenced by the implications of Darwinism, he also drew for much of his topographical detail on Alfred Russel Wallace's *The Malay archipelago* (London: Richard Clay & Sons, 1869). 'It was his favorite bedside companion. He had an intense admiration for those pioneer explorers – "profoundly inspired men" as he called them – who have left us a record of their work; and of Wallace, above all, he never ceased to speak in terms of enthusiasm. . . . *The Malay Archipelago* had been his intimate friend for many years' (quoted by Sherry, *op. cit.*, p. 142, from R. Curle, 'Joseph Conrad: ten years later', *Virginia Quarterly Review*, 10 (1934), p. 413). On Conrad and Darwinism, see A. Hunter, *Joseph Conrad and the ethics of Darwinism* (London: Croom Helm, 1983), and R. O'Hanlon, *Joseph Conrad and Charles Darwin* (Edinburgh: Salamander Press, 1984).

[3] F. L. M'Clintock [*sic*: McClintock altered the orthography of his name in 1860], *The voyage of the 'Fox' in the Arctic seas: a narrative of the discovery of the fate of Sir John Franklin and his companions* (London: John Murray, 1859).

[4] G. L. Griest, *Mudie's Circulating Library and the Victorian novel* (Bloomington: Indiana University Press, 1970), p. 21.

sublimity and terror of the polar regions to place organized exploration in a new perspective.[5]

But at the same time, with peculiar prescience, Conrad saw the contrast between this world of action and discovery, and that of the lecture theatre and library. Already (as Huxley feared would happen), education had come to be controlled by 'persons of no romantic sense for the real, ignorant of the great possibilities of active life; with no desire for struggle, no notion of the wide spaces of the world – mere bored professors, in fact, who were not only middle-aged but looked to me as if they had never been young. And their geography was very much like themselves, a bloodless thing with a dry skin covering a repulsive armature of uninteresting bones'.[6]

And so it is that in the academic histories of our subject the names of McClintock, even of Franklin, do not appear: certainly not that of Conrad himself. The first sentence of Walter Freeman's *History of modern British geography* tells us that 'A hundred years ago the geographer of popular conception and widespread adulation was an explorer of pathless African jungles or Arctic snows.' But the exploits of these 'solid hunks of British manhood', as he calls them, are condescendingly consigned to a prehistory which has little to do with the achievements of successive 'new geographies'. Hunt, Hillary, Fuchs do not appear; even Scott is rapidly eclipsed by descriptions of Herbertson's regional schemata and Mackinder's geopolitical speculations.[7]

How has this happened? Why is it so fashionable to denigrate or ignore our traditional concern with the farthest corners of the earth? In this chapter I wish briefly to consider some aspects of the role and achievements of fieldwork, exploration and discovery within geography, and to show how they must remain central to our discipline, if it is not to lapse permanently (with its practitioners) into the dry and bloodless thing from which Conrad suffered a century ago.

FIELDWORK

Insofar as our tradition has survived at all in British geography it is in the

[5] C. C. Loomis, 'The Arctic sublime', in U. C. Knoepflmacher and G. B. Tennyson, eds, *Nature of the Victorian imagination* (Berkeley: University of California Press, 1977), 95–112.

[6] Joseph Conrad, *op. cit.* (note 1), p. 17.

[7] T. W. Freeman, *A history of modern British geography* (London: Longman, 1980), pp. 1, 69.

form of 'fieldwork'.[8] No one denies the heuristic value in education of direct confrontation with nature, through organized field classes and individual projects. It sprang in Europe from Pestalozzi's innovations, which so impressed Ritter, and in Britain from Patrick Geddes's repulsion at the 'vast and reckless destruction of natural and historic beauty by the railway and manufacturing age'.[9] The camera obscura in Geddes's Outlook Tower in Edinburgh, attracting scholars as diverse as Herbertson and Reclus, came to symbolize a form of observation and a definition of subject matter which has governed field studies since that time.

S. W. Wooldridge repeatedly defined this educational commitment. 'Field work for the geographer consists essentially in comparing the map with the actual ground. . . . For the geographer the ground is the primary document. . . . The field for us is the primary source of inspiration and ideas and inspires a great part of both the matter and the method of our subject.'[10] In a later paper he suggested that 'The object of field teaching . . . is to develop an eye for country. . . . The fundamental principle is that the ground not the map is the primary document. . . . From this first principle I pass to a second, that the essence of training in geographical field work is the comparison of the ground with the map.'[11]

While doubtless derived from his own topographic studies in southeast England, Wooldridge's objectives seem curiously limited, even by the standards of the times. They have the merit of ease of execution, and – let it be said – of satisfaction to the participants. There can be few products of our educational system who have not carried out some form of observation, recording and interpretation in the field, be it surveys of drumlins or questionnaires in supermarkets. Such work has been institutionalized by the Field Studies Council at its field centres, and Wooldridge himself was a notable supporter of such work (see plate 16).[12]

[8] C. Board, 'Field work in geography, with particular emphasis on the role of land-use survey', in R. J. Chorley and P. Haggett, eds, *Frontiers in geographical teaching* (London: Methuen, 1965), 186–214.

[9] P. Geddes, 'A great geographer: Elisée Reclus, 1830–1905', *Scottish Geographical Magazine*, 21 (1905), 490–496, 548–551; reference on p. 495. Also P. Boardman, *The worlds of Patrick Geddes: biologist, town planner, re-educator, peace-warrior* (London: Routledge & Kegan Paul, 1978).

[10] S. W. Wooldridge, *The spirit and significance of field work* (London: Council for the Promotion of Field Studies, 1948), p. 2.

[11] S. W. Woodridge, 'The status of geography and the role of field work', *Geography*, 40 (1955), 73–83; reference on pp. 78–79. Reprinted in Wooldridge's *The geographer as scientist* (London: Nelson, 1956), 66–79; reference on p. 73.

[12] See also his *Field Studies Council: retrospect and prospect* (London: Field Studies

But for many, fieldwork meant much more than map interpretation and annotation. It is instructive to examine the work of the Le Play Society in this context, for this gives insights into both the strengths and the weaknesses of the fieldwork movement.[13] Le Play himself was interpreted for a British audience by Patrick Geddes, who adopted the formula of 'Place – Work – Folk' as an agenda for study through the method of direct observation. Founded in 1930, the Society had the support of many prominent geographers – Mackinder, Herbertson, Fleure, Stamp, Beaver, Edwards, Estyn Evans. Stamp led its first field excursion, in 1932, to Solcava, Yugoslavia. 'The members put up with rather primitive conditions, and made a full survey: the physical setting, the ethnography, the settlement, the high farms, the village, its life and sociological structure, the school and the Church; statistical details of the population were collected, and a vegetation survey was made.'[14] Similar projects were undertaken in Russia, the Carpathians and Montenegro, and later in Poland. 'The great country houses were still functioning. . . . The peasants were very friendly.' The vodka would flow, and the members of the party became involved in such festivals as that of Our Lady of the Green Leaves, in which they joined.

Though efforts were made to systematize the survey procedures,[15] it is clear that such projects promoted the education and enjoyment of the participants rather than inspiring original contributions to knowledge. This was necessarily so, given the absence of precise objectives, the brief periods involved in the studies, and the cultural and language barriers encountered

---

Council, 1960). Wooldridge was much supported in this enthusiasm by G. F. Hutchings and C. C. Fagg; see Fagg's 'The history of the regional survey movement', *South Eastern Naturalist*, 33 (1928), 71–94; and Hutching's 'The geographer as a field naturalist', *School Nature Study*, 44 (1949), 33–38, and 'Geographical field teaching', *Geography*, 47 (1962), 1–14. Fagg of course is best known for his work on dry valleys on the English chalk, but he also practised psychoanalysis and his presidential address to the Croydon Natural History and Scientific Society in 1923 was entitled 'The significance of the Freudian psychology for the evolution theory' (*Proceedings and Transactions of the Croydon Natural History and Scientific Society*, 9 (1923), 137–164). Fagg had originally intended to give this address on the study of the Croydon Survey Area and the recently completed 'Surface Utilization Survey', but decided that it would be too difficult. Between 1923 and 1933 he corresponded with Freud, initially about evidence of sexual fetishism in Darwin's geological writings. It would be of great interest to pursue Freud's connection with Fagg, and also with that other field naturalist Arthur Tansley, who resigned his lectureship in the Cambridge Botany School to go to Vienna to study under Freud.

[13] S. H. Beaver, 'The Le Play Society and field work', *Geography*, 47 (1962), 226–239.

[14] E. J. Russell, *The Le Play Society* (Oxford: The Le Play Society [?], 1960), pp. 6–7.

[15] C. C. Fagg and G. F. Hutchings, *An introduction to regional surveying* (Cambridge: Cambridge University Press, 1930).

in the areas and communities studied. Not surprisingly the most successful component was land-use mapping, a relatively unambiguous and objective procedure, taken up on a massive scale by L. Dudley Stamp in the Land Utilization Survey of Britain.[16]

It is a paradox that, in spite of the east European excursions of the Le Play Society, the fieldwork movement came to be increasingly confined to local areas at home. Partly this was necessitated by political and economic conditions in the thirties and forties, but it was also strongly reinforced by those senior geographers whose own research activities were regionally restricted. Wooldridge, for example, in his address to the Institute of British Geographers at Cambridge in 1951, argued that 'the human geography of Somerset is more interesting and in many ways more significant than that of, say, Somaliland, and though one should wish both to be studied, it is the former, in general, which is neglected.' He went on to make the extraordinary observation that while the practical problems of studying Somerset were admittedly less, 'the intellectual difficulties are incomparably greater.'[17] Not content with this, he attacked what he called 'the cult, or rather the disease, of . . . "otherwheritis"', seeing it as 'a real peril to the well being of geography. . . . At its worst it can only lead to a species of egotistic, impressionistic geography, a congeries of travellers' tales, agreeably titillating, no doubt . . . but making no claim whatever to scholarship'.[18]

Behind the hyperbole, Wooldridge here was signalling a retreat from one of the traditional occupations of geography (fortunately only imperfectly heeded), as well as a hardening of the division between the geographers of Kensington Gore and those of the universities. The ascent of Everest in 1953 and the first crossing of Antarctica in 1957–1958 symbolized the breadth of the gulf which had opened within the discipline. For whatever its educational advantages, 'fieldwork' at home was but a pale substitute for active exploration overseas.

Carl Sauer had a broader view. He was convinced that 'Geography is first of all knowledge gained by observation, that one orders by reflection and reinspection the things one has been looking at, and that from what one has

[16] C. Board, *op. cit.* (note 8).

[17] S. W. Wooldridge, 'Reflections on regional geography in teaching and research', *Transactions and Papers of the Institute of British Geographers 1950, Publication* 16 (1952), 1–11, reference on p. 7. Reprinted in his *The geographer as scientist*, *op. cit.*. (note 11), 51–65; reference on p. 59.

[18] *Ibid.*, references on pp. 7 and 61 respectively. D. L. Linton (among others) tried to bridge the gap, but only by making true research the chronological successor to exploration and discovery: *Discovery, education and research* (Sheffield: University of Sheffield, 1946).

experienced by intimate sight come comparison and synthesis. In other words, the principal training of the geographer should come, wherever possible, by doing field work.'[19] For him it was a *learning* experience, often with local people as guides, rather than a *recording* one: indeed, 'the more energy that goes into recording predetermined categories, the less likelihood there is of exploration.'[20] The fieldworker must go where the evidence requires: at the end of the day 'he will wish that he had been in the field longer and more often, to secure the observations which he needs.'[21]

EXPLORATION AND DISCOVERY

To enquire why academic geography failed to cope, at least in its public pronouncements and its fieldwork, with the subject's commitments, we need to examine more closely what is involved in exploration and discovery. It is a commonplace to say that discovery represents an event, exploration a process, but this neglects the profound complexity of the activities included in these terms.[22] Consider the literature on the discovery of America. Columbus set sail with fixed objectives: he explored the western ocean, he

[19] C. O. Sauer, 'The education of a geographer', *Annals of the Association of American Geographers*, 46 (1956), 287–299; reference on pp. 295–296.

[20] *Ibid.*, p. 296.

[21] C. O. Sauer, 'Forword to historical geography', *Annals of the Association of American Geographers*, 31 (1941), 1–24; reference on p. 14. Sauer's views have been treated recently in a manner that might leniently be termed uncharitable. Gould finds the message of the 1956 paper 'shabby, parochial and unintelligent' and denounces what he is pleased to term the 'bumbling amateurism and antiquarianism' which led to the 'tip-heap' of Sauer's life work: P. Gould, 'Geography 1957–1977: the Augean period', *Annals of the Association of American Geographers*, 69 (1979), 139–151. Nor can I say that I am much enlightened by Pred's remark that 'the distinction made between "field work" and other more everyday observations and experiences is but one manifestation of a general unwillingness to accept the fact that our "professional" lives are not in dichotomous opposition to one another, but dialectically interrelated': A. Pred, 'From here and now to there and then: some notes on diffusions, defusions and disillusions', in M. D. Billinge, D. J. Gregory and R. L. Martin, eds, *Recollections of a revolution: geography as spatial science* (London: Macmillan, 1984), 86–103; reference on pp. 91–92. Pred's insight (if I read him aright) that here is not there, yesterday cannot be tomorrow, and if you are doing one thing you cannot be doing another has, I believe, been apprehended by the fieldworkers among us for some time. Certainly the life-paths of most explorers would have been abruptly terminated in short order without this knowledge. I am glad to think that the insights of physical geography are in general somewhat more profound. For more on Sauer's own field work, see R. C. West, *Carl Sauer's fieldwork in Latin America* (Ann Arbor: University Microfilms International, 1979).

[22] For an encyclopaedic disussion of the activities encompassed by these terms, see D. Boorstin, *The discoverers* (London: Dent, 1983).

came upon land, and he misconstrued it as Asia. In what sense then could he be said to have 'discovered' America when he had no concept of it, and indeed, as O'Gorman has argued, before that concept had been invented?[23] What of the inadvertent contacts of the Northmen with shores they likewise could not possibly recognize as those of a world in some sense newer than Iceland and Greenland?[24] Had Columbus indeed reached Asia, on the other hand, could this in itself have constituted a discovery, given the vast amount of knowledge already available about this continent, which indeed had stimulated his expeditions?

These questions raise profound issues in the historiography of exploration from the fifteenth to the nineteenth centuries, and in the understanding of the development of geographical science.[25] They are issues I do not propose to consider here on a technical level: rather let me clarify the nature of these activities by reference to actual examples.

First, and most simply, there is the inadvertent journey, often undertaken after some catastrophe under conditions of great hardship. When Bligh lost his ship to Fletcher Christian off Aitutaki in April 1789, for example, he made his way with 18 other persons in a 23-foot open boat across the west Pacific to the East Indies. He reached civilization in Timor, after passing through the Great Barrier Reef and round Cape York, 41 days after a brief and disastrous landing at Tofua, Tonga, a distance of 5,800 km, with no fatality other than the sailor murdered on the beach at Tofua. When she was cast away from the *Bounty*, Bligh's boat had but 150 lb of bread, about 30 lb of salt pork, 6 quarts of rum, 6 bottles of wine and 28 gallons of water. It was a month before they were sufficiently close to land to catch boobies by hand, and it was these, each divided into nineteen parts, and eaten bones, blood, entrails and all, that kept them alive.[26]

Likewise, after the destruction of the *Endeavour* at the end of October 1915 (having been beset by ice for seven months), Shackleton spent over five months drifting on the ice, waiting for it to melt, before finally reaching Elephant Island in open boats. The distance was about 1,800 km from the point where the ship had first been frozen in. Incredibly, Shackleton then set sail in a boat the same size as Bligh's for the 1,300 km crossing to South

[23] Edmundo O'Gorman, *The invention of America: an inquiry into the historical nature of the New World and the meaning of its history* (Bloomington: Indiana University Press, 1961); also L. Olschki, 'What Columbus saw on landing in the West Indies', *Proceedings of the American Philosophical Society*, 84 (1941), 633–659.

[24] C. O. Sauer, *Northern mists* (Berkeley: University of California Press, 1968).

[25] W. E. Washburn, 'The meaning of "discovery" in the fifteenth and sixteenth centuries', *American Historical Review*, 68 (1962), 1–21.

[26] G. Mackaness, ed., *A book of the Bounty* (London: Dent, 1938).

Plate 1 Europe encounters the Pacific

From M. L. A. Milet-Mureau, ed., *A voyage round the world, in the years 1785, 1786, 1787, and 1788* (London: J. Johnson, 3 volumes, 1798), volume 3, p. 195.

Plate 2 Recording nature in New Guinea

From P. Sonnerat, *Voyage à la Nouvelle Guinée, dans lequel on trouve la description des lieux, des observations physiques et morales, et des détails relatifs à l'histoire naturelle dans le règne animal et le règne vegetal* (Paris: Ruault, Libraire, 2 volumes, 1776), volume 1, frontispiece. Sonnerat made the unfortunate mistake in volume 2 of figuring penguins, which he allegedly found roaming the shores of New Guinea (the only tropical penguins are those of the eastern Pacific); it transpired that he bought them at the Cape. The road to objectivity in distant parts of the earth was not an easy one.

Plate 3 Reinhold and Georg Forster at Tahiti during Cook's second voyage
From J. C. Beaglehole, ed., *The voyage of the Resolution and Adventure 1772–1775* (Cambridge: Hakluyt Society, 1961), p. xxix.

Plate 4 Professor Newton
From A. F. R. Wollaston, *Life of Alfred Newton*
(London: John Murray, 1921).

Plate 5 Henry Guillemard during his expedition to South Africa
From F. H. H. Guillemard, 'The years that the locusts have eaten'
(typescript, 7 volumes, 1927–1932).

Plate 6 Henry Guillemard (second from the left) in the presence of the Tunku Jerawi (seated), during the voyage of the *Marchesa* in the East Indies

From the Guillemard Collection of Photographs, formerly housed in the Department of Geography, Cambridge University, and now in the Cambridge University Library.

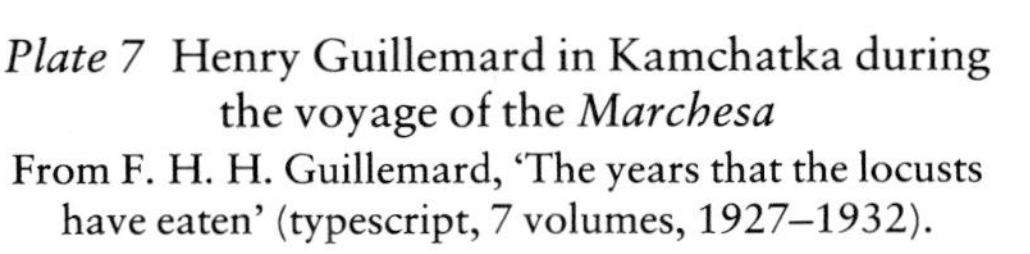

Plate 7 Henry Guillemard in Kamchatka during the voyage of the *Marchesa*
From F. H. H. Guillemard, 'The years that the locusts have eaten' (typescript, 7 volumes, 1927–1932).

Plate 8 Henry Guillemard in Montenegrin dress
From F. H. H. Guillemard, 'The years that the locusts have eaten' (typescript, 7 volumes, 1927–1932).

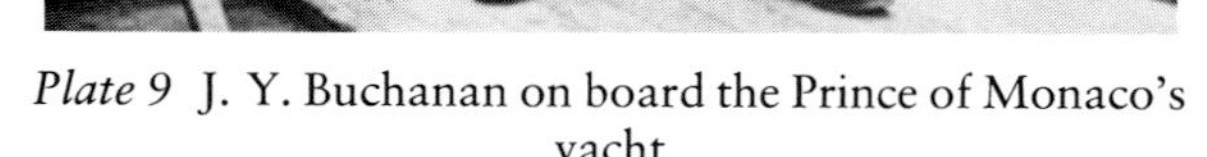

Plate 9 J. Y. Buchanan on board the Prince of Monaco's yacht
From a photograph in the Department of Geography, Cambridge University.

Plate 10 J. Y. Buchanan: 'a humour all his own'
From a photograph in the Department of Geography, Cambridge University.

La Corona Coronas Havana Cigars 1/6 From all tobacconists

CLOSING CITY PRICES—See Page Twenty-seven.

LATE NIGHT FINAL

Evening Standard

TO-MORROW'S WEATHER — MAINLY FAIR.

No. 33,011. LONDON, TUESDAY, JUNE 3, 1930. ONE PENNY.

MARTELL for AGE and QUALITY

UNDERGRADUATE SHOOTS TUTOR DEAD, WOUNDS DETECTIVE, DIES IN HOSPITAL

AMAZING DRAMA AT CAMBRIDGE.

SENIOR TUTOR AT KING'S COLLEGE KILLED DURING INTERVIEW WITH AN UNDERGRADUATE.

FOUR SHOTS HEARD.

From Our Special Correspondent.

CAMBRIDGE, Tuesday.

Mr. A. F. R. Wollaston, senior tutor at King's College, Cambridge, was fatally shot through the heart to-day.

Det.-sergt. Willis, of the Cambridge police, was also shot at and wounded.

An undergraduate, Mr. D. N. Potts, who is alleged to have turned a revolver on himself, was

BIG SWEEP CONFUSION: DIOLITE NUMBER RIDDLE.

CONFLICTING REPORTS IN CALCUTTA.

F or P?

CALCUTTA, Tuesday.

EXTRAORDINARY confusion exists in Calcutta about the numbers which have been drawn in the Derby Sweepstake.

It was announced yesterday on good authority that P3000 had drawn one of the Diolite numbers.

[This number is held by three young bricklayers at Hoylake (Cheshire). See story on Page Three.]

But this horse appears on another member's list as F3000, and the member is certain that his version is correct.

Many lists of numbers, held by various members, do not coincide.

The policy of the club in ensuring secrecy seems to have been successful.

A FREE MARKET FOR MANY BRITISH GOODS.

Dr. Earle Page's Message on Country Party Policy.

Dr. Earle Page, Leader of the Australian Country Party and former Federal Treasurer, states in reply to an inquiry cabled to him by Viscount Rothermere and Lord Beaverbrook that the policy of Tariff Reform and Empire Preference advocated by the party includes free entry for British imports into Australia in many lines.

Dr. Page adds (says a Reuter message from Sydney to-day) that the object of this tariff policy is to lower Australian costs of production and thus expand her industries.

On May 26 Dr. Page proposed at a conference of the Country Party that British manufacturing machinery should be admitted free of charge in return for protection against foreign competition for Australian produce, including wheat and meat, in the British market and for preference in Govern-

Plate 11 The murder of Mr Wollaston
From the *Evening Standard*, 3 June 1930.

Plate 12 Prince Kropotkin
From G. Woodcock and I. Avakumović, *The anarchist prince*
(London: T. V. Boardman, 1950).

Plate 13 Elisée Reclus
From the *Geographical Journal*, 26 (1905), opp. p. 338.

Plate 14 Education
Engraving by François Kupka, in Elisée Reclus, *L'Homme et la terre* (Paris: Libraire Universelle, 6 volumes, 1905–1908), volume 6, p. 433.

Plate 15 The Empire in India
Engraving by François Kupka, in Elisée Reclus, *L'Homme et la terre* (Paris: Libraire Universelle, 6 volumes, 1905–1908), volume 5, p. 81.

Plate 16 S. W. Wooldridge engaged in fieldwork on the Fernhurst Anticline
From a photograph in the Department of Geography, King's College, London.

Plate 17 Charles Darwin in 1853
From the chalk by Samuel Laurence at Down House.

Plate 18 Thomas Henry Huxley
Reproduced by permission of Sir Andrew Huxley.

Georgia, which he accomplished in two and a half weeks. Landing on a bleak and uninhabited shore, he immediately walked across the island to reach the whaling settlement. 'Not a life lost and we have been through Hell', he wrote to his wife, prudently omitting to say that anxious thought had been given to the growing necessity of eating the cook.[27]

The final journey was most harrowing of all. It began in the east Pacific on 20 November 1819, when the *Essex* was destroyed by a whale just south of the equator in longitude 119°W, thus originating the legend of Moby Dick. The small boats with the crew took a month to drift the 2,500 km to uninhabited Henderson Island. Two of them sailed on, narrowly missing the settled island of Pitcairn nearby. The cabin boy, Owen Coffin, was selected by lot to be killed for food (he did not object, saying he could hardly be worse off than he already was), and another man who died was also cannibalized. One of the boats was finally rescued after 4,000 km close to the Juan Fernandez Islands, in 33°45′S, 81°′W, after 93 days.[28]

My second category is that of the planned journey, of which the aim is simply to proceed between known points, with no suggestion of adding to knowledge other than through traversing unfamiliar routes. A classic of its kind is the journey made by the young Francis Younghusband, who in 1886 had been exploring the Chang-pai-shan on the Russo-Chinese border in Manchuria. He was only 23 and was clearly unused to the rough frontier life (he records in his diary dinner with Colonel Sokolowski at Ninguta where 'the drinking was terrible' – 'backwards and forwards, first one and then another' – beginning with vodka and sardines, then into dinner at which each person had six bottles of wine and beer; the glasses were filled indiscriminately with Crimean claret, beer, port, sherry, Guinness's stout, and yet more vodka).[29] Undaunted, he left Peking on 4 April 1887 to walk to India. Travelling mostly by night because of the heat, he crossed the Gobi, travelled through the Altai and Turfan to Urumchi, onwards to Kashgar and Yarkand, and finally crossed the Himalaya with immense difficulty by the Muztagh Pass. When he finally walked into the mess in Srinagar, after a

[27] M. Fisher and J. Fisher, *Shackleton* (London: Barrie, 1957), p. 395. The plight of the cook goes unrecorded in this book.

[28] R. Gibbings, ed., *Narrative of the wreck of the whale-ship Essex of Nantucket which was destroyed by a whale in the Pacific Ocean in 1819, told by Owen Chase First Mate, Thomas Chappel Second Mate and George Pollard Captain of the said vessel* (London: Gold Cockerel Press, 1935); T. F. Heffernan, *Stove by a whale: Owen Chase and the Essex* (Middletown: Wesleyan University Press, 1981).

[29] G. Seaver, *Francis Younghusband: explorer and mystic* (London: John Murray, 1952), p. 51. I recall a very similar experience a few years ago in the Sikhote Alin, north of Vladivostok, at an enormous meal of boiled bear and vodka; Guinness's stout is apparently no longer to be had in that part of the world, however.

journey of seven months, the political agent suggested he take a bath. The proposal rather shook Younghusband: he had deliberately just done exactly that.[30]

A most significant point about this exploit was that Younghusband made no systematic observations at all. He greatly enjoyed it, was deeply affected by the strangeness and beauty of the lands through which he passed, but nevertheless was astonished and embarrassed by his reception at the Royal Geographical Society when he returned. 'Geologists had wanted to know if I have observed the rocks; botanists, if I had collected the flowers; glaciologists, if I have observed the motions of the glaciers; anthropologists, if I have measured the people's skulls; ethnologists, if I had studied their languages; cartographers, if I had mapped the mountains.' Poor Younghusband: he had not thought to do any of these things – it would have been so easy and so very interesting, he admitted with hindsight, and he 'would have liked to have been able to satisfy these men thirsting so keenly for knowledge'. He left with the Society's gold medal (for his 'topographical notes'), but 'deeply repentant'.[31]

Clearly it was no longer enough, to earn the title of geographer, simply to make a dashing ride to Khiva, as Colonel Fred Burnaby had done,[32] or to be thrown by the local emir down a pit filled with vipers, which had been the lot of Colonel Stoddart in Bokhara.[33] The systematic and organized accumulation of knowledge during planned investigations became the central objective and this forms my third category. There was knowledge to be gained which, as Younghusband discovered too late, was more than purely topographic. Banks and the Forsters on Cook's first and second

[30] *Ibid.*, p. 91.

[31] *Ibid.*, p. 96.

[32] F. G. Burnaby, *A ride to Khiva* (London: Cassell, 1876); M. Alexander, *The True Blue: the life and adventures of Colonel Fred Burnaby, 1842–1885* (London: Rupert Hart-Davis, 1957).

[33] An episode with an obvious fascination for me; it is best told by Fitzroy Maclean, *A person from England and other travellers to Turkestan* (London: Jonathan Cape, 1958). After his sojourn in the pit Stoddart and his companion Captain Conolly were publicly beheaded, the latter dying with the memorable line, 'Stoddart, we shall see each other in Paradise.' In 1976, while briefly in Bokhara, I tried to find the pit but (perhaps fortunately) without success: doubtless such incidents, which so appalled his countrymen (I first read of Stoddart's fate on the other side of the world, in a logwood cutters' newspaper published in Belize in 1846), were commonplace at that time in the emir's domains. The 'person from England' who went to rescue the pair was the egregious Joseph Wolff; when he finally escaped to Erzerum it took a friend five days to pick the vermin off his body. Once back in England he not surprisingly never travelled again: with his wife he lies buried in the country churchyard of Ile Brewers in Somerset.

voyages knew it well. Increasingly no respectable expedition could ignore the systematic collection of data in the field sciences: all the early Everest expeditions made geological maps and recorded temperatures; Scott died dragging 20 kg of rock samples back from the Pole.[34]

It will be apparent that my emphasis on achievement through research reflects my own experience of fieldwork and exploration and has largely governed the interpretation offered here. I resolved on going to Cambridge to get to know the tropical world: Cambridge of course was ideally suited to make this possible, both in attitudes and resources.[35]

While the stimulus of the formal courses was astonishing in its breadth,[36] true education came with travel and exploration. In my first long vacation I took a train from Charing Cross station to Calcutta, a journey the central part of which was performed by deck passage between Basra and Bombay.[37] The summer was spent simply coming to terms with a sub-continent I have often since revisited, mainly in Bihar and West Bengal. The following year I was in the diamond country and mangrove estuaries of Sierra Leone, having taken passage in the steerage from Liverpool, not without some opposition from the Elder Dempster Line. Graham Greene's *Journey without maps* and *The end of the affair* (later in Central America, of course, it was *The power and the glory* and Aldous Huxley's *Beyond the Mexique Bay*) helped open my eyes to what it was possible to see and to know. And finally – at last with an expedition – to the Orinoco headwaters in eastern Colombia, a journey which almost made me an ethnologist,[38] and

[34] There is need for a comprehensive study of the ways in which science was incorporated with exploration at the end of the nineteenth century. Certainly at first, and notably in the case of Captain Scott, it caused great dissension between those who actually made the observations and collected the specimens and those who interpreted them. See my 'Social amenities and scientific explorations: geography and geographers in late Victorian London', in R. McLeod and S. Forgan, eds, *Scientific London* (forthcoming).

[35] My mentor there was the noted tropical geographer B. H. Farmer. Fortunately when I applied for admission he was engaged in fieldwork in Ceylon; he often tells me that had he been in Cambridge I would never have got in.

[36] What physical geographer would now find it necessary to read (and enjoy) Previté-Orton's *Shorter Cambridge mediaeval history* for a supplementary course? And what a difference it made to stand on the walls of Constantinople a few weeks later and imagine the Turk at the gate. Some of my contemporaries evidently enjoyed Cambridge geography less. B. T. Robson, for example, clearly went to the same lectures as I did, but evidently heard different things: see his 'A pleasant pain', in M. D. Billinge, D. J. Gregory and R. L. Martin, eds, *op. cit.* (note 21), 104–116.

[37] My mother was appalled. I was fortunate that the only thing I lost on this voyage was my watch, though off Muscat it was (as Wellington might have said) a damned close-run thing. One did learn a lot, however, about the Arabs, travelling thus.

[38] D. R. Stoddart, 'Myth and ceremonial among the Tunebo Indians of eastern Colombia',

introduced me to the massive scholarship of the *Handbook of South American Indians* and thus to Sauer, Métraux and Lévi-Strauss. It was entirely by chance that my next expedition, three weeks after that had ended, was to the coral reefs of British Honduras, and the shock of this unknown world, its plants and animals, extraordinary beauty and great tragedy, determined the course of my life. Ever since then I have been in the tropics, once, twice, sometimes three times a year,[39] trying to understand the reef environment around the world, always with the growing realization that I could never know enough. And always in the background too has been Sauer's warning that 'field time is your most precious time – how precious you will know only when its days are past.'[40] But I also know, again with Sauer, that 'one of the finest experiences of youth is to go where none of your kind has been, to see and learn to make some sense out of what has not been known to any of us.'[41]

SCIENCE AND DISCOVERY

It is thus the quest for systematic and objective knowledge that distinguishes the traveller, inadvertent or otherwise, from the scientific explorer and discoverer.[42] But one can go further, and draw a broad analogy between the structure of these activities and the nature of discovery itself. If it is to be other than a trivial activity, discovery must be seen as having theoretical significance. Facts are given meaning by their theoretical context, rather than by naive apprehension: it is the mind, not the eye, that sees, as Hanson reminds us in his *Patterns of discovery*.[43] Indeed, as O'Gorman argued, it is precisely this which defines the problem of who 'discovered' America.[44] One might argue that one of the weaknesses of the field-study movement in education was its failure adequately to recognize that observation is itself a

Journal of American Folklore, 75(1962), 147–152, and (with J. D. Trubshaw), 'Colonization in action in eastern Colombia', *Geography*, 47 (1962), 47–53.

[39] Colleagues point out that this could also be said of Cambridge.

[40] C. O. Sauer, *op. cit.* (note 19), p. 298.

[41] *Ibid.*, p. 296.

[42] For some recent trends, see D. R. Stoddart, 'Geography and contemporary exploration', *Geographical Journal*, 146 (1980), 169–173.

[43] N. R. Hanson, *Patterns of discovery: an inquiry into the conceptual foundations of science* (Cambridge University Press, 1958), p. 6: 'People, not their eyes, see. Cameras, and eye-balls, are blind.'

[44] Edmundo O'Gorman, *op. cit*, (note 23).

theory-laden procedure. The distinction is thus to do with the meaning of knowledge: between those who see facts as in some unambiguous way the building blocks of explanation, used in the inductive generation of general laws, and those who see them as only achieving significance through their theoretical contexts. Huxley in his *Physiography* certainly believed in the former view, though perhaps particularly as a pedagogic device.[45] Darwin knew that it was the latter which held the key to science. 'How odd it is', he wrote in 1861, 'that anyone should not see that all observation must be for or against some view if it is to be of any service.'[46] And he told Lyell that 'without the making of theories I am convinced there would be no observation.'[47]

And it is this distinction which explains why the practitioners of field observation came to be held in such low regard by the New Geographers of the universities and colleges. It was increasingly realized that a piece of granite or a box of butterflies, gained at whatever cost, made no necessary contribution to knowledge in themselves: one might as well, as Darwin said, go into a gravel pit and count the pebbles and record their colours.[48] The response to this realization should, of course, have been the opposite. For just as philosophers of science may use the map and the act of mapping as an analogue of the structure of theories and of theory-making,[49] so the acts of geographical exploration and discovery can be seen as the practical expression of a common view of knowledge. It is strange that this interpretation, which would serve to bind geography's classical concerns with the needs of a developing discipline, has not been more widely recognized. Discovery becomes the test of a prediction, exploration the process by which that test is carried out.

In its most simple form this was outlined by William Bunge in his *Theoretical geography* in 1966,[50] a book which had great impact at the time but which it is now becoming fashionable to decry. He posed the following

[45] See chapter 9.

[46] C. R. Darwin to Henry Fawcett, 18 September 1961, in F. Darwin and A. C. Seward, eds, *More letters of Charles Darwin: a record of his work in a series of hitherto unpublished letters* (London: John Murray, 2 volumes, 1903), volume 1, p. 195.

[47] C. R. Darwin to C. Lyell, 1 June 1860, in F. Darwin, ed., *The life and letters of Charles Darwin, including an autobiographical chapter* (London: John Murray, 3 volumes, 1887), volume 2, p. 315.

[48] C. R. Darwin, *op. cit.* (note 46). Many of our colleagues have made their reputations by doing precisely this, however.

[49] S. Toulmin, *The philosophy of science* (London: Hutchinson, 1953), pp. 105–139.

[50] W. Bunge, *Theoretical geography* (Lund: C. W. K. Gleerup, 2nd edition, 1966), in chapter 8, 'The meaning of spatial relations', which does not occur in the 1st edition of 1962.

question: 'Consider an idealized topographic [map] sheet. What will the adjacent sheet be like?'[51] It is a question which epitomizes the explorer's quest, and which involves the prediction and confirmation of what he will find. In answer to it, Bunge hazarded the general proposition that 'The farther in space away from the known location the less confident the predictions.'[52] And he pointed directly, if in the language and the concepts of his time, to the implications of such a view for the nature of exploration:

> Geographers predict a village because geographic theory demands it. If geographic theoreticians provide geographic experimenters, including explorers, with the most likely sites of the missing Maya sites, compute their k's, their rank-size rule, estimate the number of hamlets, compute their average space and so forth are they not predicting? To place our finger on the map of Yucatán and say, 'There', is no less impressive a prediction than an astronomer pointing in the Heavens to a missing planet.[53]

But Bunge's readers remained entrenched in their laboratories: not for them the jungles of the Petén and Quintana Roo. Waldo Tobler made a lone but inconclusive foray into this area in his attempt to predict housemound locations in Anatolia, though his predictions were not tested in the field.[54] The archaeologists proved more receptive, most notably in studies of the Maya site of Lubaantun, Belize, though in large degree the point was lost: batteries of locational techniques, including Thiessen polygons and nearest-neighbour statistics, were used to *interpret* known sites rather than to *discover* new ones, as Bunge had suggested.[55]

The example is not perhaps of great significance: what is important is to recognize that exploration and discovery, far from being simply the activities of 'solid hunks of manhood' in a past colonial age, are central in concept as well as execution to our whole endeavour. My interpretation of the need for exploration and discovery differs from most others generally

[51] *Ibid.*, p. 246.
[52] *Ibid.*, p. 246.
[53] *Ibid.*, p. 247.
[54] W. Tobler and S. Wineburg, 'A Cappadocian speculation', *Nature, Lond.*, 213 (1971), 39–41.
[55] See, for example, J. Marcus, 'Territorial organization of the lowland Classic Maya', *Science*, 180 (1973), 911–916; N. D. C. Hammond, 'Locational models and the site of Lubaantun: a classic Maya centre', in D. L. Clarke, ed., *Models in archaeology* (London: Methuen, 1972), 757–800; N. D. C. Hammond, 'The distribution of late Classic Maya major ceremonial centres in the Central Area', in Hammond, ed., *Mesoamerican archaeology: new approaches* (London: Duckworth, 1974), 313–334.

accepted, since these have almost without exception concentrated on the personal characteristics of those involved in exploration. Thus Leonard Darwin[56] singled out determined perseverance and great personal courage as the prime requirements, together with a capacity for organization and for friendship. Sir Vivian Fuchs[57] listed endurance, judgement, emotional stability and again determination. But these qualities are the necessary rather than the sufficient conditions for explanation: they answer the question 'how?' rather than 'why?' a man is a successful explorer and contributor to knowledge. As generalizations they fail to satisfy, even though on an individual level, as in the cases of Burton, Livingstone and Fremont, they may help us to understand a personal achievement better.[58] Fuchs I believe comes closer when he adds curiosity and imagination (though specifically excluding 'the man who lives by theories'), for these

[56] L. Darwin, 'On the personal characteristics of great explorers', *Scottish Geographical Magazine*, 28 (1912), 134–146.

[57] V. Fuchs, 'The qualities of an explorer', *Geographical Magazine*, 36 (1963), 205-215.

[58] For such interpretations, see F. M. Brodie, *The devil drives: a life of Sir Richard Burton* (London: Eyre & Spottiswoode, 1967); T. Jeal, *Livingstone* (New York: G. P. Putnam's Sons, 1973); and A. Rolle, 'Exploring an explorer: psychohistory and John Charles Fremont', *Pacific Historical Review*, 51 (1982), 135–163. For a general discussion of biography in this field, see W. B. Fairchild, 'Explorers: men and motives', *Geographical Review*, 38 (1948), 414–425. Burton in particular deserves more attention, if only for his aphorism 'Geography is good, but gold is better.' The first European to see the Kaaba in Mecca (which he mapped by pacing) and to penetrate Medina, the leader of an expedition with Speke to search for the source of the Nile in 1858–1859, founder's medallist of the Royal Geographical Society in the year of *The origin of species*, participant in the debate with Speke at the British Association in Bath in 1864, after five years of vicious enmity, which led to Speke's dramatic suicide, Burton has likewise largely disappeared from the history of geography, as opposed to that of exploration. Yet one might make a case for his inclusion quite apart from his speculations on African geography. The Terminal Essay of his *A plain and literal translation of the Arabian nights' entertainments* (Benares: Kama-shastra Society, 10 volumes, 1885) supplies a regional delineation of sodomitical practices which long pre-dates anything Herbertson wrote on temperature and rainfall; Burton called the areas involved the 'Sotadic Zone'. It was moreover based upon a lifetime of original research that had begun with his report on buggery in Karachi, prepared at the age of 24 and a constant source of scandal for the rest of his life. Fawn Brodie in her magnificent biography writes that 'an extraordinary amount of energy went into his "field research" and speaks of 'the intensity and almost frantic quality of his searching, especially on matters sexual' (*op. cit.*, p. 336). In a piece published posthumously in 1894 he wrote that 'Discovery is mostly my mania': Brodie ranks him among explorers with Livingstone, Stanley, Baker and Speke, and among the scientists of his time with Darwin, Galton, Lyell, Frazer, Flinders Petrie, Evans, Sayce and T. H. Huxley (*op cit.*, p. 16). It would be a confident subject indeed which could afford to exclude such a remarkable scholar (he knew 35 languages and published over 50 books) from its history. Not surprisingly, Burton's wife burned everything she could lay her hands on of his papers at his death (a curious collection of letters is to be found, however, in Trinity College, Cambridge). They are both buried in a stone tomb built in

qualities can be read as the unstructured counterpart of the process of knowing. The only alternative to the view suggested here – Overton's 'theory of exploration'[59] – sees the answer as a response to a demand for new resources generated by economic conditions. While doubtless applicable to the opening of countries like Canada, Australia and New Zealand in the nineteenth century, this explanation seems to me to be too closely rooted in specific historical and geographical conditions, and to lack the generality required for wider understanding.

OCEANS OF IGNORANCE, ISLANDS OF KNOWLEDGE

It was David Brewster who recorded the famous words of Newton, looking back on his achievements at the end of his life: 'I know not what I may appear to the world, but to myself I seem to have been only like a boy playing on the sea-shore, and diverting myself in now and then finding a smoother pebble or a prettier shell than ordinary, whilst the great ocean of truth lay all undiscovered before me.'[60] For a geographer especially it is a most attractive metaphor, more particularly for someone like myself whose work has concentrated on coral island beaches. Serendipitous discovery undoubtedly enlivens more systematic study, and even occasionally yields treasure of a sort, though I fear that the wealth of the pirates still eludes me.[61]

---

the form of an Arab tent, complete with battery-operated camel bells, at Mortlake in Surrey. There can have been few more bizarre sights in recent years than that of the elegant figure of the director of the Royal Geographical Society, Sir Laurence Kirwan, leading a small party of Burton's disciples through the cemetery on 2 July 1974, with a ladder, and climbing into the tomb through the high rear window (it was the first step in its renovation). It is pleasing to know that the Society at least does not forget its medallists. See L. P. Kirwan, 'Meditations on the Burton mausoleum at Mortlake', *Geographical Journal*, 141 (1975), 49–54.

[59] J. D. Overton, 'A theory of exploration', *Journal of Historical Geography*, 7 (1981), 53–70.

[60] D. Brewster, *Memoirs of the life, writings, and discoveries of Sir Isaac Newton* (Edinburgh: Thomas Constable, 2 volumes, 1855), volume 2, p. 407.

[61] But see my 'Destruction of Maya remains by shoreline erosion, Belize sand cays, Central America', in M. L. Schwartz, ed., *Proceedings of the Commission on Coastal Environment Field Symposium, Shimoda, Japan* (Bellingham: Western Washington University, 1980), 159–168, for what one can find on beaches. Sadly, my stay on uninhabited Suwarrow Atoll in the central Pacific in 1981 – where treasure has frequently been found and men murdered in the search for it – yielded nothing other than sediment samples and botanical specimens.

Yet I prefer a different picture: of knowledge as an island in a sea of ignorance, for the more the island expands the longer its shoreline will be with the unknown.[62] This alone supplies the justification for our subject's continuing commitment to exploration and discovery, and simultaneously explains why this commitment need never end.

It is, too, a matter for pride that our history contains such a record of achievement at the farthest ends of the earth.[63] Let us salute with Conrad 'men great in their endeavour and in hard-won successes of militant geography; men who went forth each according to his lights and with varied motives . . . but each bearing in his heart a spark of the sacred fire'.[64] If that spark ever dies, then our geography will indeed have become a dry and bloodless thing.

[62] The image is not original. I noted it 20 years ago, attributed it to Rodberg and Weisskopf (1957), and cannot now trace the citation.

[63] Told most recently by I. Cameron, *To the farthest ends of the earth* (London: Macdonald & Jane's, 1980).

[64] Joseph Conrad, *op. cit.* (note 1), p. 31.

8

Darwin's Impact on Geography

At a time when many sciences are re-examining the impact of biological thinking, and particularly of Charles Darwin's writings, on their methods and theoretical foundations,[1] geographers have been strangely silent, and the Darwin centenary in geographical circles passed almost unremarked.[2] It is, in fact, strange that Darwin's name is not prominent in either of Hartshorne's volumes on geographic methodology, where only passing reference is made to the impact of the life sciences on geography.[3] Whereas the centenaries of the deaths of Humboldt and Ritter on 6 May and 28 September 1859 were commemorated by geographers, the first publication

[1] B. J. Loewenberg, 'Darwin, Darwinism and history', *General Systems*, 3 (1958), 7–17. For recent reviews, see B. J. Loewenberg, 'Darwin and Darwin studies, 1959–63', *History of Science*, 4 (1965), 15–54; E. Mayr, 'The nature of the Darwinian revolution', *Science, N.Y.*, 176 (1972), 981–989; and J. C. Greene, 'Reflections on the progress of Darwinian studies', *Journal of the History of Biology*, 8 (1975), 243–273.

[2] A meeting at the Royal Geographical Society for the Darwin Centenary did not consider Darwin's contribution to scientific thought: Sir C. G. Darwin, 'Darwin as a traveller', *Geographical Journal*, 126 (1960), 129–136. See also note 63. A paper entitled 'Ch. Darwin's influence on the progress of science in geography', by A. Malicki, was announced for presentation at the XI International Congress of the History of Science, Warsaw, 1965, but no abstract of this paper was published (*Sommaires*, XI Congrès International d'Histoire des Sciences, Cracow, 24–29 August 1965, 2 volumes, 594 pp.). There is an early treatment of the geographical content of Darwin's own writings by Giovanni Marinelli, 'Carlo Roberto Darwin e la geografia', *Atti dell'Istituto Veneto de Scienze, Lettere ed Arti*, ser. 5, 8 (1882), 1279–1321. Marinelli treats Darwin's coral-reef work at length, and in analysing the geographical nature of Darwin's other writings he concludes that their principle was essentially chorological. The paper is reprinted in *Rivista de Filosofia Scientifica*, 2 (1882–1883), 385–410, and in *Scritti minori de Giovanni Marinelli*, volume 1: *Metodo e storia della geografia* (Firenze: Tipografia de M. Ricci, 1908), 99–141. See also Willi Ule, 'Darwin's Bedeutung in der Geographie', *Deutsche Rundschau für Geographie und Statistik*, 31 (1909), 433–443.

[3] R. Hartshorne, *The nature of geography, a critical survey of current thought in the light of the past* (Lancaster, Pa.: Association of American Geographers, 1939) and R. Hartshorne, *Perspective on the nature of geography* (Chicago: Rand McNally, for Association of American Geographers, 1959).

of *On the origin of species* on 24 November in the same year remained unnoticed.

Much of the geographical work of the past hundred years, however, has either explicitly or implicitly taken its inspiration from biology, and in particular from Darwin. Many of the original Darwinians, such as Hooker, Wallace, Huxley, Bates and Darwin himself, had been actively concerned with geographical exploration, and it was largely facts of geographical distribution in a spatial setting which provided Darwin with the germ of his theory. This chapter seeks to trace the broad lines of the biological impact on geography since 1859, to assess in what respects this impact was Darwinian and, equally important, what essential features of Darwin's thought were ignored by geographers.

It is important to recall that Darwin's theory was not simply one of 'evolution', a word which did not appear in *The origin* until the fifth edition in 1869, but concerned a mechanism whereby random variations in plants and animals could be selectively preserved, and by inheritance lead to changes at the species level. In geography, however, Darwinism was interpreted primarily as evolution, in the sense of a 'continuous process of change in a temporal perspective long enough to produce a series of transformations'.[4] It was in this sense that many natural and social scientists welcomed evolution from about 1860 onwards. Darwin, however, was primarily concerned with the mechanism of the change or, as *The origin* was subtitled, 'the preservation of favoured races in the struggle for life'. This element of struggle was applied in a deterministic way, particularly in human geography, at about the same period of time. The crux of Darwin's theory, the randomness of the initial variations,[5] passed almost unnoticed. In both physical and human geography, supposedly Darwinian ideas were applied in an eighteenth rather than a nineteenth century manner, and geographers were still applying essentially Newtonian views of causation well into the twentieth century. Why the central theme of Darwin's work was thus neglected is, therefore, a fundamental problem in the history of ideas. Finally, the Darwinian revolution gave fresh impetus to concepts of biological origin which date back to Ritter and before, and the subsequent development of ecology led to new insights in some branches of geographical thinking.

[4] R. Scoon, 'The rise and impact of evolutionary ideas', in S. Persons, ed., *Evolutionary thought in America* (New Haven: Yale University Press, 1950), 4–42; reference on p. 5.

[5] Samuel Butler neatly phrased the issue: 'To me it seems that the "Origin of Variation", whatever it is, is the only true "Origin of Species,"' in *Life and habit* (London: Trubner, 1878), reference on p. 263.

In this chapter four themes are taken to be especially significant contributions to geographical thought from biology and, particularly, from Darwin.[6] They are: (1) the idea of change through time; (2) the idea of organization; (3) the idea of struggle and selection; (4) the randomness or chance character of variations in nature. Each of these four themes is examined in turn from the geographical point of view, to determine in what sense such views were biological or Darwinian in origin, how geography reacted to them, and what geographical insights they stimulated.

TIME AND EVOLUTION

The first part of the nineteenth century, culminating in Lyell's *Principles of geology*,[7] saw the breakdown of the medieval view of the age of the earth, just as the Copernican revolution had revised ideas on its position in the universe four centuries earlier. The expansion of physical geography towards the end of the century drew on a double inspiration: that of the early geologists from Hutton to Lyell, and that of evolutionary biology, which was itself dependent on the earlier breakdown of restrictive cosmological ideas.

The strongest and most explicit impact of evolution was in the study of landforms, a field in which Darwin had worked during the *Beagle* years, when he formed his theory of the transformation of fringing reefs into barrier reefs and then into atolls by the slow operation of subsidence of their foundation through time.[8] The initial deduction and subsequent develop-

[6] No attempt is made to cover more general issues, such as the influence of Darwin's work on classification and taxonomy, with the resulting emphasis in geography on 'genetic classification', or such fundamentally biological fields as zoogeography, on which both Darwin and Wallace worked. For commentary on these, see particularly D. B. Grigg, 'The logic of regional systems', *Annals of the Association of American Geographers*, 55 (1965), 465–491, and P. J. Darlington, Jr, 'Darwin and zoogeography', *Proceedings of the American Philosophical Society*, 103 (1959), 307–319.

[7] C. Lyell, *Principles of geology: being an attempt to explain the former changes of the earth's surface, by reference to causes now in operation* (London: John Murray, 3 volumes, 1830–1833). See also C. C. Gillispie, *Genesis and geology: a study in the relations of scientific thought, natural theology, and social opinion in Great Britain, 1790–1850* (Cambridge: Harvard University Press, 1951).

[8] D. R. Stoddart, 'Coral islands, by Charles Darwin, with an introduction, map and remarks', *Atoll Research Bulletin*, 88 (1962), 1–20, and C. R. Darwin, *The structure and distribution of coral reefs* (London: Smith, Elder, 1842). For a fuller consideration, see D. R. Stoddart, 'Darwin, Lyell, and the geological significance of coral reefs', *British Journal for the History of Science*, 9 (1976), 199–218.

ment of this theory, as Gruber[9] observed, closely resembles the later development of Darwin's biological ideas, and it could serve as the archetype for the 'cyclic' ideas later developed in geomorphology. Huxley[10] himself published on the new subject of 'physiography' in the 1870s, but it was Davis who took evolution as his inspiration in the idea of the geographical cycle. Earlier workers, faced with the bewildering complexity of landforms, had sought to reduce them to order by nominal classification, much as had Linnaeus in taxonomy, but the failure to supply any unifying principle to such study reduced it to cataloguing.[11] In his first paper on the development of landforms, however, Davis referred to a 'cycle of life', and used such terms as birth, youth, adolescence, maturity, old age, second childhood, infantile features and struggle to emphasize the analogy of an organism undergoing a sequence of changes in form through time.[12] The power of evolutionary thinking to bring diverse facts into new meaningful relationships fascinated Davis. Writing of the cycle in 1900, he stated that:

> in a word it lengthens our own life, so that we may, in imagination, picture the life of a geographical area as clearly as we now witness the life of a quick growing plant, and thus as readily conceive and as little confuse the orderly development of the many parts of a landform, its divides, cliffs, slopes, and water course, as we now distinguish the cotyledons, stems, buds, leaves, flowers, and fruit of a rapidly maturing annual that produces all these forms in appropriate order and position in the brief course of a single summer. The time is ripe for the introduction of these ideas. The spirit of evolution has been breathed by the students of the generation now mature all through their growing years, and its application in all lines of study is demanded.[13]

[9] H. E. Gruber and V. Gruber, 'The eye of reason: Darwin's development during the *Beagle* voyage', *Isis* 53 (1962), 186–200.

[10] T. H. Huxley, *Physiography, an introduction to the study of nature* (London: Macmillan, 1877); and see chapter 9.

[11] See, for example, the writings of Elisée Reclus, *The earth: a descriptive history of the phenomena of the life of the globe*, edited by A. H. Keane (London: J. S. Virtue, 1886).

[12] W. M. Davis, 'Geographic classification. Illustrated by a study of plains, plateaus and their derivatives', *Proceedings of the American Association for the Advancement of Science* (1884), 428–432. The linkage between evolution and landforms was not, of course, original: Archibald Geikie had already lectured to the Royal Geographical Society on 'Geographical evolution', *Proceedings of the Royal Geographical Society*, 1 (1879), 422–443.

[13] W. M. Davis, 'The physical geography of the lands', *Popular Science Monthly*, 57 (1900), 157–170, reprinted in W. M. Davis, *Geographical essays*, edited by D. W. Johnson (Boston: Ginn, 1909), 70–86; reference on pp. 85–86. See also W. M. Davis, 'The relations of the earth sciences in view of their progress in the nineteenth century', *Journal of Geology*, 12 (1904),

Throughout his working life, Davis emphasized this theme of orderliness and development through time, which he termed evolution, but it is perhaps significant that he took his illustrations not from the species or the population, but from the individual. So successful was Davis in promoting this view that in his hands geomorphology became more the study of the origin of landforms than of landforms themselves[14] and was thus readily channelled into the restricted field of denudation chronology.

Darwin was, of course, deeply influenced by Lyell's *Principles*,[15] and the two distinct components of Lyell's uniformitarianism, gradualism and actualism, are implicit in *The origin*. It has been argued that a strict uniformitarianism had no place for progression or transmutation of species,[16] and Lyell himself emphatically rejected their mutability in the early editions of the *Principles*.[17] But in the sense of excluding catastrophic explanation, Huxley was certainly correct in his view that 'consistent uniformitarianism postulates evolution as much in the organic as the inorganic world,'[18] and that 'the *Origin of Species* is the logical sequence of the *Principles of Geology*.'[19]

669–687, especially p. 675; 'The physical factor in general geography', *The Educational Bi–monthly*, 1 (1906), 112–122; 'The geographical cycle', *Geographical Journal*, 14 (1899), 481–504, and *Essays*, *op. cit.*, 249–278, especially pp. 249 and 254; 'Peneplains and the geographical cycle', *Bulletin of the Geological Society of America*, 23 (1922), 589–598; especially pp. 594–595.

[14] C. O. Sauer, 'The morphology of landscape', *University of California Publications in Geography*, 2 (1925), 19–54; see p. 32.

[15] As early as 1835 Darwin wrote to W. D. Fox that 'I am become a zealous disciple of Mr Lyell's views. . . . I am tempted to carry parts to a greater extent even than he does.' C. R. Darwin to W. D. Fox, Lima, July 1835, in Francis Darwin, ed., *Life and letters of Charles Darwin, including an autobiographical chapter* (London: John Murray, 3 volumes, 1887), reference in volume 1, p. 263.

[16] W. F. Cannon, 'The uniformitarian–catastrophist debate', *Isis*, 51 (1960), 38–55; R. Hooykaas, *Natural law and divine miracle: the principle of uniformity in geology, biology and theology* (Leiden: E. J. Brill, 1963), pp. 93–101.

[17] See Gavin de Beer's comment: 'Lyell used the principle of uniformitarianism to prove that evolution was impossible because evolution involved progressionism and progressionism involved catastrophism and catastrophism must be rejected. Darwin used uniformitarianism to show that simple, existing causes produced and directed evolution, and there there was no link between catastrophism and progressionism. . . . The supreme paradox was, therefore, that Darwin used Lyell's methods to show that Lyell's views on biology were wrong.' G. de Beer, *Charles Darwin: evolution by natural selection* (London: Thomas Nelson, 1963), reference on p. 104.

[18] T. H. Huxley, 'On the reception of the "Origin of Species"', in F. Darwin, ed., *op. cit.* (note 15), 179–204; reference on p. 109.

[19] T. H. Huxley, 'The coming of age of "The Origin of Species"', in T. H. Huxley, *Science and culture, and other essays* (London: Macmillan, 1882), 310–324; reference on p. 315.

Uniformitarianism in geology, and subsequently in biology, involved, as Hutton clearly saw, the need for time in excess of that allowed by theology. In a famous passage on the Alps, Hutton described the continuous mantle of waste from the mountaintops to the sea: 'throughout the whole of this long course, we may see some part of the mountain moving some part of the way. What more can we require? Nothing but time.'[20] Once the reality of small but cumulative variations was established in biology, a similar conclusion followed. Time became one of Darwin's chief requirements, to the extent that he refused to accept Lord Kelvin's apparent demonstration of the youth of the earth based on estimated rates of cooling and the second law of thermodynamics.[21] And it was Kelvin, not Darwin, who was later shown to be wrong. When Davis in 1899, therefore, wrote his paper on the cycle of erosion, with the trinity of factors, structure, process and time, it was time which he singled out as 'the one of the most frequent application and of a most practical value'[22] in landform study. The key to the cyclic view in geomorphology in fact lies in systematic, irreversible change of form through time, and from this derives the biological analogy of ageing used by Davis, Johnson and their school. Davisian geomorphology was deductive, time-oriented and imbued with mechanistic notions of causation, deriving it uniformitarianism from Lyell and its theme of change through time at least partly from a simplified view of evolution.

Closely similar views were being proposed at about the same time in plant geography and particularly in ecology. Hooker, among the founders of the subject, was an explorer in his own right; later workers such as Shelford, Cowles and Tansley were members of professional geographic bodies; and one man, Clements, occupied in plant ecology a position similar to that of

[20] J. Hutton, *Theory of the earth, with proofs and illustrations* (London and Edinburgh: printed for Messrs Cadell, Junior & Davies, London, and William Creech, Edinburgh, 2 volumes, 1795), references in volume 2, p. 329.

[21] W. Thomson, Baron Kelvin of Largs, 'On the secular cooling of the earth', in W. Thompson and P. G. Tait, eds, *Treatise on natural philosophy* (Oxford: Clarendon Press, 1867), volume 1 (all published), 711–727; and W. Thomson, Baron Kelvin of Largs, '"The doctrine of uniformity" in geology briefly refuted', *Popular Lectures and addresses*, volume 2: *Geology and general physics* (London: Macmillan, 1894), 6–9. Darwin admitted to being 'greatly troubled at the short duration of the world according to Sir W. Thomson'. C. R. Darwin to James Cross, 31 January 1869, in F. Darwin and A. C. Seward, eds, *More letters of Charles Darwin: a record of his work in a series of hitherto unpublished letters* (London: John Murray, 2 volumes 1903), reference in volume 2, p. 163. For a historical treatment of the problem of time, see F. C. Haber, *The age of the world: Moses to Darwin* (Baltimore: Johns Hopkins University Press, 1959). especially pp. 265–290.

[22] W. M. Davis, 'The geographical cycle', *Geographical Journal*, 14 (1899), 481–504, and *Geographical Essays*, *op. cit.* (note 13), 249–278; reference on p. 249.

Davis in geomorphology. In soil science also, similar naive views of evolution as change through time were emphasized by Marbut and his school, in introducing the ideas of Dokuchaiev and Sibirtsev into the English literature.[23] Both conceptually, and in the imagery employed, plant ecologists and pedologists followed Davis's biological analogy for change through time. Clements emphasized succession as the 'universal process of formation development . . . the life-history of the climax formation'.[24] The conceptual similarity to geomorphology was seized on by Cowles, a botanist trained in physiography by Salisbury and Chamberlin, who brought Davisian geomorphology and Clementsian ecology together in 'physiographical ecology' following fieldwork in the Chicago area on the coincidence of plant formations and physiographic units.[25] Plant ecologists and geomorphologists both adopted terms such as infancy, youth, maturity and old age to describe development through time.

As in geomorphology, the time-framework has proved too restrictive for later investigations. Whittaker, in his thorough review of the climax concept, perceptively summarizes Clements's contribution to plant ecology, but he could well have been speaking of Davis's role in the development of geomorphology or Marbut's in pedology:

> It was the great contribution of Clements to have formulated a system, a philosophy of vegetation, which has been a dominating influence on American ecology as a framework for ecological thought and investigation. . . . Some negative aspects of Clements' system are . . . the superficial verbalism, the tendency to fit evidence by one means or another into the philosophical structure, the thread of non-empiricism which runs through his thought and work. . . . The Clementsian system had a certain symmetry about it, it was a fine design if its

[23] For Dokuchaiev's views, see K. D. Glinka, *Treatise on soil science*, 4th edition (Jerusalem: Israel Program for Scientific Translations, for the National Science Foundation, Washington, D.C., 1963), p. 188; and on the American school, C. F. Marbut, 'Soils of the Great Plains', *Annals of the Association of American Geographers*, 13 (1923), 41–46.

[24] F. E. Clements, *Plant succession, an analysis of the development of vegetation* (Washington: Carnegie Institution, 1916), reprinted in F. E. Clements, *Plant succession and indicators: a definitive edition of Plant succession and Plant indicators* (New York: Hafner, 1963), reference on p. 3.

[25] H. C. Cowles, 'The causes of vegetative cycles', *Botanical Gazette*, 51 (1911), 161; 'The causes of vegetative cycles', *Annals of the Association of American Geographers*, 1 (1911), 3–20, reference on p. 3. For Cowles's substantive work, see H. C. Cowles, 'The physiographic ecology of Chicago and vicinity: a study of the origin, development and classification of plant societies', *Botanical Gazette*, 31 (1901), 73. Cowles's teacher, Salisbury, was not himself an advocate of the cycle of erosion concept, but he used it as a teaching device.

premises were granted; and for its erection Clements may rank as one of the truly creative minds of the field.[26]

In the social sciences the development of a time-perspective awaited that of a historical tradition, especially the emergence of a concept of prehistory in the 1830s. That evolutionary ideas were in the air by 1859 is shown by the appearance of Sir Henry Maine's *Ancient law* as early as 1861. E. B. Tylor's *Early history of mankind* (1865) and *Origin of civilization* (1870), and the Duke of Argyll's *Primeval civilization* (1869) set a fashion in the developmental interpretation of prehistory and ethnology which dominated the subject until Malinowski's functional reinterpretation in the 1920s. In social anthropology also, McLennan's *Primitive marriage* (1865), Frazer's *Golden bough* (1890), Westermarck on religion, and Lewis Morgan, Durkheim and Lévy-Bruhl on social structures, established an evolutionary position over a period of decades which dominated thinking in these subjects until the reaction in the twentieth century.[27] A few workers, some of them influential in geography, maintained a developmental framework, ranging from the 'unilinear' school of White to the 'multilinear' evolution of Steward.[28] Childe[29] influenced historical interpretation of technological development, particularly among English geographers; the botanist Geddes's work on cities influenced early ideas on urban geography;[30] Taylor

[26] R. H. Whittaker, 'A consideration of climax theory: the climax as a population and pattern', *Ecological Monographs*, 23 (1953), 41–78; see also C. C. Nikiforoff, 'Reappraisal of the soil', *Science*, N.Y. 129 (1959), 186–196.

[27] I. Goldman, 'Evolution and anthropology', *Victorian Studies*, 3 (1959), 55–75; D. G. MacRae, 'Darwinism and the social sciences', in S. A. Barnett, ed., *A century of Darwin* (London: William Heinemann, 1958), 296–312; E. R. Leach, 'Biology and social anthropology: the current status of the biological analogy', *Cambridge Review*, 85 (1964), 248–251; F. S. C. Northrop, 'Evolution in its relation to the philosophy of nature and the philosophy of culture', in S. Persons, ed., *op. cit.* (note 4), 44–83; R. W. Gerard, C. Kluckholn and A. Rapoport, 'Biological and cultural evolution: some analogies and explorations', *Behavioral Science*, 1 (1956), 6–34. For popular views, see H. R. Hays, *From ape to angel: an informal history of social anthropology* (New York: Alfred A. Knopf, 1958); G. E. Daniel, *The idea of prehistory* (London: Watts, 1962); and L. Newson, 'Cultural evolution: a basic concept for human and historical geography', *Journal of Historical Geography*, 2 (1976), 239–255.

[28] L. White 'Evolutionary states, progress and the evolution of cultures', *Southwestern Journal of Antrhopology*, 3 (1947), 165–192; J. H. Steward, *Theory of culture change: the methodology of multilinear evolution* (Urbana: University of Illinois Press, 1955).

[29] V. G. Childe, *Man makes himself* (London: Watts, 1936); V. G. Childe, *Social evolution* (London: Watts, 1951).

[30] P. Geddes, *Cities in evolution: an introduction to the town planning movement and to the study of civics* (London: Williams & Norgate, 1915); T. G. Taylor, *Urban geography: a study of site, evolution, pattern and classification in villages, towns and cities* (London: Methuen, 1st edition 1949, 2nd edition 1951), see pp. 7–9, 421–423.

applied developmental principles to the study of race and culture history;[31] and Beaver and others attempted to introduce cyclic ideas into the interpretation of economic landscapes[32] though with little success.

Change through time has been a dominant theme in much human geography, particularly in the work of the Berkeley School on the settlement of the American southwest and other areas. Here Sauer's influence has been dominant, and it is interesting that he himself studied at Chicago under the physiographer Salisbury and the plant ecologist Cowles.[33] Another pupil of Cowles, who also studied under Davis and published in both geomorphology and vegetation geography, was the historical geographer Ogilvie, who carried their emphasis on time into regional studies.[34] The influence of plant ecology and the historical viewpoint was also clear, both in concept and language, in Whittlesey's idea of sequent occupance in the development of landscapes.[35]

The history of these narrowly evolutionary views in geography, however, resembled that in social anthropology: the early and enthusiastic application of time-frameworks to the data, and then a retreat from a developmental to a functional approach, or to a much modified and refined evolutionary interpretation. Primarily, however, geographers interpreted

[31] T. G. Taylor, *Environment and race: a study of the evolution, migration, settlement and status of the races of man* (London: Humphrey Milford, Oxford University Press, 1927); T. G. Taylor, *Environment and nation: geographical factors in the cultural and political history of Europe* (Toronto: University of Toronto Press, 1936); T. G. Taylor, *Environment, race and migration. Fundamentals of human distribution: with special sections on racial classification; and settlement in Canada and Australia* (Toronto: University of Toronto Press, 1937).

[32] S. H. Beaver, 'Technology and geography', *Advancement of Science*, 18 (1961), 315–327; H. Bobek, 'Die Hauptstufen der Gesellschafts- und Wirtschaftsentfaltung in geographischer Sicht', *Die Erde*, 90 (1959), 259–298; H. Carol, 'Stages of technology and their impact upon the physical environment: a basic problem in cultural geography', *Canadian Geographer*, 8 (1964), 1–9. The time dimension is also emphasized in a different way by the Swedish school of human geography, for example by T. Hägerstrand, 'The propagation of innovation waves', *Lund Studies in Geography, Series B, Human Geography*, 4 (1952), 3–19, and R. L. Morrill, 'Simulation of central place patterns over time', *Lund Studies in Geography, Series B, Human Geography*, 24 (1962), 109–120.

[33] J. Leighly, 'Introduction', in J. Leighly, ed., *Land and life: a selection from the writings of Carl Ortwin Sauer* (Berkeley and Los Angeles: University of California Press, 1963), 1–8.

[34] A. G. Ogilvie, 'The time-element in geography', *Transactions and Papers of the Institute of British Geographers*, 18 (1952), 1–16; see p. 6. Similar views were expressed by H. C. Darby, 'On the relations of geography and history', *Transactions and Papers of the Institute of British Geographers*, 19 (1953), 1–11, but E. W. Gilbert resisted the introduction of 'scientific' or 'evolutionary' ideas into historical geography, in 'What is historical geography?', *Scottish Geographical Magazine*, 148 (1932), 129–136.

[35] D. Whittlesey, 'Sequent Occupance', *Annals of the Association of American Geographers*, 19 (1929), 162–165.

the biological revolution in terms of change through time: what for Darwin was a process became for Davis and others a history. This was powerfully reinforced not only when geology burst through theological restrictions on time, but also when man himself was found to have a history going back into antiquity. For a time 'evolution' implied little more than the idea of change, development and 'progress', and Darwin was in spite of himself seen as its author.

ORGANIZATION AND ECOLOGY

Darwin's second major contribution to geography was the idea of the interrelationships and connections between all living things and their environment, developed in Haeckel's new science of ecology.[36] In the third chapter of *The origin*, Darwin had been impressed by the 'beautiful' and 'exquisite' adaptation and interrelationships of organic forms in nature, and the theme of ecology is implicit if unstated in many of his writings.[37] 'How infinitely complex and close-fitting', he wrote in *The origin*, 'are the mutual relations of all organic beings to each other and to the physical conditions of life.'[38] This was the theme of ecology and, while Clements in America was forcing vegetation into a time-framework, European workers in general were more concerned with community structures and functions, culminating in Tansley's idea of the ecosystem. Perhaps Darwin's most significant contribution to ecological thinking, however, was to include man in the living world of nature. This had been becoming inevitable, with Boucher de Perthes's work on the Somme gravels after 1837, but it was in 1859 that the importance of finds of ancient man was formally recognized, in Prestwich's paper to the Royal Society.[39] Darwin deliberately left the implications of *The origin* for the history of man unspoken in his first edition, but he was

[36] E. Haeckel, 'Entwicklungsgang und Aufgaben der Zoologie', *Jenaische Zeitschrift*, 5 (1869), 353.

[37] R. C. Stauffer, 'Ecology in the long manuscript version of Darwin's *Origin of Species* and Linnaeus' *Economy of Nature*', *Proceedings of the American Philosophical Society*, 104 (1960), 235–241; and P. Vorzimmer, 'Darwin's ecology and its influence upon his theory', *Isis*, 56 (1965), 148–155.

[38] C. R. Darwin, *On the origin of species by means of natural selection: or, the preservation of favoured races in the struggle for life* (London: John Murray, 1859).

[39] J. Prestwich, 'On the occurrence of flint-implements, associated with the remains of extinct mammalia, in undisturbed beds of a late geological period', *Proceedings of the Royal Society*, 10 (1860), 50–59, read 26 May 1859. Published in full with amended title in *Philosophical Transactions of the Royal Society*, 150 (1861), 277–317.

disappointed by Lyell's reluctance to draw the obvious conclusions in *The antiquity of man* four years later.[40] The theological difficulties were considerable for, although it was possible to reinterpret the biblical account of time, fundamental Christian beliefs such as the fall of man and original sin hinged on specific details of special creation; if the creation proved a myth, what of the theology? In spite of Darwin's own reticence, his theory soon took the popular title of 'the ape theory', and controversy in the 1860s centred around the problem of man's ancestry rather than of variation and selection.[41] Huxley's *Man's place in nature* in 1863, for example, dealt not with man's ecological status, but with his relationship with the apes;[42] man had emphatically become a subject for scientific speculation. Darwin himself, in the *Expression of the emotions in man and animals* (1868) and in *The descent of man* (1871), went further by treating modern man on the same level as other living things.

Haeckel used the term 'ecology' in 1869, and from about 1910 'human ecology' was used for the study of man and environment, not in a deterministic sense, but for man's place in the 'web of life' or the 'economy of nature'. Park's statement of the scope of human ecology[43] deals with concepts taken from plant and animal ecology. The simplicity of human ecology as a methodological framework when stated in purely biological terms was echoed by Barrows in his presidential address to the Association of American Geographers in 1923.[44] Barrows's address, perhaps hastening the expulsion of geomorphology from geography in the United States, aroused considerable animosity and little positive support among geographers, and the sociologists themselves gradually moved away from Park's position.[45]

[40] C. Lyell, *The geological evidence of the antiquity of man, with remarks on theories of the origin of species by variation* (London: John Murray, 1863). See C. R. Darwin to J. D. Hooker, 24 February 1863, in F. Darwin, ed., *op. cit.* (note 15) p. 9.

[41] A. Ellergård, 'Darwin and the general reader: the reception of Darwin's theory of evolution in the British periodical press, 1859–1872', *Acta Universitatis Gothoburgensis: Göteborgs Universitets Arsskrift*, 64 (1958), 1–394.

[42] T. H. Huxley, *Evidence as to man's place in nature* (London: Williams & Norgate, 1863).

[43] R. E. Park, 'Human ecology', *American Journal of Sociology*, 42 (1936), 1–15. See also R. D. McKenzie, 'The ecological approach to the study of the human community', *American Journal of Sociology*, 30 (1924), 287–301, and 'The scope of human ecology', *Publications of the American Sociology Society*, 20 (1926), 141–154.

[44] H. H. Barrows, 'Geography as human ecology', *Annals of the Association of American Geographers*, 13 (1923), 1–14.

[45] L. F. Schnore, 'Geography and human ecology', *Economic Geography*, 37 (1961), 207–217; and R. J. Chorley, 'Geography as human ecology', in R. J. Chorley, ed., *Directions in geography* (London: Methuen, 1973), 155–169.

Ecology has become, moreover, increasingly empirical in method, and in doing so it has run counter to, and ultimately has superseded, the synthetic geographical tradition of explanation by analogy, which attempted to understand the complexity and interrelationships of phenomena by reference to the even greater complexity of living organisms.[46] This is a theme which may be traced to classical writings and medieval scholasticism, and is in no sense Darwinian in origin, but after Darwin such treatment lost its more extreme metaphysical implications and became more directly biological in expression. The organism analogy is explicit in both classical plant ecology and pedology, particularly in Clements's writings. For him, the plant community is 'a complex organism, or superorganism, with characteristic development and structure';[47] 'As an organism the formation arises, grows, matures, and dies. . . . Each climax formation is able to reproduces itself.'[48] Clements believed that 'this concept is the "open sesame" to a whole new vista of scientific thought, a veritable *magna carta* for future progress.'[49] Similarly, Shaler, Whitney and others interpreted the functional interrelationships in soils as a phase in the 'higher estate of organic existence'.[50] Later workers, however, failed to demonstrate the discrete existence of organic unity in either soil or vegetation formations,[51] and the organic analogy in physical geography never carried the influence which it did in other branches of the subject.[52]

The idea of the organic unity of the earth can best be traced to Ritter: 'The earth is one . . . all its parts are in ceaseless action and reaction on each other. . . . The earth is therefore . . . a unit, an organism of itself it has its own cosmical life; it can be studied in no one of its parts.'[53] To him the

[46] For commentary on the geographic relevance of the concepts of community and ecosystem, see W. B. Morgan and R. P. Moss, 'Geography and ecology: the concept of the community and its relationship to environment', *Annals of the Association of American Geographers*, 55 (1965), 339–350; and D. R. Stoddart, 'Geography and the ecological approach: the ecosystem as a geographic principle and method', *Geography*, 50 (1965), 242–251.

[47] F. E. Clements, *op. cit.* (note 24), p. 3.

[48] F. E. Clements and V. E. Shelford, *Bio-ecology* (New York: John Wiley, 1939), p. 24.

[48] *Ibid.*, p. 24.

[50] N. S. Shaler, 'The origin and nature of soils', *12th Annual Report*, U.S. Geological Survey (1890–1891), part 1, 213–345.

[41] See, for example, C. C. Nikiforoff, 'Reappraisal of the soil', *Science*, N.Y. 129 (1959), 186–196; and T. Kira, H. Ogawa and K. Yoda, 'Some unsolved problems in tropical forest ecology', *Proceedings, Ninth Pacific Science Congress*, 4 (1962), 124–134.

[52] But see more extreme writings of, for example, C. Strickland, *Deltaic formation, with special reference to the hydrographic processes of the Ganges and Brahmaputra* (Calcutta and London: Longmans, Green, 1940), especially chapter 2, 'The sea in pregnancy'.

[53] C. Ritter, *Comparative geography*, translated by W. L. Gage (Edinburgh and London: William Blackwood & Sons, 1865), pp. 64, 65.

earth was a 'living work from the hand of a living God, [with] a close and vital connection, like that between body and soul, between nature and history.'[54] For both Humboldt and Ritter, unity, harmony and interdependence of parts constituted the organic analogy. Half a century later Vidal de la Blache reached similar conclusions and acknowledged his debt to Ritter both at the earth and at the regional level.[55]

SELECTION AND STRUGGLE

Although the limitations of organic analogies require little demonstration, the problem of the effect of environment on man leads into the difficult fields of environmental influence, selection and adaptation.[56] Most pre-Darwinian writers on the effects of environment were content to look for cause–effect relationships, without enquiring too closely into process, and this theme was taken up by Ratzel in the first volume of *Anthropo-Geographie* and later by his students Miss Semple and Demolins. To the French school, imbued with Vidal's notions of harmony and interrelationship, this was too rigid a framework for analysis, but in America Davis attempted to carry simplistic ideas of causality into the definition of geography itself: 'Any statement is of geographical quality if it contains a relation between some inorganic element of the earth on which we live, acting as a control, and some element of the existence, or growth, or behaviour or distribution of the earth's organic inhabitants, serving as a response.'[57] This suggestion gained little support among geographers, who

[54] C. Ritter, *The comparative geography of Palestine and the Sinaitic Peninsula, translated and adapted to the use of biblical students* by W. L. Gage (Edinburgh: T. & T. Clark, 4 volumes, 1866), reference in volume 2, p. 4.

[55] P. Vidal de la Blache, 'Le principe de la géographie générale', *Annales de Géographie*, 5 (1896), 129–142; 'Des caractères distinctifs de la géographie', *Annales de Géographie*, 22 (1913), 289–299. The North American case has been examined by J. S. Duncan, 'The superorganic in American cultural geography', *Annals of the Association of American Geographers* 70 (1980), 181–198. For a fuller consideration, see chapter 11.

[56] G. Tatham, 'Environmentalism and possibilism', in T. G. Taylor, ed., *Geography in the twentieth century: a study of growth, fields, techniques, aims and trends* (London: Methuen, 1951), 128–162; L. Febvre with L. Bataillon, *A geographical introduction to history* (London: Kegan Paul, Trench, Trubner & Co., 1925), being the translation of *La Terre et l'évolution humaine: introduction géographique à l'histoire* (Paris: La Renaissance de Livre, 1922); F. Ratzel, *Anthropo-geographie oder Grundzüge der Anwendung der Erdkunde auf die Geschichte* (Stuttgart: J. Engelhorn, 1882).

[57] W. M. Davis, 'An inductive study of the content of geography', *Bulletin of the American Geographical Society*, 38 (1906), 67–84, and *Geographical essays*, *op. cit.* (note 13), 3–22; reference on p. 8. Also W. M. Davis, 'Systematic geography', *Proceedings of the American Philosophical Society*, 41 (1902), 235–259.

realized that no science can take as its field of study a specific relationship rather than a body of data, for if it did the statement of its aims would presuppose the existence of the relationship itself.[58]

If causal relationship provided an unsound methodological principle, however, the problem of environmental influence remained.[59] Fleure, who came deeply under the influence of Darwinism in 1892–1895, stressed the need for the physiographical study of environmental effects on man, and in his typology of human regions (regions of difficulty, of effort, of increment) came close to applying Darwinian ideas of natural selection through environmental influence to human groups.[60] The study of physiological effects, however, has become a specialized branch of biology outside geographical competence, and geographers have generally restricted themselves to the inference of causation from covariance on a coarser scale. Huntington particularly took up the problem of natural selection, environmental influences and human population on a world scale, and Taylor explored the same theme in a series of studies of race, peoples, states and towns, emphasizing their development through time under the influence of environmental factors.[61] The questions which these determinists raised were posed in so gross a manner that they could only invite the grossest answers; most geographers realized this, and neither Taylor nor Huntington gained full academic acceptance. The questions which they asked could not be meaningfully answered in geographical terms, and the whole determinist–possibilist controversy, 'unreal and futile' as Hartshorne

[58] C. O. Sauer, *op. cit.* (note 14), pp. 51–52.

[59] A. P. Brigham, 'Problems of geographic influence', *Annals of the Association of American Geographers*, 5 (1915), 3–25; C. R. Dryer, 'Genetic geography: the development of the geographic sense and concept', *Annals of the Association of American Geographers*, 10 (1920), 3–16.

[60] H. J. Fleure, 'Geography and the scientific movement', *Geography*, 22 (1937), 178–188; H. J. Fleure, 'The later development in Herbertson's thought: a study in the application of Darwin's ideals', *Geography*, 37 (1952), 97–103; H. J. Fleure, 'Human regions', *Scottish Geographical Magazine*, 35 (1919), 94–105, revised from 'Régions humaines', *Annales de Géographie*, 26 (1917), 161–174; H. J. Fleure, 'Geography and evolution', *Geography*, 34 (1949), 1–9.

[61] E. Huntington, *Mainsprings of civilisation* (New York: John Wiley, 1945); E. Huntington, 'Geography and natural selection', *Annals of the Association of American Geographers*, 14 (1924), 1–16; T. G. Taylor, *Environment and race*, *op. cit.* (note 31); T. G. Taylor, *Environment and nation*, *op. cit.* (note 31); T. G. Taylor, *Urban geography*, *op. cit.* (note 30). On the general theme of the extension of Darwinian ideas in North American geography, see D. R. Stoddart, 'Darwin's influence on the development of geography in the United States, 1859–1914', in B. W. Blouet, ed., *The origins of academic geography in the United States* (Hamden: Archon Books, 1981), 265–278.

termed it, moved on to a philosophical rather than an empirical level.[62]

Darwin's theory was, of course, one of natural selection rather than of evolution, and basic to his thesis was the idea, taken from Malthus, that populations tended to expand at a geometric rate and thus outstrip resources: 'Thus, from the war of nature, from famine and death, the most exalted object of which we are capable of conceiving, namely, the production of the higher animals, directly follows,' wrote Darwin in the last paragraph of *The origin*, 'There is grandeur in this view of life.'[63] Such views in turn influenced social thinking, particularly in America, where Spencer's idea of the survival of the fittest and Darwin's of the struggle for life were used to justify laissez-faire in politics and economics, particularly in the 'Social Darwinism' of Sumner.[64] Hofstadter has shown how the geologists Ward and Shaler denied the value of unrestricted competition in social life, and how they and the Russian geographer Kropotkin stressed cooperation and mutual aid in social development.[65] The idea of social selection was often somewhat crudely phrased, especially in geographical writing. Thus Turner's frontier hypothesis and especially Roosevelt's book

[62] R. Hartshorne, *Perspective on the nature of geography*, *op. cit.* (note 4), p. 57; A. F. Martin, 'The necessity for determinism, a metaphysical problem confronting geographers', *Transactions and Papers of the Institute of British Geographers*, 17 (1951), 1–11; for a review of the whole group of issues around environmental influence and determinism, see G. R. Lewthwaite, 'Environmentalism and determinism: a search for clarification', *Annals of the Association of American Geographers*, 56 (1966), 1–23.

[63] C. R. Darwin, *op. cit.* (note 38), p. 560.

[64] Herbst has attempted to trace the dichotomy in American geography between physical and human studies to the early influence of Social Darwinism on, for example, Davis's definition of the nature of the subject. See J. Herbst, 'Social Darwinism and the history of American geography', *Proceedings of the American Philosophical Society*, 105 (1961), 538–544. J. A. Campbell and D. N. Livingstone have recently reinterpreted the deterministic emphasis of late-nineteenth- and early-twentieth-century American geography in Lamarckian rather than Darwinian terms. The point is an interesting one, but it remains the case that dominant scholars at the time saw their own intellectual indebtedness as being directly to Darwin, however much these views may have undergone a Lamarckian transformation in the process. See J. A. Campbell and D. N. Livingstone, 'Neo–Lamarckism and the development of geography in the United States and Great Britain', *Transactions of the Institute of British Geographers*, n.s. 8 (1983), 267–294; and D. N. Livingstone, 'Natural theology and neo-Lamarckism: the changing context of nineteenth-century geography in the United States and Great Britain', *Annals of the Association of American Geographers*, 74 (1984), 9–28. See also H. Cravens, *The triumph of evolution: American scientists and the heredity–environment controversy, 1900–1941* (Philadelphia: University of Pennsylvania Press, 1978).

[65] R. Hofstadter, *Social Darwinism in American thought* (Philadelphia: University of Pennsylvania Press, 1944; revised edition, Boston: The Beacon Press, 1955). The decline of the more extreme Social Darwinism in America closely paralleled the eclipse of Spencer's evolutionary philosophy by the pragmatism of William James and John Dewey in the later

on *The winning of the west* both took the naive view that frontier conditions selected all that was pioneering and democratic in a society, which then itself took on the pioneer spirit.[66] It is interesting that, except in the idea of competition and its implications, Darwinism had little effect in classical equilibrium economics, and both the implications of random variation and of development through time are relatively recent innovations.[67]

It was in political geography, however, that ideas of struggle and selection on a national level were most significant. The political usage made of the organic view of the state, and the ideas of struggle and *Lebensraum*, brought the subject into intellectual disgrace in the 1930s, as Troll[68] has outlined, and modern political geography is at pains to dissociate itself from any kind of organic analogy.

RANDOMNESS AND CHANCE

This review of biological ideas in geography has demonstrated that 'Darwinism' or 'evolution' was almost always interpreted by geographers either in the sense of change through time or of social struggle and selection. In both cases the application has been largely deterministic: it has in fact

---

years of the nineteenth century. See expecially P. S. Wiener, *Evolution and the founders of pragmatism* (Cambridge: Harvard University Press, 1949). For a reappraisal of Hofstadter's thesis, see P. F. Boller, Jr, *American thought in transition: the impact of evolutionary naturalism* (Chicago: University of Chicago Press, 1969); C. E. Russett, *Darwin in America: the intellectual response 1865–1912* (San Francisco: W. H. Freeman, 1976); and R. C. Bannister, *Social Darwinism: science and myth in Anglo-American social thought* (Pennsylvania: Temple University Press, 1979).

[66] F. J. Turner, *The frontier in American History* (New York: Henry Holt, 1920). For commentary see R. H. Block, 'Frederick Jackson Turner and American geography', *Annals of the Association of American Geographers*, 70 (1980), 31–42; W. Coleman, 'Science and symbol in the Turner frontier hypothesis', *American Historical Review*, 72 (1966), 22–49; and D. H. Burton, 'Theodore Roosevelt's Social Darwinism and views on imperialism', *Journal of the History of Ideas*, 26 (1965), 103–118.

[67] T. Veblen, 'Why is economics not an evolutionary science?', *Quarterly Journal of Economics*, 13 (1898) 373–397; J. J. Spengler, 'Evolutionism in American economics, 1800–1946', in S. Persons, ed., *op. cit.* (note 4), pp. 202–266; A. Alchian, 'Uncertainty, evolution, and economic theory', *Journey of Political Economy* (1950), 211–221; and, on development economics, W. W. Rostow, *The stages of economic growth: a non–communist manifesto* (Cambridge: Cambridge University Press, 1960).

[68] C. Troll, 'Geographic science in Germany during the period 1933–1945: a critique and justification', *Annals of the Association of American Geographers*, 39 (1949), 99–137.

been said that simple geographical determinism, in its picture of causality, was one of the last fields of operation of the Newtonian world-view in the twentieth century. Any discussion of the biological impact on geographical thinking must hinge on this central question: why was Darwinism, a theory for the selection of randomly occurring variants, interpreted in a deterministic and not a probabilistic sense?[69] Why was chance omitted in geography? The question is of more than historical interest, for a century after *The origin* geographers are beginning to recognize the importance of stochastic processes in geographic change.[70]

The problem is more remarkable because the study of random processes in the nineteenth century was by no means limited to Darwinian biology: indeed, Merz has written that 'the study of this blind chance in theory and practice is one of the greatest scientific performances of the nineteenth century.'[71] In the natural sciences Laplace laid the foundations of probability theory at the beginning of the century, and the theme was subsequently taken up by Adolphe Quetelet in the social sciences and used by Buckle in a work with which Darwin was certainly familiar.[72] The new

[69] R. A. Fisher goes so far as to state that 'Darwin's chief contribution, not only to Biology but to the whole of natural science, was to have brought to light a process of which contingencies *a priori* improbable, are given, in the process of time, an increasing probability, until it is their non-occurrence rather than their occurrence which becomes highly improbable.' See 'Retrospect of the criticisms of the theory of natural selection', in J. Huxley, A. C. Hardy and E. B. Ford, eds, *Evolution as a process* (London: Allen & Unwin, 1954), 84–98; reference on p. 91.

[70] For an early statement on probability in geography, see J. Brunhes, 'Du caractère propre et du caractère complexe des faits de géographie humaine', *Annales de Géographie*, 33 (1913), 1–40, and translation in 'The specific characteristics and complex characters of the subject-matter of human geography', *Scottish Geographical Magazine*, 29 (1913), 304–322, 358–374: 'Every truth concerning the relations between natural surroundings and human activities can never be anything but approximate; to represent it as something more exact than that is to falsify it' (pp. 362–363). 'All biological relations, all oecological truths, are, and can be, nothing more than statistical truths' (p. 364). For recent substantive work, see, for example, T. Hägerstrand, *op. cit.*, and R. L. Morrill, *op. cit.* (note 32).

[71] J. T. Merz, *A history of European thought in the nineteenth century*, volume 2 (Edinburgh: William Blackwood, 1928), p. 264. For a historical review, see E. Nagel, 'Principles of the theory of probability', *International Encyclopedia of Unified Science*, volume 1, no. 6 (Chicago: University of Chicago Press, 1939).

[72] P. S. de Laplace, *Théorie analytique des probabilités* (Paris: Mme Ve. Courcier, Imprimeur-Librairie pour les Mathématiques, 1812); P. S. de Laplace, *Essai philosophique sur les probabilités* (Paris: Bachelier, Successeur de Mme Ve, Courcier, Librairie pour les Mathématiques, 1814); L. A. J. Quetelet, *Sur l'homme et le développement de ses facultés: ou essai de physique sociale* (Bruxelles: L. Hauman et Compe., 2 volumes, 1836): J. Herschel, 'Quetelet on probabilities', *Edinburgh Review*, 42 (1850), 1–57; H. T. Buckle, *History of civilisation in England*, volume 1 (London: J. W. Parker, 1857), and Darwin's comments in F.

kinetic theory of gases developed by Herapath, Clausius and Clerk Maxwell was appearing at the same time as *The origin*.[73] Boltzmann extended statistical conceptions in mechanics; and in biology itself Darwin's work stimulated a long series of statistical studies, from Galton and Pearson to Fisher and Haldane. Why, then, in such an intellectual atmosphere, was the geographical interpretation so deterministic?[74]

Part of the answer lies in Darwin himself. Darwin's theory made a clear distinction between the way in which evolution was effected and the course of evolution itself: geography seized on the latter and ignored the former. Darwin began with the idea of the selection of 'chance' variations, which are 'no doubt' governed by laws.[75] These laws Darwin failed to discover, and in time he came to emphasize chance less and less, and by the last edition of *The origin* he was thinking of directional variation in a Lamarckian sense. Nowhere does he use the word 'random' and in the fourth chapter of *The origin* he states that the use of the word 'chance' is 'wholly incorrect'.[76] Although he undoubtedly believed that unfavourable variations could be as numerous as favourable ones, this became less clear with each successive edition. Darwin's difficulty was this: whereas his theory explained adaptation in nature by variation and natural selection, he could not, before the discovery of Mendel's work on genetics, offer any explanation of the basic variation, but the very facts of adaptation, which provided his strongest evidence, and which natural selection explained, had long been

Darwin, ed., *op. cit.* (note 15), volume 2, pp. 110 and 386. One may of course argue that these earlier workers used statistical analysis as a tool to overcome error and incompleteness in our perception of the world, rather than recognizing that the real world is itself subject to chance. See the comments by M. D. B. Hesse on C. C. Gillispie, 'Intellectual factors in the background of analysis of probabilities', in A. C. Crombie, ed., *Scientific change, symposium on the history of science, University of Oxford, 9–15 July 1961* (London: William Heinemann, 1963), 430–453, comments on pp. 471–476.

[73] J. Clerk Maxwell, 'Illustrations of the dynamical theory of gases, part I', *London, Edinburgh and Dublin Philosophical Magazine and Journal of Science*, 19 (1860), 19–32, read 21 September 1859.

[74] F. Lukermann, in an interesting discussion, has drawn attention to the dependence of the French school of human geography on the work of French statisticians and natural scientists in the nineteenth century, from Laplace to Cournot and later Henri Poincaré, and the intellectual milieu in which they worked. The French possibilists thus form an exception to the generalizations in this paragraph. See F. Lukermann, 'The "Calcul des probabilités" and the École Française de Géographie', *Canadian Geographer*, 9 (1965), 128–137.

[75] C. R. Darwin to J. D. Hooker, 23 December 1856, in F. Darwin, ed., *op. cit.* (note 15), volume 2, p. 87.

[76] C. R. Darwin, *op. cit.* (note 38), p. 138. Huxley, of course, interpreted even 'chance' variations deterministically, *op. cit.* (note 18), pp. 199–201.

accounted for by the church in terms of design.[77] Early nineteenth-century theology in England was a curious mixture of revelation and natural theology, exemplified in the Bridgewater Treatises and by William Paley. Paley wrote, for example, in 1802, that:

> There cannot be a design without a designer; contrivance without a contriver; order without choice; arrangement, without any thing capable of arranging; subserviency and relation to a purpose, without that which could intend a purpose; means suitable to an end, and executing their office in accomplishing that end, without the end ever having been contemplated, or the means accommodated to it. Arrangement, disposition of parts, subserviency of means to an end, relation of instruments to an use, imply the presence of intelligence and mind.[78]

Lacking a mechanism for variation, and shaken by the theoretical objections in Jenkin's *North British Review* article in 1867,[79] Darwin changed his ground. Although maintaining privately that 'the old argument of design . . . fails [and that] there is no more design . . . than in the course which the wind blows,'[80] he still had doubts: the thought of the eye made him cold all over, the sight of feathers in a peacock's tail made him sick.[81] To Asa Gray he wrote in distress in 1860 on the problem of evil and the question of design: 'I am inclined to look at everything as resulting from designed laws, with the details, whether good or bad, left to the working of what we may call chance.'[82] But the effort to reconcile the unreconcilable was a failure. 'A dog might as well speculate on the mind of Newton. . . . The more I think the more bewildered I become.'[83]

Darwin therefore abandoned the fundamental issue of random variation, on which both the natural theologians and the exponents of revealed

[77] See on this theme A. Ellergård, *op. cit.* (note 41).

[78] W. Paley, *Natural theology; or, evidences of the existence and attributes of the deity collected from the appearances of nature* (London: printed for R. Faulder, 1802), p. 12. For a full analysis, see D. N. Livingstone, *op. cit.* (note 64).

[79] See the anonymous article by Fleeming Jenkin, 'The origin of species', *North British Review*, 92 (1867), 277–318, and discussion by P. Vorzimmer, 'Charles Darwin and blending inheritance', *Isis*, 54 (1963), 371–390.

[80] C. R. Darwin, *The autobiography of Charles Darwin 1809–1882*, edited by Nora Barlow (London: Collins, 1958), p. 87.

[81] C. R. Darwin to Asa Gray, April 1860, in F. Darwin, ed., *op. cit.* (note 15), volume 2, p. 296. Perhaps he recalled Sturmius's remark, quoted by Paley, 'that the examination of the eye was a cure for atheism', in W. Paley, *op. cit.* (note 78), p. 35.

[82] Darwin to Asa Gray, 22 May 1860, in F. Darwin, ed., *op. cit.* (note 15), volume 2, p. 312.

[83] *Ibid.*

religion could unite, and concentrated on descent and on selection. If descent could be demonstrated, then the argument from design would appear much less plausible than that from evolution,[84] whereas selection could be demonstrated, for example in pigeons, on a purely empirical level. Darwin thus outflanked his opponents and deflected them from his most serious weakness, but at the same time he laid himself open to the charge of plagiarism and lack of originality. After all, *evolution* was not new, only Darwin's mechanism was, yet before Mendel Darwin could only defend the former, not the latter.[85] Darwinism in the sense of development of evolution through time was seized on by geographers as a unifying principle to subsume vast quantities of otherwise discrete and apparently unrelated data: the clarity and order which this interpretation revealed had a remarkable effect on the progress of the sciences. But called Darwinism or not, it omitted Darwin's central theme. Mendel's work, and particularly the statistical treatment of heredity by Sir Ronald Fisher in *The genetical theory of natural selection*, gave Darwinists the weapons they needed; but Fisher's book appeared 70 years after Darwin's, and by that time the 'evolutionary' impact in geography and other sciences had been made.

CONCLUSION

Biological influences in geography during the past century, therefore, although often claiming descent from evolution or from Darwin, have been interpreted in ways which at times subtly and at times blatantly diverge from Darwin's actual philosophy. The major themes of change through time, of selection and struggle, and of the interrelatedness of things (the organic analogy, and later ecology) are all present in Darwin's writings, specifically in the eleventh, fourth and third chapters of *The origin of species*, but the unique contribution of Darwin's theory, that of random variation, was, for religious and scientific reasons, neglected in geographical

[84] A. Ellergård, *op. cit.* (note 41).

[85] D. Fleming, 'The centenary of the *Origin of Species*', *Journal of the History of Ideas*, 20 (1959), 437–446. On the general theme of Darwin's theological difficulties, see M. Mandelbaum, 'Darwin's religious views', *Journal of the History of Ideas*, 20 (1958), 363–378; and D. Fleming, 'Charles Darwin, the anaesthetic man', *Victorian Studies*, 4 (1961), 219–236. For an alternative interpretation, deriving ideas of randomness from natural theology, see W. F. Cannon, 'The bases of Darwin's achievement: a revaluation', *Victorian Studies*, 5 (1961), 109–134, and for a different view of Fleming's thesis, see J. A. Campbell, 'Nature, religion and emotional response: a reconsideration of Darwin's affective decline', *Victorian Studies*, 18 (1974), 159–174.

circles. It is interesting that methods which incorporate randomness are now being increasingly used by geographers.

The discussion of these four themes demonstrates that geographical thinking in the past hundred years has cut across biological thinking, incorporating some ideas into the corpus of thought derived by Hartshorne and Hettner from Kant and Humboldt, but neglecting others. Even in their most extreme statement, however, these themes never came to dominate geographical thinking, which, by concentrating on the interdependence of phenomena on the earth's surface, evolved a rationale of its own. In this, however, Darwin's influence can still be distinguished, in the impact which he made on the nineteenth-century world-view. Darwin established a sphere of scientific enquiry free from *a priori* theological ideas, and freed natural science from the arguments of natural theology. With the publication of *Essays and reviews* in 1860,[86] theology itself began to turn away from science, and to acknowledge that in this field the Bible was no authority. Darwin, by empirical argument and inductive method, thus dismissed teleology as a live issue in scientific explanation,[87] and though similar arguments persisted in vitalist biology they were gradually reduced by the expansion of knowledge. Darwin, furthermore, sealed the acceptance of uniformitarianism and law in science, and completed the dismissal of providential interference and catastrophism in scientific writing. And finally, and in this he was alone, Darwin established man's place in nature, both in Huxley's sense and in Haeckel's, and in so doing made man a fit object for scientific study. Modern geography is inconceivable without these

[86] F. Temple and others, *Essays and reviews* (London: John W. Parker, 1860).

[87] The role of teleology in Darwin's thought is notoriously difficult to assess, especially in the later editions of *The origin* as Darwin shifted his ground over mechanism. These, together with the much-quoted last paragraph, have led to the argument that Darwinian evolution was in fact of a teleological nature. The situation is complicated by the curious reaction in some theological circles, which saw in this intepretation a way out of the crises which the publication of *The origin* had caused. A reviewer has drawn my attention to G. Himmelfarb's account of this in *Darwin and the Darwinian revolution* (London: Chatto & Windus, 1959), pp. 325–329. Asa Gray, among others, was acutely aware of the teleology issue, and his attempt to interpret *The origin* teleologically led to growing estrangement from Darwin himself. Gray saw natural selection as purposive, 'and to most minds Purpose will imply Intelligence'; A. Gray, 'Relation of insects to flowers', *Contemporary Review*, 41 (1882), 609. This quotation is taken from the elegant treatment of Gray's position in 'A theist in the age of Darwin', chapter 18, pp. 355–83, in A. H. Dupree, *Asa Gray 1810–1888* (Cambridge: Belknap Press of Harvard University Press, 1959).

general advances, but their elaboration belongs to the study of intellectual history, not to that of geographical thought.[88]

[88] In preparing this chapter the following discussions have been valuable: C. C. Gillispie, *Genesis and geology: a study in the relations of scientific thought, natural theology, and social opinion in Great Britain, 1790–1850* (Cambridge: Harvard University Press, 1951); A. Ellergård, *Darwin and the general reader* (note 44); L. Eiseley, *Darwin's century: evolution and the men who discovered it* (London: Victor Gollancz, 1959); Gertrude Himmelfarb, *Darwin and the Darwinian revolution* (London: Chatto & Windus, 1959); J. C. Greene, *Darwin and the modern world view: the Rockwell Lectures, Rice University* (Baton Rouge: Louisiana State University Press, 1961); Jacob Bronowski, 'Introduction', in M. Banton, ed., *Darwinism and the study of society* (London: Tavistock Publications, 1961); and the writings of Darwin himself, particularly *The origin of species* (1859), the *Life and letters* (1888), and the *Autobiography* (1958).

The quantity of Darwin scholarship in recent years has, of course, been enormous (note 1). Connoisseurs of erudition should not miss R. B. Freeman, 'Darwin's negro bird-stuffer', *Notes and Records of the Royal Society of London*, 33 (1978), 83–86, and R. Colp, Jr, 'Charles Darwin's coffin, and its maker', *Journal of the History of Medicine*, 35 (1980), 39.

9

'That Victorian Science'

The third quarter of the nineteenth century was a critical period in the development of geography in Britain. It saw the establishment of the subject in the older universities, the institution of universal elementary education in which geography played an important part, and the organization of a national system of school examinations which served to impose uniform standards on an expanding discipline. It also saw the gradual movement of opinion in the Royal Geographical Society and the British Association for the Advancement of Science away from an exclusive concern with exploration towards a new emphasis on 'scientific geography'.[1] In retrospect these developments can be seen as part of the revolution in thought inspired by Darwin's *Origin of species*, and, though the main lineaments of its effects on geography are now well known,[2] there have hitherto been few detailed analyses of how its impact was felt.

During this period, no work was more influential in the transformation of geography at all levels than Thomas Henry Huxley's *Physiography*, first published in 1877.[3] This work defined for a generation the way in which the earth's physical features were studied in Britain; it defined also the nature of school education during a period of rapid extension to all parts of the country and all sectors of the population. In this chapter I investigate the nature of its impact, in the context of the Darwinian revolution, and I show how its popularity and ultimate transformation and decline resulted less

[1] R. Strachey, 'Address to the Geographical Section of the British Association at Bristol, August 26, 1875', *Proceedings of the Royal Geographical Society*, 20 (1876), 79–89.

[2] See chapter 8.

[3] T. H. Huxley, *Physiography: an introduction to the study of nature* (London: Macmillan, 1877). Huxley's book is not mentioned by R. Hartshorne in his *The nature of geography* (Lawrence, Kansas: Association of American Geographers, 1939) nor in his *Perspective on the nature of geography* (Chicago: Rand McNally, 1959). For general studies of Huxley, see C. Bibby, *T. H. Huxley: scientist, humanist and educator* (London: Watts, 1959), and J. G. Paradis, *T. H. Huxley: man's place in nature* (Lincoln: University of Nebraska Press, 1978), and for an insight into what drove him on, the comment by Beatrice Webb in her diary entry for 6 May 1887: 'he is a leader of men. I doubt whether science was pre-eminently the bent of

from the intrinsic quality of its idea than from the educational pressures of the day. This is, then, partly the history of an idea and partly the history of a 'peculiarly Victorian science';[4] but it is also a study of the way in which the development of ideas is influenced by other, more practical concerns.

When Humboldt died, and *The origin* was published, in 1859, the comparative philosophy of an earlier generation still dominated the content and method of physical geography, as could well be seen in Herschel's article on the subject published in that year in the eighth edition of the *Encyclopaedia Britannica*.[5] After Humboldt's *Cosmos*, the most popular text was Mary Somerville's *Physical geography*, first published in 1848.[6] These works aimed to demonstrate the interconnections of phenomena and the continuity and completeness of knowledge, rather than to adduce

---

his mind. He is truth-loving, his love of truth finding more satisfaction in demolition than in construction. He throws the full weight of thought, feeling, will, into anything that he takes up. . . . When he talks to man, woman, or child he seems all attention and he has, or rather had, the power of throwing himself into the thoughts and feelings of others and responding to them. And yet they are all shadows to him: he thinks no more of them and drops back into the ideal world he lives in. For Huxley, when not working, dreams strange things: carries on lengthy conversations between unknown persons living within his brain. There is a strain of madness in him; melancholy has haunted his whole life. "I always knew that success was so much dust and ashes. I have never been satisfied with achievement". None of the enthusiasm for what is, or the silent persistency in discussing facts; more the eager rush of the conquering mind, loving the fact of conquest more than the land of the conquered. And consequently his achievement has fallen far short of his capacity. Huxley is greater as a man than as a scientific thinker. The exact opposite might be said of Herbert Spencer': B. Webb: *My apprenticeship* (London: Longmans, Green, 1946), p. 28.

[4] The phrase (which supplies my title) is from A. J. Meadows, *Science and controversy: a biography of Sir Norman Lockyer* (London: Macmillan, 1972), p. 227.

[5] J. F. W. Herschel, 'Physical geography', *Encylopaedia Britannica*, 8th edition, 17 (1859), 569–647; reprinted as J. F. W. Herschel, *Physical geography* (Edinburgh: Adam & Charles Black, 1861). The 8th edition was the last to carry a separate article on physical geography. The *Encyclopaedia Britannica* never carried a separate article on physiography, though part of the subject was treated as 'Physiographical geology' under 'Geology' in later editions.

[6] M. Somerville, *Physical geography*, (London: John Murray, 1848) 2 volumes. For discussion of Mrs Somerville see E. C. Patterson, 'Mary Somerville', *British Journal for the History of Science*, 4 (1969), 311–339; J. N. L. Baker, 'Mary Somerville and geography in England', *Geographical Journal*, 111 (1948), 207–222; M. Sanderson, 'Mary Somerville: her work in physical geography', *Geographical Review*, 64 (1974), 410–420. It is one of the ironies of academic life that whereas the fellows of Somerville College, Oxford (founded in her memory in 1879) show a very proper concern for the state of her grave in Naples (*The Times*, 5 February 1980), they decline to admit to the college undergraduates interested in the subject about which she wrote. Since Mrs Somerville was a woman, moreover, the Royal Society was prepared only to admit her bust and not her person; it now stands in the restaurant at Carlton House Terrace.

principles or explore problems: hence Mackinder's little-remembered remark that the *Cosmos* 'helped to delay the advance of science'.[7] In the years immediately following the appearance of the *The origin*, physical geography, in common with other earth and life sciences, was transformed. It became necessary to show how things had come to be as they are, rather than simply to describe how they mesh together in apparent harmony at the present day. New data from scientific exploration, notably from the *Challenger* expedition of 1873–1876,[8] were combined with new methods, notably of measurement and experiment, exemplified by Tyndall's work on the nature of ice and the glaciers of the Alps.[9] There was thus a need for a new framework, rejecting the implicit order and harmony of the old teleological philosophy, emphasizing laws and methods, function and morphology, in the new light of post-Darwinian nature science.

HUXLEY AS EDUCATOR

Thomas Henry Huxley, who had himself taken part in one of the mid-century voyages of scientific exploration, was appointed a lecturer in natural history at the School of Mines in London, and naturalist to the Geological Survey, in 1854. He thus began a career of research and education, first at Jermyn Street and after 1872 in Exhibition Road, associated with the new science complex in South Kensington. It was a position he used mainly to promote Darwin's ideas, in both biology and education. Ultimately he held the chair of biology in the Normal School of Science from 1881 to his retirement in 1885.

Huxley began giving evening lectures to working men at Jermyn Street in 1857. At first these were concerned with the animal kingdom, and his first book, *Man's place in nature* in 1862, incorporated much material from these lectures. But their scope gradually widened to cover the races of man (in 1864), zoogeography (in 1871), and, increasingly, elementary geology and the science of landforms. His most famous lectures, 'On a piece of chalk' and 'What is to be learned from a piece of coal', were given to audiences of working men in Norwich and Leicester in 1868 and 1870. Gradually these lectures came to comprise what he later termed

[7] H. J. Mackinder, 'Address to the Geography Section', *Report of the 65th Meeting of the British Association for the Advancement of Science* (1895), 738–748; reference on pp. 742–743.

[8] For example, C. Wyville Thomson, *Nature, Lond.*, 14 (1876), 197.

[9] J. Tyndall, *Glaciers of the Alps* (London: Macmillan, 1860).

'physiography'.[10]

At this time, too, Huxley was defining a philosophy for general education, in which the earth and life sciences played a major role. In one of his most powerful addresses, 'A liberal education; and where to find it', given at the South London Working Men's College in 1868, he attacked the emphasis on classical learning in English schools and argued that:

> We need what, for want of a better name, I must call Physical Geography. What I mean is that which the Germans called 'Erdkunde'. It is a description of the earth, of its place and relation to other bodies; of its general structure, and of its great features – winds, tides, mountains, plains: of the chief forms of the vegetable and animal world, of the varieties of man.[11]

Before the Liverpool Philomathic Society the following year he developed the argument:

> let every child be instructed in those general views of the phenomena of nature . . . that is to say, a general knowledge of the earth, and what is on it, in it, and about it. If anyone who has had experience of the ways of young children will call to mind their questions, he will find that so far as they can be put into any scientific category, they come under the head of 'Erdkunde'. The child asks, 'What is the moon, and why does it shine?' 'What is this water, and where does it run?' 'What is the wind?' 'What makes this waves in the sea?' 'Where does this animal live, and what is the use of that plant?'[12]

Huxley's intention was thus primarily pedagogical and his justification practical: he likened each person's life to a game of chess played according to the laws of nature, success or failure being a function of knowledge. Education for him was 'learning the rules of this mighty game', and must hence primarily be concerned with 'the chess board . . . the world'.[13]

[10] L. Huxley, *Life and letters of Thomas Henry Huxley* (London: Macmillan, 2 volumes, 1900); see volume 1, p. 137.

[11] T. H. Huxley, 'A liberal education; and where to find it', in *Science and education* (London: Macmillan, 1893), p. 108.

[12] *Ibid.*, p. 123.

[13] *Ibid.*, pp. 82–83. For an analysis of Huxley's thought on this question and the influence of J. S. Mill, see O. Stanley, 'T. H. Huxley's treatment of "Nature"', *Journal of the History of Ideas*, 18 (1957), 120–127.

ORIGINS OF THE *PHYSIOGRAPHY*

It was in this context that the lectures which subsequently became the *Physiography* were given. In 1869 Huxley lectured on 'Elementary physical geography' at the London Institution, and the following year on 'Elements of physical science' to women at the South Kensington Museum. The text of the first course is still extant;[14] it comprised 12 lectures given between 12 April and 28 June, and closely resembled the later book in plan. In his first lecture, Huxley stated clearly that he was concerned with young people, not with adults; the lectures themselves differ somewhat from the printed syllabus,[15] and towards the end of the course Huxley compressed much material on the solar system, climate and oceanography to compensate for his more leisurely treatment of denudation in earlier lectures. The term 'physiography' does not appear in the printed syllabus or the manuscript text of the 1869 course, but it forms the sub-title for the printed syllabus of the course given in 1870.[16] No separate manuscript text of the 1870 lectures survives, but the syllabus closely follows that for 1869, and it is likely that the same notes were used for both courses.

Apart from a proposal to issue a *Physiography* in ten small volumes,[17] Huxley took no action to publish the lectures for several years. They were finally put in publishable form by an assistant, and appeared as *Physiography: an introduction to the study of nature* in November 1877. The book was an immediate success: it sold 3,386 copies in six weeks, 13,000 over three years, was frequently reprinted, and went into a revised edition some 30 years after its first appearance, long after Huxley's death.[18]

Before considering its content and the reasons for its success, we must first discuss its title. Huxley at first believed he had invented the term,[19] and

[14] Huxley Papers, volume 64, ff. 1–205; volume 65, ff, 1–197.

[15] Huxley Papers, volume 63, ff. 1, 22–33; volume 64, ff. 206–208.

[16] Huxley Papers, volume 63, ff. 9–21; volume 64, ff. 206–207. R. A. Gregory was therefore quite mistaken in stating that Huxley lectured about the Thames and only adopted the title for the book after 'Physiography' had been instituted as an examination subject by the Department of Science and Art in 1876. In view of Huxley's close connection with the department, the borrowing was certainly the other way about. See R. A. Gregory, 'Physiography and physical geography', *Nature, Lond.*, 63 (1900), 207–208. The word is from the Greek φῦσις nature and γραφῖα description.

[17] Huxley Papers, volume 65, ff. 198–199.

[18] T. H. Huxley and R. A. Gregory, *Physiography: an introduction to the study of nature*, rev. edition (London: Macmillan, 1904).

[19] L. Huxley, *op. cit.* (note 10), volume 1, p. 476, though he was aware of its prior use in Rosenbusch's sense: *op. cit.*, (note 3), p. vi (see note 36 below).

he is still generally credited with doing so.[20] In the year of publication, however, his attention was drawn to an earlier book with the same title, by Eugène Cortambert, published in Paris in 1836.[21] This is, however, an extremely rare work,[22] and Huxley was almost certainly unaware of its existence earlier. The term 'physiography' had, nevertheless, been in use for over a century. Kant used it in 1788, in the sense of *Erdbeschreibung*, for the systematic rather than the chorological sciences,[23] and Humboldt in a famous footnote to his *Flora Fribergensis*[24] distinguished similarly between:

> *Geognosia*, or *Erdkunde*, the chorological sciences [Hartshorne[25] translates 'Geognosia' as both 'geography' and 'geognosy']; *Historia telluris*, or *Erdgeschichte*, the chronological or historical sciences; and *Physiographia*, or *Naturbeschreibung* [translated by Hartshorne[26] as 'nature study'], the systematic sciences of zoology (*Zoognosia*), botany (*Phytognosia*), and geology (*Oryctognosia*).

Humboldt repeats these distinctions in an identical footnote in the *Cosmos* 50 years later, but his text gives a fuller discussion of the divisions of knowledge, omitting the term *Physiographia*.[27]

[20] J. Challinor, *Dictionary of geology*, 3rd edition (Cardiff: University of Wales Press, 1967), p. 111; M. Gary, R. McAfee, Jr, and C. L. Wolf, eds, *Glossary of geology* (Washington: American Geological Institute, 1972), p. 538.

[21] P. E. F. Cortambert, *Physiographie: description générale de la nature pour servir d'introduction aux sciences géographiques* (Paris: Picquet, Kilian, 1836).

[22] The book is in neither the British Library nor the Library of Congress; no copy has been traced in the United States. There is a copy in the Bibliothèque Nationale, Paris, and an incomplete one in the Musée National d'Histoire Naturelle, Paris, which Dr M.-H. Sachet first located for me. The final page number of the book is misprinted 336 instead of 636, and this accounts for the misleading, if technically correct, entry in *Bibliographie de la France*, 25 (1836), 33 (and similarly in the catalogue of the Bibliothèque Nationale). Since the Museum copy lacks the final pages and did not agree with these entries, this initially caused some confusion.

[23] Dr Manfred Büttner has in preparation a paper on Kant's concept of *Physiographie*.

[24] F. A. von Humboldt, *Flora Fribergensis specimen plantas cryptogamicas praesertim subterraneas exhibens* (Berolini: H. A. Rottmann, 1793), pp. ix–x. The note is repeated in A. von Humboldt, *Cosmos: entwurf einer physischer Weltbeschreibung* (Stuttgart: J. C. Cotta'schen Buchhandlung, 1845), volume 1 (1845), pp. 486–487. It is discussed by R. Hartshorne (1939), *op. cit.* (note 3), pp. 124–126; R. Hartshorne (1959), *op. cit.* (note 3), pp. 173–182; and G. Tatham, 'Geography in the nineteenth century', in T. G. Taylor, ed., *Geography in the twentieth century* (London: Methuen, 1951), p. 51. A translation is given by R. Hartshorne, 'The concept of geography as a science of space, from Kant and Humboldt to Hettner', *Annals of the Association of American Geographers*, 48 (1958), 97–108; see p. 100.

[25] R. Hartshorne (1958), *op. cit.* (note 24), p. 100.

[26] *Ibid.*, p. 100.

[27] A. von Humboldt, 'Begrenzung und wissenschaftliche Behandlung einer physischen

Further, as Mill[28] pointed out, the term physiography was in common use in Scandinavia during the second half of the eighteenth century, and may well have been introduced by Linnaeus, who died in 1778. A Physiographic Society (Physiographiska Sallskapet) was founded in Lund on 2 December 1772, and survives as the Kunglinga Fysiografiska Sallskapet i Lund, which published four numbers of a journal, *Phyiographiska Sallskapets Tidskrift*, in 1837–1838.[29] Shortly after the Lund society was founded, a counterpart in Stockholm (Physiographisk Salliskapets i Stockholm) published three parts of a *Physiographisk Sallskapets Handlingar* in 1776: this is the first printed usage of the term so far traced.[30] Rather later, a physiographic society in Oslo (Physiographiske Forening i Christiania) published a journal, *Magazin for Naturvidenskaberne*, which included physiography in its title from 1828 to 1924; it now appears in two series as specialist botanical and zoological journals.[31] It seems likely that 'physiographic' was used in Scandinavia at this time as a general synonym for 'geographic' or 'topographic' in society titles. A similar society at Uppsala, formed in 1758, named Kosmografiska Sallskapet, is generally regarded as the first Swedish geographical society, and further illustrates the interchangeability of terms at this time.[32] There is clearly scope for further study of early physiographic societies in eighteenth-century Scandinavia, and into the possibility that the term was given currency by Linnaeus himself.[33]

The continental usage of physiography in the sense of description of nature was given wider currency in Britain by J. H. Green in his 1840

Weltbeschreibung', *op. cit.* (note 24), volume 1, pp. 49–78. He distinguishes *physische Weltbeschreibung* (physical description of the universe); *physische Erdbeschreibung* (physical geography); *Naturgeschichte* (descriptive natural history); *specielle Länderbeschreibung* (regional geography); and *vergleichende Geographie* or *Erdbeschreibung* (general geography *sensu lato*). There is a large exegetical literature on this terminology and Kant's.

[28] H. R. Mill, 'The word Physiography', *Nature, Lond.*, 63 (1900), 231.

[29] This journal is not in C. D. Harris and J. Fellman, 'International list of geographical serials', *University of Chicago, Department of Geography Research Paper* 138 (1971), though the *Tidskrift*'s successors (*Handlingar, Förhandlingar, Årsbok*) are (p. 138). These more recent volumes have been mainly biological, but the *Tidskrift* included topography, geology and biology. There is a set in Cambridge University Library (Q340.7.c.2.1).

[30] This too is not in the Harris and Fellmann list (note 29); there is a set in the British Museum Library (961.k.9).

[31] Also not in the Harris and Fellmann list (note 29).

[32] H. Richter, 'Geografiens historia i Sverige intill år 1800', *Lychnos-Bibliotek* (Uppsala: Almqvist & Wiksell), 17(1) (1959), 1–287; see p. 151. Richter regards the Uppsala society as Sweden's first geographical society. I am indebted to Professor W. R. Mead for this reference.

[33] Professor W. R. Mead tells me that the word was apparently not in use in Finland, for example, at that time.

Hunterian Oration. In this he distinguished 'physiography', which aimed 'to enumerate and delineate the effects and products of nature, as they appear' from 'sensible experience', from 'physiology' (theory of nature) and 'physiogony' (history of nature).[34] These three terms, identically defined, had previously been used by Coleridge in his essay on 'Henry More's theological works'.[35] In the narrow sense of the study of the external characteristics of objects, 'physiography' was also used at a later date by Rosenbusch in his treatise on mineralogy.[36]

But even before Huxley's lectures, 'physiography' was in common use in the English-speaking world in a closely comparable sense. Dana in 1863 made 'Physiographical geology' the first major section of his influential *Manual of geology*, and gave a succinct definition:

> Physiography, which begins where geology ends, – that is, with the adult or finished earth, – and treats (1) of the earth's final surface arrangements (as to its features, climates, magnetism, life, etc.), and (2) its systems of physical movements and changes (as atmospheric and oceanic currents, and other secular variations in heat, magnetism, etc.).[37]

Under this head Dana treated relief, ocean currents and temperature,

[34] J. H. Green, *Vital dynamics: the Hunterian Oration before the Royal College of Surgeons in London, 14th February 1840* (London: William Pickering, 1840), p. 101.

[35] S. T. Coleridge, 'Henry More's theological works', in H. N. Coleridge, ed., *The literary remains of Samuel Taylor Coleridge* (London: William Pickering, 1834), volume 3, 156–167; reference on p. 158.

[36] H. Rosenbusch, *Mikroskopische Physiographie der petrographischen wichtigen Mineralien: ein Hülfsbuch bei mikroskophischen Gesteinsstudien* (Stuttgart: 1st edition, 1873; 2nd edition, 1885).

[37] J. D. Dana, *Manual of geology: treating of the principles of the science* (Philadelphia: T. Bliss, 1863), p. 2. Whether Huxley knew this book before 1869, or even in 1877, cannot be definitely determined. It is cited in the *Physiography, op. cit.* (note 3), p. 313, but since Huxley's lecture notes were prepared for the press by an assistant this is not evidence that Huxley himself was familiar with the *Manual.* There was a copy of this edition in the working library which Huxley presented to the Royal College of Science in 1887. I am grateful to Mrs Pingree for a copy of the relevant entry in the manuscript 'Catalogue of Books presented by Professor Huxley to Science and Education Library'. Mrs Pingree has determined that the volume is no longer in the collection, which is now held in the archives of Imperial College. Dana did, however, present a copy to the Royal College of Science, and this is still in the Library, Department of Geology, Imperial College. It is possible that Huxley may have seen this copy, even if he acquired his own at a later date. He could also have seen the word used in the *American Journal of Science*: see C. C. Parry, 'Physiographical sketch of that portion of the Rocky Mountain Range, at the headwaters of South Clear Creek, and east of Middle Park; with an enumeration of the plants collected in the district, in the summer months of 1861', *American Journal of Science*, 83 (1862), 231–243; 84, 249–261, 330–341.

atmospheric movements and temperature, and the distribution of vegetation; but he also distinguished 'Dynamical geology', and the study of processes, as a distinct sub-division. By the time that Huxley's book was published (though admittedly several years after his lectures), the term 'physiography' was in almost common usage,[38] and much credit for this must certainly to be given to Dana.

STRUCTURE OF THE *PHYSIOGRAPHY*

Huxley's book, and the 1869 lectures, begin with the Thames at London Bridge. Working from the local and familiar to the unfamiliar, Huxley dealt with springs, rainfall and climate, water chemistry, denudation, glacial erosion, marine erosion, earth movements and volcanicity, deposition in the ocean and the formation of rocks, the geology of the Thames basin, and finally the earth as a planet, its movement and the seasons, and its place in the solar system. The long lecture on *Bathybius*, the existence of which had been strikingly disproved by Buchanan during the *Challenger* expedition, was omitted from the book, which did, however, include up-to-date material from Powell's *Exploration of the Colorado River of the west*, published in 1875, as well as Hayden's observations on the Yellowstone geysers, data on rainfall and sediment yield in Britain from the Rivers Pollution Commission, and new profiles of the North Atlantic from the voyage of the *Challenger*.

The *Physiography* exemplified Huxley's view of science as 'nothing but *trained and organized common sense*':[39] in education, for him, 'the fundamental principle was to begin with observational science, facts collected; to proceed to classificatory science, facts arranged; and to end with inductive science, facts reasoned upon and laws deduced.'[40] The

[38] E.g. J. Geikie, *The Great Ice Age and its relation to the antiquity of man* (London: W. Isbister, 1874), pp. xxi, 176, 519. Even so neither 'physiography' nor 'geomorphology' were used in J. W. Powell, *Exploration of the Colorado River of the west* (Washington: Government Printing Office, 1875), or C. E. Dutton, 'Tertiary history of the Grand Canyon District', *Monographs of the United States Geological Survey*, 2 (1882). Physiography is used without definition for the description of topographic features in I. C. Russell, 'Geological history of Lake Lahontan', *Monographs of the Unites States Geological Survey*, 11 (1885), p. 36, and it appears (but only in the index) in G. K. Gilbert, 'Lake Bonneville', *Monographs of the United States Geological Survey*, 1 (1890).

[39] T. H. Huxley, 'On the educational value of the natural history sciences' (first published 1854), in *Science and education* (London: Macmillan, 1893), p. 45.

[40] L. Huxley, *op. cit* (note 10), volume 1, p. 309; also T. H. Huxley (1893), *op. cit.* (note 39), pp. 52–53.

organizing principle in Huxley's science was causality, and the idea of causal chains and interconnections of causes supplied a theme which held together the diverse phenomena of the physical world. He explained his philosophy in the introduction to the book:

> the application of the plainest and simplest processes of reasoning to any of these phenomena suffices to show, lying behind it, a cause, which will again suggest another; until, step by step, the conviction dawns upon the learner that, to attain even an elementary conception of what goes on in his parish he must know something about the universe; that the pebble he kicks aside would not be what it is and where it is unless a particular chapter of the earth's history, finished untold ages ago, had been exactly what it was.[41]

Huxley's genius, and a reason for the success of the book, was to link this mode of explanation directly with a child's own experience, thus inverting the normal approach of physical geography texts of the day, and supplying an organizing principle which many of them lacked. 'I do not think', he wrote dismissively,

> that a description of the earth, which commences, by telling a child that it is an oblate spheroid, moving round the sun in an elliptical orbit; and ends, without giving him the slightest hint towards understanding the ordnance map of his own country; or any suggestion as to the meaning of the phenomena offered by the brook which runs through his village, or the gravel pit whence the roads are mended; is calculated either to interest or instruct.[42]

This emphasis on local study was both revolutionary and timely, for it coincided with the start of the movement to transform geographical

[41] T. H. Huxley (1877), *op. cit.* (note 3), pp. vii–viii. The same principle was emphasized by C. L. Morgan, 'Physiography', *Geographical Magazine*, n.s. 2 (5) (1878), 241–254. Huxley constantly emphasized his deterministic view of causality. In his *Essays upon controverted questions* (London: Macmillan, 1892), p. 247, he wrote that 'The fundamental axiom of scientific thought is that there is not, never has been, and never will be any disorder in nature. The admission of the occurrence of any event which was not the logical consequence of the immediately antecedent events, according to these definite, ascertained, or unascertained rules which we call the "laws of nature", would be an act of self-destruction on the part of science.' And in his *American Addresses* (London: Macmillan, 1877), pp. 2–3, he argued that 'it has ceased to be conceivable that chance should have any place in the universe, or that events should depend upon any but the natural sequence of cause and effect. . . . the chain of natural causation is never broken'. Such views were pedagogically attractive, but as an acceptable philosophy of science were already outmoded in Huxley's lifetime.

[42] T. H. Huxley (1877), *op. cit.* (note 3), p. vii.

education in which Huxley himself played a major part. There is doubtless truth as well as hyperbole in his allegation that in spite of a great deal of what he contemptuously termed 'Jewish history and Syrian geography' in the schools, he doubted 'if there is a primary school in England [in 1868] in which hangs a map of the hundred in which the village lies'.[43] Thus he intended his treatment of the Thames only as an example of what might be done for any local river. 'It is easy, for example, to make the Medway, the Severn, and the Forth, or the Clyde the starting point of our studies of nature.'[44]

RECEPTION

The book was immensely popular: within six months its circulation had become 'enormous'.[45] John Morley wrote to Huxley to say that 'Your *Physiography* is worth silver and gold, and I have predicted to Macmillan that it will have a great and perpetual sale. There is nothing at all like it in

[43] T. H. Huxley (1893), *op. cit.* (note 11), p. 87. Huxley here and elsewhere closely echoes the divine and antiquary Thomas James (1809–1863), who in 1857 in an anonymous review of a series of books on the history and antiquities of Northampton wrote: 'Geography should begin from the school-walls. "Which side of this room does the sun rise on?" – "Does Churchlane run west or north?" – "Whither does the brook flow that rises on Squash-hill?" In this way the young scholar would in time be brought to comprehend the round world and his own position on it; and probably with some clearer perception of the truth and relation of things than if he had begun by rote, "The earth is a terraqueous globe, depressed at the poles, consisting of", etc. etc. But we are all taught on the contrary plan. We begin at the wrong end. . . . We start from the Equator instead of High-street, and the result is the lamentable fact, that even educated men are strangers in their own country. . . . perhaps there is not a schoolroom in England where a county map is to be found hung up on the wall'. See *Quarterly Review*, 101 (1857), 1–56; reference on pp. 1–2. Huxley's robust agnosticism gave considerable offence and may have hindered the ready acceptance of his views. The Duke of Argyll wrote to Tyndall that 'He [Huxley] writes as if every believer in Christianity were no better than the blackbeetle beneath his feet': Dowager Duchess of Argyll, ed., *Duke of Argyll: autobiography and memoirs* (London: John Murray, 1906), volume 2, p. 526.

[44] J. W. Judd, 'Huxley's physiography', *Nature, Lond.*, 17 (1878), 178–180; see p. 179. My own copy of the *Physiography*, purchased in Cambridge, has marginal substitutions of 'Cam' for 'Thames'. Two translations of Huxley's book appeared. In the first, *Physiographie: introduction a l'étude de la nature* (Paris: G. Baillière, 1882), the Thames was replaced by the Seine throughout; in the second, *Physiographie: eine Einleitung in das Studium der Natur* (Leipzig: F. Brockhaus, Internationale wissenschaftliche Bibliothek, 1884), the Elbe and the Weser were the replacements. Dr M.-H. Sachet and Dr G. Scheer located copies of these translations for me in Paris and Darmstadt, respectively.

[45] F. Galton, *Journal of the Royal Geographical Society*, 48 (1878), p. cxxviii. For Galton, its publication was the main event of the year in geographical education.

the way of making nature ... interesting and intelligible to young people. ... Such a book is a real service to the human race.'[46] Green found it 'one of the best books I have read for many a long day'.[47] The Rev. E. Hale at Eton College referred his class to the *Physiography* and 'never found that they had any difficulty in understanding it'.[48] Morley's son preferred reading it before bedtime to a novel, and Leonard Huxley later recalled that it was 'vitalising to a child's mind, and left a lasting stimulus in quite a personal way',[49] Joseph Hooker had to provide successive copies for half a dozen sons as they became old enough to appreciate it, since each appropriated the copy then available for himself.[50]

In contrast to the unity of Huxley's approach, the physical geographies of the time were too often 'a collection of information, too abstract, too incoherent, too wide, and too superficial at the same time, to be of any use in education'.[51] Mary Somerville's *Physical geography*, dating from 1848, appeared in a sixth edition, revised by Bates, shortly before she died, aged 92, in 1872; a seventh edition appeared in the same year as the *Physiography*, and it was the last.[52] It was a book without illustrations, which gave detailed descriptions of the topography of each continent, of minerals, landforms and vegetation, and of the distribution of plants, animals and man. It included no local study, and while vastly more attractive than some of the purely topographic texts,[53] it must have been largely outside a child's personal experience. Ansted's *Physical geography* of 1867 began with the solar system, and was organized round the Aristotelian themes of earth, air, water, fire and life. It was unillustrated and compendious: the list of contents alone occupied 28 pages.[54] Arnold Guyot's *Physical geography*, published as late as 1873, belonged to a

[46] J. Morley to T. H. Huxley, 14 December 1877: Huxley Papers, General Letters, volume 23, ff. 36–37. The book was also favourably reviewed in *Athenaeum*, 30 (1878), 449.

[47] A. H. Green to T. H. Huxley, 22 January 1878: Huxley Papers, Scientific and General Correspondence, volume 17, f. 115.

[48] E. Hale, *Proceedings of the Royal Geographical Society*, n.s. 8 (1886), 203.

[49] L. Huxley, 'Home memories', *Nature, Lond.*, 115 (1925), 698–702, reference on p. 701; J. Morley to T. H. Huxley, *op. cit.* (note 46).

[50] J. D. Hooker to T. H. Huxley, 27 December 1877, and to Lord Redesdale, 25 January 1899, in L. Huxley, ed., *Life and letters of Sir Joseph Dalton Hooker, O.M., G.C.S.I* (London: John Murray, 2 volumes, 1918), volume 2, pp. 327, 433.

[51] P. Kropotkin, 'The teaching of physiography', *Geographical Journal*, 2 (1893), 350–359; reference on p. 350.

[52] M. Somerville, *Physical geography*, 7th edition (London: John Murray, 1877); see also note 6.

[53] Some were so dull that authors put them into verse to make them more palatable: A. Mackay, *A rhyming geography for little boys and girls* (Edinburgh: Bell & Bradfrute, 1873).

[54] D. T. Ansted, *Physical geography* (London: W. H. Allen, 1867).

pre-Darwinian tradition of teleological explanation. Probably only Keith Johnson's *Handbook of physical geography*, published in the same year as Huxley's book, approached the latter in organization and coherence.[55] In this sense, Huxley's *Physiography* represented a radically new departure, which it would be quite misleading to evaluate in terms of its scientific content in the light of subsequent developments in the earth sciences.[56]

ADVANTAGES OF HUXLEY'S APPROACH

Huxley's immediate aim in 1869 and 1877 was didactic: for almost a quarter of a century both subject and method were enormously successful in popular education. What contributed to this success, other than the inherent interest of the subject matter itself?

First, Huxley's philosophy, derived from Mill rather than from Kant and Whewell, appealed directly to sensory experience: its common-sense justification stemmed from a position of rational empiricism which stressed facts rather than theory or hypothesis. This had great advantages in both learning and teaching, but was particularly vulnerable to changes in the accepted philosophy of science. In elementary education even weak teachers would turn to the discipline of learning facts inherent in Huxley's approach and believe that they were teaching the scientific method. Indeed there were some, such as Fitch, who believed that this disciplinary aspect of geography ('training the mind') was sufficient justification in itself for the subject being taught.[57] From the student's point of view, also, science could be seen as a finite body of factual propositions which could readily be learned, rather than a less intelligible abstract calculus of theory which needed to be understood. Second, Huxley's emphasis[58] on concrete objects and his scepticism over book learning was in accord with a long tradition of teaching by object-lessons in elementary education. His lectures on a piece of chalk and a piece of coal were perhaps the best examples of this; and it

[55] A.H. Guyot, *Physical geography* (London: Sampson, Low 1873); K. Johnston, *Handbook of physical geography* (Edinburgh: W. & A. K. Johnston, 1877).

[56] An example of such anachronistic judgement is the statement that Huxley's *Physiography* was 'old fashioned in outlook, belonging to the compromise school that grew out of the marine *versus* fluvial controversy': R. J. Chorley, A. J. Dunn and R. P. Beckinsale, *The history of the study of landforms or the development of geomorphology* (London: Methuen, 1964), volume 1, p. 595.

[57] J. G. Fitch, 'Geography and the learning of facts', *Lectures on teaching* (Cambridge: University Press, 1881), pp. 344–369.

[58] T. H. Huxley (1893), *op. cit.* (note 11), p. 127.

was an approach developed by Charles Kingsley with great success in a series of lectures on *Town geology*[59] to the Chester Natural History Society in 1872. His lecture titles were: 'The soil of the field'; 'The pebbles in the street'; 'The stones in the wall'; 'The coal in the fire'; 'The lime in the mortar'; and 'The slates on the roof'. The method had the merits of immediacy, of relevance to daily life, and of appealing to a wide spectrum of intellectual ability. Ravenstein subsequently attempted to introduce the object-lesson method more formally into geography,[60] and it was readily made the basis of geographical and geological museums.

The object-lesson method was reinforced by a second mid-nineteenth-century educational reform movement, the 'science of common things'. Huxley himself criticized this as too often descending to pure information devoid of organizing principles, but his own contribution of the causal principle helped remedy this deficiency.[61] This common-sense empiricism was strongly reinforced in physiography and geography by two other concerns: experimentation and local studies. A characteristic of Huxley's book, and even more of some later texts, was an emphasis on experimental demonstration. Partly this resulted from the treatment of glaciation: John Tyndall, professor of natural philosophy at the Royal Institution from 1853 to 1887, had measured the movement of ice in Alpine glaciers and experimented with the behaviour of ice under stress. His experimental demonstrations at the Royal Institution of phenomena such as regelation, as well as of the nature of clouds, of the colour of the sea and the mechanism of Icelandic geysers had become famous. In 1872 his book *The forms of water*[62] was almost a catalogue of experiments, and had in fact originated as demonstrations to a juvenile audience. Unfortunately only the better schools could emulate such experiments, but even to describe them gave some indication of the scientific method.

Local study, or *Heimatskunde*, had become popular in German education since the time of Pestalozzi. It was a method particularly appropriate to geography,[63] and formed the basis of Huxley's use of the Thames basin

[59] C. Kingsley, *Town geology* (London: Strahan, 1872).

[60] E. G. Ravenstein, 'Object lessons in geography', *Proceedings of the Royal Geographical Society*, n.s. 6 (1884), 674–675.

[61] D. Layton, *Science for the people: the origins of the school science curriculum in England* (London: Allen & Unwin, 1973), p. 112.

[62] J. Tyndall, *The forms of water in clouds and rivers, ice and glaciers* (London: Henry King, 1872).

[63] J. S. Keltie, 'Report to the council of the Royal Geographical Society', in Royal Geographical Society, *Report of the proceedings of the Society in reference to the improvement of geographical education* (London: John Murray, 1886), pp. 39–41; E. G. Ravenstein, 'The aims and methods of geographical education', *ibid.*, 163–176, discussion, 176–181, see p. 166.

round which he developed his lecture plan. The method was also used in America by Guyot, where 'the geography of the school yard' soon became 'a popular fad'.[64] As will be seen, physiography itself soon abandoned Huxley's emphasis on local studies and fieldwork in its increasing concern with elementary science and astronomy.[65] It is also true, as Kropotkin argued, that it was a method better adapted to 'some remote village, lost in the Black Forest' 50 years before, and was less appropriate to schools in the great industrial cities.[66] When it disappeared from physiography, however, it left too important a field to be completely abandoned. Patrick Geddes, a student of Huxley's in 1874–1878, resurrected it as 'nature study', a field accepted in the Scottish school curriculum in 1899 and in England in 1900.[67] Geddes's assistant, A. J. Herbertson, transferred the idea to geography as regional or local survey, both in teaching and research. The Le Play Society catered for both, and specialist societies such as the Liverpool and District Regional Survey Association were founded.[68] Cowham's excursions from Croydon to Godstone formed one of the first formal attempts to establish the school field trip as a method in geography teaching.[69] In abandoning local studies, therefore, physiography lost one of its most persuasive practical concerns, which soon became a major theme in British geography.

THE EDUCATIONAL CONTEXT

Physiography was published at an opportune time. In 1860 the Department of Science and Art had begun its scheme of school examinations, and in 1870 the Elementary Education Act made elementary education compul-

[64] J. W. Redway, 'What is physiography?', *Education Review*, 10 (1895), 352–363; reference on p. 362.

[65] C. Lapworth, 'The relations of geology', *Scottish Geographical Magazine*, 19 (1903), 393–417. Grenville Cole noted in *Nature* that 'the rock is removed out of its place, and the waters wear the stones, while the eye of the pupil, made for wonder, remains fixed upon the printed page', *Nature Lond.*, 90 (1912), 159.

[66] P. Kropotkin, *op. cit.* (note 51), p. 350.

[67] P. Geddes, 'Nature study and geographical education', *Scottish Geographical Magazine*, 18 (1902), 525–536. Also P. Boardman, *Patrick Geddes: maker of the future* (Chapel Hill: University of North Carolina Press, 1944), pp. 234, 265–276.

[68] The society was founded by P. M. Roxby in 1918. See W. Hewitt, *The Wirral Peninsula: an outline regional survey* (London: Hodder & Stoughton, 1922); and P. Geddes, 'Huxley as teacher', *Nature, Lond.*, 115 (1925), 740–743, reference on p. 743.

[69] J. H. Cowham, *The school journey: a means of teaching geography, physiography, and elementary science* (London: Westminster School Book Depot, 1900).

sory throughout the country: by 1876 this had added one million new places in elementary schools.[70] There was, too, an increasing concern with scientific education. In 1868 a Parliamentary Select Committee on Scientific Instruction, after hearing evidence from the Rev. J. G. Cromwell, who urged that no school should get a government grant if it did not teach physical geography, recommended 'That elementary instruction in drawing, in physical geography, and in the phenomena of nature, should be given in elementary schools'.[71] Nor was this movement confined to elementary education. In 1870 the Headmasters' Conference was organized as an association of public schools, and in 1873 the Oxford and Cambridge Schools' Examination Board was established, with the first Higher Certificate examinations the following year. In 1873, also, the first university extension lectures began, organized at Cambridge, and this developed into a movement, in which geographers and geologists played a major part, which carried higher education into industrial towns throughout the country.[72] Mackinder later recalled that in three years he travelled 30,000 miles giving extension lectures, chiefly on physiography.[73]

The role of the earth sciences in this educational revolution was, in retrospect, astonishing. The Department of Science and Art began examining in physical geography in 1864,[74] with D. T. Ansted as the examiner. By

[70] D. Layton, *op. cit.* (note 61).

[71] Select Committee on Scientific Instruction, *Report* (1868), pp. viii, 207. Much of this movement of reform came from alarm at German educational advancement. See G. Haines IV, 'German influence upon English education and science, 1800–1866', *Connecticut College Monograph* 6 (1957), 1–107; 'German influence upon scientific instruction in England, 1867–1887', *Victorian Studies*, 1 (1958), 215–244; and 'Essays on German influence upon English education and science 1850–1919', *Connecticut College Monograph* 9 (1960), 1–88.

[72] H. J. Mackinder and M. E. Sadler, *University extension, past, present, and future* (London: Cassell, 1891). See also C. Bibby, 'Thomas Henry Huxley and university development', *Victorian Studies*, 2 (1958), 97–116; E. Welch, *The peripatetic university: Cambridge Local Lectures 1873–1973* (Cambridge: Cambridge University Press, 1973); and N. A. Jepson, *The beginnings of English University adult education – policy and problems. A critical study of the early Cambridge and Oxford University Extension Lecture monuments between 1873 and 1907, with special reference to Yorkshire* (London: Michael Joseph, 1973).

[73] H. J. Mackinder, 'Geography as a pivotal subject in education', *Geographical Journal*, (1921), 376–384; reference on p. 378. R. D. Roberts, secretary of the London University extension scheme, had been university lecturer in geology at Cambridge, 1884–1886, and wrote *An introduction to modern geology* (London: John Murray, University Extension Manuals, 1893). See also R. D. Roberts, *Eighteen years of university extension* (Cambridge: Cambridge University Press, 1891), and B. B. Thomas, 'R. D. Roberts and adult education', *Harlech Essays* (Cardiff: University of Wales Press, 1933), pp. 1–35.

[74] Previously physical geography had been included with nautical astronomy, steam and navigation under the general subject of navigation.

1867 over 10 per cent of all papers written were in this subject, and by 1871 over 20 per cent. In 1873 physical geography accounted for 15,238 papers, or 27 per cent of the total number of papers in all subjects (65,577). By comparison, in that year mathematics attracted less than 8,000 applicants, animal physiology less than 7,000, inorganic chemistry 4,000, and applied mechanics only 567. No other academic subject remotely rivalled physical geography in its popularity from 1868, when it took the lead, to 1876. At its peak, Ansted had 14 assistants to help with the marking.[75]

Then, by a minute dated 15 August 1876, it was resolved to terminate examinations in physical geography and to substitute, as a new subject, physiography.[76] The last examinations in physical geography were held in 1878, the first in physiography in 1877, the year of publication of Huxley's book. Ansted resigned as examiner, and was replaced by Norman Lockyer, later professor of astronomical physics at the Royal College of Science, and the geologist J. W. Judd. The immediate effect of the change was catastrophic, presumably because of the unfamiliar nature of the new subject, and at least at first from the lack of textbooks. The number of papers taken fell from about 15,000 a year to less than 5,000 (see figure 4); not until 1886 did the total again exceed 10,000. Meanwhile the total number of papers in all subjects was increasing rapidly: from 65,000 in 1877 to 128,000 in 1887, and to over 200,000 in 1892. Though physiography increased slowly (it reached 22,000 in 1891), it rarely topped 10 per cent of the total number of papers, in contrast to the 26 per cent exceeded by physical geography in earlier days.[77]

Partly this was simply not a result of the change in title, but of a change in the function of the subject. Huxley himself had seen physical geography (he had not then adopted the term physiography) as 'the peg upon which the greatest quantity of useful and entertaining scientific information can be suspended',[78] and Judd, more picturesquely, saw physiography as 'the

[75] Department of Science and Art, *Annual Reports*.

[76] Department of Science and Art, *24th Annual Report* (1877), pp. 2–3. The reasons were various, one being that physical geography was more an academic than a practical subject and hence not properly within the remit of the Department of Science and Art. It was also possible for student teachers to obtain successive grants from both the Education Department and the Department of Science and Art, to do virtually the same subject and qualify after passing very similar examinations; and hence it was decided to make the Science and Art Department's subject quite different from that of the Education Department to prevent this abuse. See R. A. Gregory, *op. cit.* (note 16), p. 208. The fullest account of the Department of Science and Art, which I did not see until after the present paper was written, is H. Butterworth, 'The Science and Art Department 1853–1900', unpublished Ph.D. thesis, University of Sheffield (1968).

[77] Department of Science and Art, *24th Annual Report* (1877).

[78] T. H. Huxley (1893), *op. cit.* (note 11), p. 109.

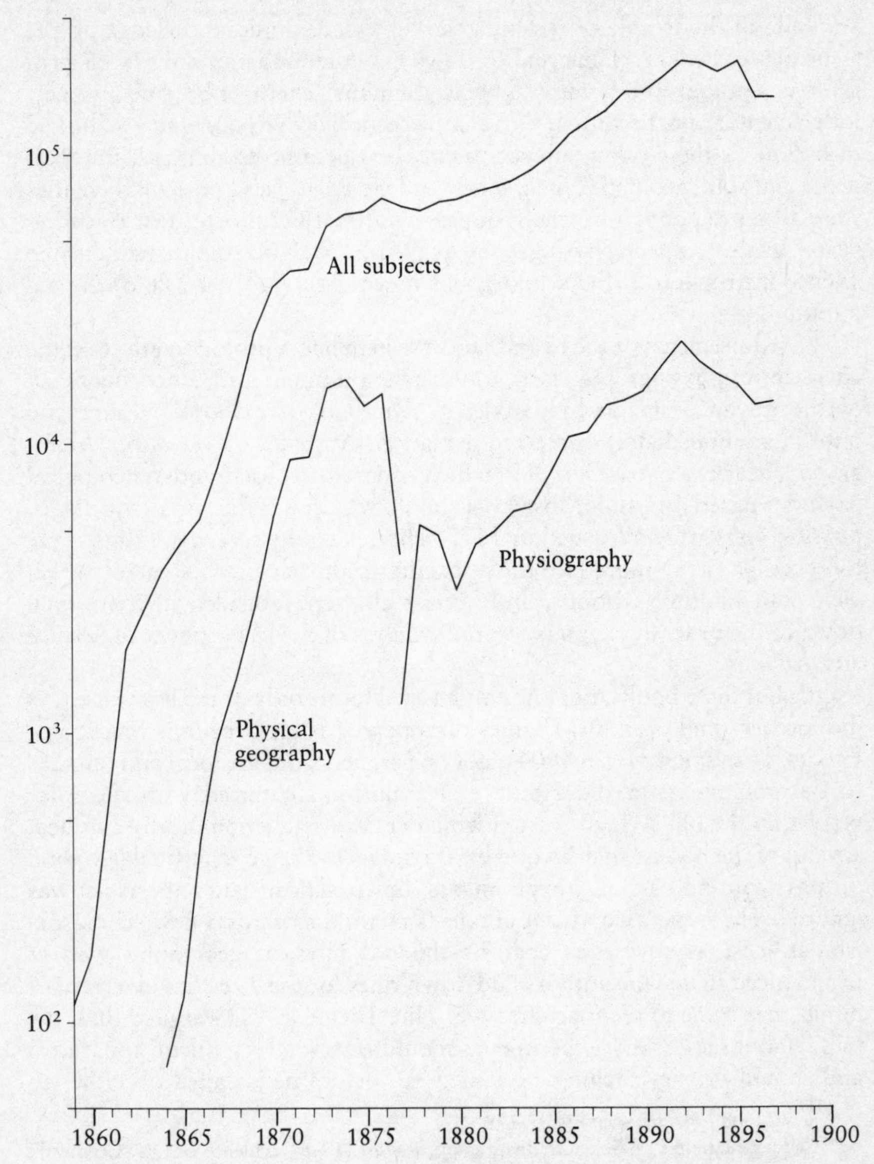

Figure 4 Physical geography and physiography in the examinations of the Department of Science and Art, 1864–1897.
Source: Department of Science and Art, *Annual Reports.*

vestibule of the temple of natural science'.[79] It required knowledge of 'the principles of heat and light, and the laws of matter and motion' as well as of geology, zoology and botany.[80] Hence in many schools it became a vehicle for elementary instruction in the sciences, especially physics, and was bound to decline as these other subjects themselves became established. Practical plane and solid geometry, magnetism and electricity, and pure mathematics were all more popular than physiography after 1877, in contrast to earlier years. In this respect physiography played a catalytic role in establishing science instruction in the schools, but it was necessarily a role of limited duration.

The requirements of the examination system had a profound effect on the character of physiography itself, to the extent ultimately of cancelling many of the advances made by Huxley's own book. Textbook writers and publishers immediately perceived a market. Ansted's *Elements of physiography*, Mackay's *Physiography* (which adopted an obscurantist theological position based on the Mosaic chronology), and Prince's *Elements of physiography* were all issued in 1877. Though coolly received,[81] they were followed by many more (see the appendix to this chapter), some of which went into multiple editions, and almost all were modelled directly, even down to the order of chapters, on the syllabus of the Department of Science and Art.

Of all of these books, not one emphasized local study or the home area, as did Huxley (and even the Thames disappeared from Gregory's revision of Huxley's *Physiography* in 1904); not one argued from the local and familiar to the unfamiliar, as did Huxley; and almost all started with the solar system, in the old style of Ansted which Huxley had emphatically avoided. Several of the books, such as that by Davies in 1897, were particularly good on physical and chemical experiments, but practical field experience was ignored. The trend away from Huxley's method towards a new scholasticism at least as severe as that of the old physical geography was so pronounced that some authors laid down rules for the direct involvement of pupils that came to seem progressive. Thus Dickie in 1893 argued that 'the pupil should *see* every experiment, should *handle* every bit of apparatus, and should *hear* everything explained . . . before he is called on either to

[79] J. W. Judd, *op. cit.* (note 44), p. 179.

[80] Select Committee on Scientific Instruction, *Report* (1868), evidence of J. G. Cromwell, p. 207.

[81] Even Ansted was roughly handled in *Nature*: 'A student had better not be taught at all than be taught in this manner, and, in fact, a student of average intelligence . . . if he does not doubt the competence of his teacher, will have no more to do with him'. *Nature, Lond.*, 18 (1878), 563–564.

read or record it in words.'[82] Books which attempted a broader view (such as Gee's in 1895, which gave a more literary treatment, with quotations from Humboldt, Charles Kingsley and David Livingstone, with greater geographical content) were not well received and did not reach a second edition.[83]

This trend resulted directly from the examination questions set by Lockyer and Judd. Lockyer was an astronomer,[84] and clearly a much stronger character than the geologist Judd,[85] who might have been expected to have retained some emphasis on field science. The change in emphasis became generally apparent. 'Many teachers and authors of textbooks', wrote Gregory, 'suppose that Physiography is synonymous with Physical Geography. That this idea is not held by Professors Judd and Lockyer, the examiners in Physiography, is shown by . . . the questions.'[86] It was 'Professors Judd and Lockyer . . . [who] have practically determined the present form of the Honours stage in this subject'.[87] By the time that the syllabus and regulations were revised in 1894, Lockyer had transformed Huxley's original earth science into an elementary stage of general instruction in mechanics, physics and chemistry, and an advanced stage consisting almost entirely of astronomy. The former necessarily had a limited appeal as individual scientific subjects became established in their own right, and the latter was of little educational or practical significance. Though the examination requirement generated a continuous flow of

[82] H. Dickie, *Elements of physiography* (London: William Collins, 1893), p. 3.

[83] W. Gee, *Short studies in nature knowledge; an introduction to the science of physiography* (London: Macmillan, 1895).

[84] A. J. Meadows, *op. cit.* (note 4). Lockyer's emphasis was apparent from an early date; subsequently he expanded this portion to include the whole of physiography. See N. Lockyer, 'The earth's place in nature: a sketch of a branch of physiography', *Good Words for 1878*, 129–135, 233–239, 479–486, 638–645, 842–849.

[85] H. G. Wells, 'Professor Judd and the science of geology (1886–1887)', gives an unflattering portrait of Judd at this time, in *Experiment in autobiography* (London: Gollancz, 1934), volume 1, pp. 227–233. Mrs Pingree thinks this view of Judd too harsh and has drawn my attention to tributes to him on his retirement as dean of the Royal School of Mines. One speaker recalled Judd's field trips with students, 'and the enormous thirst they used to develop. Once they were followed by a squad of police, under the impression that they were a prize fight' (*Mining Journal*, 17 February 1906). Nevertheless, Lockyer's view of the syllabus clearly prevailed over what one might have expected Judd's to have been.

[86] R. A. Gregory, *Elementary physiography; being a description of the laws and wonders of nature* (London: J. Hughes, 1892), p. 2. And as the reviewer in *Nature* pointed out, it was Professors Judd's and Lockyer's 'opinion in this matter [which] is final for the student interested', *Nature, Lond.*, 47 (1892), 74.

[87] R. A. Gregory and H. G. Wells, *Honours physiography* (London: J. Hughes, 1892), p. 11.

textbooks, stimulated by each syllabus revision, many were written as purely commercial ventures (H. G. Wells and R. A. Gregory received £10 each from Hughes for their *Honours physiography*, which Wells did simply for the money),[88] and the subject attracted an increasing degree of educational antagonism which was more properly directed at Lockyer's physiography than at Huxley's.[89]

A major problem, exacerbated by the system of payment by results, was that many entered the examinations simply to gain a qualification rather than to learn about a subject, and physical geography was apparently particularly popular in this way. Huxley himself commented on students who 'work to pass, not to know. . . . They do pass, and they don't know', and in the year his book appeared he stated that 'The educational abomination . . . of the day is the excessive stimulation of young people to work at high pressure by incessant competitive examinations' – or 'book gluttony and lesson bibbing', as he called it.[90] It is doubtful whether physical geography was significantly worse than some other subjects in this respect (though it was certainly less immediately practical): the failure rate was somewhat higher than the average, but not excessively so.[91]

[88] H. Hartley, 'The life and times of Sir Richard Gregory, Bt., F.R.S. 1864–1952', *Nature, Lond.*, 171 (1953), 1040–1046; reference on p. 1041.

[89] Royal Geographical Society, *Report of the proceedings of the Society in reference to the improvement of geographical education* (London: John Murray, 1886), p. 115. The change in emphasis was not welcomed by all reviewers. One found it 'a mystery why one part of chemistry added to two parts of physics could produce "physiography"', *Nature, Lond.*, 73 (1906), 578. Some of these textbooks were appallingly bad. A review of H. C. Martin, *Notes on elementary physiography* (Manchester: John Heywood, 1891) spoke of 'scraps of information . . . themselves incorrect . . . [with] crude statements which render them ridiculous', and described the book as a whole as 'untrustworthy . . . useless . . . [and] much to be condemned', *Nature, Lond.*, 44 (1891), 589–60. W. Mather in *Physiography: an elementary textbook* (London: Marshall, 1888) tried to instil some cohesion into a fragmentary subject matter by the novel expedient of abandoning conventional division into chapters altogether, simply starting at the beginning and going on to the end.

[90] Quoted by C. Bibby, *op. cit.* (note 3).

[91] The percentage failure rates were:

	Advanced	*Elementary*
Physical geography	39.0	38.1
Physiography	36.9	36.2
Both	37.7	36.8
All subjects	32.1	33.0

Source: Department of Science and Art, *Annual Reports* (1864–1897)

But in spite of all these factors, the educational contribution of both physical geography and physiography in the later nineteenth century was impressive. During the 34 years of annual reports of the Department of Science and Art, 1864–1897, over three million examination papers were worked in all subjects: 10 per cent were in physical geography or physiography, and the latter alone accounted for close to a quarter of a million papers.[92]

PHYSIOGRAPHY AND THE NEW GEOGRAPHY

But if physiography increasingly departed from Huxley's conception in the schools, the importance of the field of study it represented was increasingly recognized at higher levels. The Royal Commission on Scientific Instruction and the Advancement of Science (the Devonshire Commission), which sat from 1870 to 1875, had Huxley as a member and Lockyer as secretary. It was primarily concerned with the establishment of subjects such as physics, chemistry, engineering, zoology and modern languages in the universities, and with the support of scientific research, but it also reviewed the lecturing at Berlin in, among other subjects, physical geography, and it received a memorial from the Royal Geographical Society on the need for chairs in geography at both Oxford and Cambridge.[93]

Under the influence of Francis Galton and D. W. Freshfield, the council of the Royal Geographical Society was moving towards greater emphasis on geography in higher education,[94] and it recognized that some change in the traditional character of the subject was itself necessary if this were to succeed. Mackinder, already active in university extension work, lectured to the Society on 'The scope and methods of geography' in 1887.[95] His 'new geography' presupposed a knowledge of physiography: 'Physiography asks of a given feature, "Why is it?" Topography, "Where is it?" Physical geography, "Why is it there?" Political geography, "How does it act on man in society, and how does he react on it?" Geology asks, "What riddle of the

[92] The figures were: total papers worked 3,083,047; physical geography 92,992; physiography 224,351. Department of Science and Art, *Annual Reports* (1864–1897).

[93] Royal Commission on Scientific Instruction and the Advancement of Science, *Third Report* (1873), pp. xxiii–xxiv. See also chapter 4.

[94] D. W. Freshfield, 'The place of geography in education', *Proceedings of the Royal Geographical Society*, n.s. 8 (1886), 698–714, discussion 714–718.

[95] H. J. Mackinder, 'On the scope and methods of geography', *Proceedings of the Royal Geographical Society*, n.s. 9 (1887), 141–160, discussion 160–174.

[96] *Ibid.*, p. 147.

past does it help to solve?"'[96] While physiography (in Huxley's sense) was thus a contributor to both geography and geology, it necessarily formed, in Mackinder's scheme, an integral part of the former. In this he was doubtless influenced by the success of physiography as an extension subject, but at that time he also found the causal framework attractive, and he made it the basis for his whole view of geography. It was an approach he later abandoned, and he came to regret the emphasis placed on physiography in departments of geography.[97] But at that time, the 'steady application' of what were seen as the 'principles of Darwinism' was thought to be the best means of recognition of the New Geography as a science.[98]

As part of the adoption of geography in the universities, a lectureship in physiography and commercial geography was established for H. R. Mill at Heriot-Watt College in 1886, and a chair of physiography for J. L. Lobley at City of London College in 1894. Thus physiography, for a time, held an influential position within geography, largely at the expense of the human aspects of the subject.[99] But as geography established its new identity, physiography again came under attack, for both internal and external reasons. The internal reason was that a mixture of 'nature study and general elementary science' was 'not truly geographical at all'; it did an 'injustice to geography' and discredited the subject.[100] With the spread of elementary education, geography no longer needed to cover all branches of elementary science itself, and the elementary nature of the instruction offered in these subjects itself detracted from the standing of geography as a discipline. The problem arose in acute form during the discussion by the University of London Board of Geographical Studies on new regulations, in 1902. Judd, long a physiography examiner with the Department of Science and Art, argued for the inclusion of elementary general science in preliminary teaching; Mackinder, seeing the danger, held out for regional geography or nothing; and Mackinder won.[101]

[97] H. J. Mackinder, 'Presidential address to the Geographical Association, 1916', *Geographical Teacher*, 8 (1916), 271–277, reference on p. 277; H. J. Mackinder, 'The human habitat', *Scottish Geographical Magazine*, 47 (1931), 321–335, reference on p. 323.

[98] J. S. Keltie in *Proceedings of the Royal Geographical Society*, n.s., 10 (1888), 53 (review of F. Darwin, ed., *The life and letters of Charles Darwin* [London: John Murray, 3 volumes, 1888]).

[99] H. J. Fleure, 'Sixty years of geography and education: a retrospect of the Geographical Association', *Geography*, 38 (1953), 231–264; reference on p. 234.

[100] R. N. Rudmose Brown, 'The province of the geographer', *Scottish Geographical Magazine*, 30 (1914), 467–479; reference on pp. 473–474.

[101] E. W. Gilbert, 'The Right Honourable Sir Halford J. Mackinder, P.C., 1861–1947', *Geographical Journal*, 127 (1961), 27–29.

The external reason was equally powerful. Huxley's conception of physiography represented a definite field of study – the old physical geography unified by a causal principle. As the new geography moved towards a more human emphasis, it gave less attention to the field of processes and landforms of the earth's surface. The gap was filled by the 'whole great new science of geomorphology',[102] with an organizing principle even more powerful than Huxley's.

PHYSIOGRAPHY AND GEOMORPHOLOGY

The origins of 'geomorphology' are surprisingly difficult to discern, and the enquiry is made more difficult by the anachronistic use of terms by almost all historians of the earth sciences:[103] the *Oxford English Dictionary* cites no use of the term before 1896. The more general term 'morphology', used by Goethe in 1817 for the science of living forms,[104] was given general circulation by Ernst Haeckel in 1866,[105] and was introduced into the earth sciences by Penck's *Morphologie der Erdoberfläche* in 1894.[106] Penck is

[102] O. J. R. Howarth, 'The centenary of Section E (Geography) in the British Association', *Scottish Geographical Magazine*, 67 (1951), 145–160; reference on p. 155 (quoting A. J. Herbertson at the British Association meeting in 1901).

[103] An example of such usage is the extraordinary statement that 'there was a general lack of interest and effort directed towards physiographic ends in Britain during the 1870s and 1880s,' ascribed to an obsession with exploration on the part of the Royal Geographical Society and to Geikie's influence at the geological Survey. This is certainly not true for physiography as then understood; and it is also not true so far as the Society is concerned. See Chorley *et al.*, *op. cit.* (note 56), p. 592.

[104] J. W. Goethe, *Zur Naturwissenschaft überhaupt besonders zur Morphologie* (Stuttgart and Tübingen: J. G. Cotta'schen Buchhandlung, 4 volumes, 1817–1823).

[105] E. Haeckel, *Generelle Morphologie der Organismen: Allgemeine Grundzüge der organischen Formenwissenschaft* (Berlin: G. Reimer, 2 volumes, 1866).

[106] A. Penck, *Morphologie der Erdoberfläche* (Stuttgart: J. Engelhorn, 2 volumes, 1894). Penck had used the term for a decade previously: see A. Penck to J. Partsch, 29 June 1884, in G. Engelmann, 'Briefe Albrecht Pencks an Joseph Partsch', *Wissenschaftliche Veröffentlichungen des Institut für Ländeskunde*, 17–18 (1960), 17–107; reference on p. 64. The cognate terms 'morphography' and 'morphometry' were also used by Penck in his *Morphologie* and earlier writings, and there are earlier usages of both terms outside the earth sciences. The word 'topology' in the restricted sense of General Berthaut ('étude raisonnée des formes topographiques') never gained general acceptance. H. M. A. Berthaut, *Topologie: étude de terrain* (Paris: Librairie Militaire R. Chapelot et Cie, 1911), reference on p. iv. The history of this usage is also uncertain. Victor Bérard referred to it as a 'mot nouveau' in his history of the Phoenicians in 1902. Bérard, following Gustav Hirschfeld, developed a 'science des sites', going further than simple topographic description, to aid in historical explanation. 'Il existe des lois *topologiques* . . .; la *topologie*, science des lieux, serait à la *topographie*, simple description des

sometimes credited with originating the derivative 'geomorphology',[107] but this is probably erroneous. McGee in 1888 wrote of the 'genetic study of topographic forms (which has been denominated geomorphology)',[108] but even before this, Emmanual de Margerie, in a review of Richthofen's *Führer für Forschungsreisende*, had referred to the field it covered 'qu'on peut appeler, a l'exemple de plusieurs savants américains, la géomorphologie';[109] he also referred to Penck's *Morphologie*, then in preparation, as a 'Manual de géomorphologie'.

The term received wide currency in Mackinder's lecture to the British Association for the Advancement of Science at Ipswich in 1895: he referred to 'what we now call geomorphology, the causal description of the earth's present relief' or 'the half artistic, half genetic consideration of the form of the lithosphere'.[110] He also coined the term 'geophysiology' to include oceanography, climatology and biogeography, with anthropogeography, which, with geomorphology, defined the content of geography. In the same year the International Geographical Congress in London had a section

lieux, exactement ce qu'est la géologie à la géographie. . . . la *topologie* nous expliquera les descriptions qui lui fornait le topographie. . . . Bref, dans le present et dans le passé, la topologie déduira les raisons des habitats humains, et réciproquement, en face d'un habitat humain, elle induira les conditions qui l'on fait naître, le genre et la période de civilisations auxquels il faut le rapporter,' V. Bérard, *Les Phéniciens et l'Odyssée* (Paris: A. Colin, 2 volumes, 1902–1903), see volume 1, pp. 6–9. Bérard's views were well received, in England. W. M. Fullerton called topology 'one of the most astonishing instruments of penetration of the past that human ingenuity has invented . . . this magic pickaxe': 'Before Homer: seapower and the Odyssey', *Cornhill*, n.s. 14 (1903), 245–259; reference on p. 251. Also G. Murray, 'The wanderings of Odysseus', *Quarterly Review*, 202 (1905), 344–370 (Murray termed Bérard a 'scientific geographer'), and the anonymous review, 'A notable work on the Odyssey', *Spectator*, 94 (1905), 855–857, in which it is noted that 'To the topographist, for instance, the site of Cambridge is a mystery; to the topologist, on the other hand, who knows the history of the Fens and of medieval trade, it is full of meaning.' These were powerful claims, and it is surprising that historical geographers, for example, seem largely unaware of Bérard's work. Historical interpretation of topography in the sense of Bérard's topology was common at that time (e.g. E. Reclus, 'La Phénicie et les Phéniciens', *Bulletin de la Société neuchâtelois de Géographie*, 12 (1900), 261–274), but when Berthaut reintroduced the term in 1911 it carried none of Bérard's associations with it, and neither usage survived.

[107] R. E. Dickinson, *The makers of modern geography* (London: Routledge & Kegan Paul, 1969), reference on pp. 102–103.

[108] W J McGee, 'The geology of the head of Chesapeake Bay', *Annual Report of the United States Geological Survey*, 7 (1888), 537–646; reference on p. 547.

[109] E. de Margerie 'Géologie', *Polybiblion Revue Bibliographique Universelle, Partie littéraire*, (2) 24 (1886), 310–330; reference on p. 315.

[110] H. J. Mackinder, 'Modern geography, German and English', *Geographical Journal*, 6 (1895), 367–379; references on pp. 373, 375.

entitled 'Geomorphology' (and Penck used the term in his paper to the meeting).[111] By 1906 a lecture course on geomorphology was being given to geologists and geographers at Cambridge University (by J. E. Marr). In American W. M. Davis, who much preferred to use 'physiography', used the new term in 1895.[112]

Davis's physiography was neither Huxley's nor Lockyer's, however: it was more accurately represented by McGee's definition of geomorphology as the 'genetic study of topographic forms'. Its organizing principle was not simple causality, but the idea of regular change of form through time, systematized in Davis's scheme of the cycle of erosion and first used in his paper on classification given in 1884.[113] Though the word 'physiography' persisted in America, it came to be synonymous with 'geomorphology' as understood in Britain and Europe,[114] whereas 'physiography' in England retained some of the catholicity of meaning of Huxley's original usage. Physiography in the narrower sense of geomorphology was, in fact, a characteristically American development.[115]

An important stage in American geographical education was the report by the Conference on Geography in 1894 to the National Educational Association's Committee of Ten.[116] In a sense this repeated the English experience of previous decades. Russell Hinman's *Eclectic physical geography* of 1889[117] covered the same ground as many English physiography texts, starting with concepts of matter and energy, dealing with the earth as a planet, the atmosphere, the sea, landforms, climate and life, and emphasizing 'proximate causes' as its explanatory theme. Perhaps unusually

[111] A. Penck, 'Die Geomorphologie als genetische Wissenschaft: eine Einleitung zur Diskussion über geomorphologische Nomenklatur', *Report of the 6th International Geographical Congress, (London 1895)* (1896), 737–747; English summary, 748–752.

[112] W. M. Davis, 'Bearing of physiography on uniformitarianism', *Bulletin of the Geological Society of America*, 7 (1895), 8–11; reference on p. 8.

[113] W. M. Davis, 'Geographic classification, illustrated by a study of plains, plateaus and their derivatives', *Proceedings of the 33rd Meeting of the American Association for the Advancement of Science (Philadelphia 1884)*, (1885), 428–432.

[114] N. M. Fenneman, 'The rise of physiography', *Bulletin of the Geological Society of America*, 50 (1939), 349–360.

[115] W. M. Davis, 'The progress of geography in the United States', *Annals of the Association of American Geographers*, 14 (1924), 159–215, reference on p. 199; J. K. Wright, 'What's American about American geography?', in *Human nature in geography* (Cambridge: Harvard University Press, 1966), 124–139, reference on pp. 131–132.

[116] C. R. Dryer, 'A century of geographic education in the United States', *Annals of the Association of American Geographers*, 14 (1924), 117–149; reference on p. 124.

[117] R. Hinman, *Eclectic physical geography* (New York: Van Antwerp; London: Sampson Low, Marston, Searle & Rivington, 1889).

in textbooks of this sort, each chapter began with a biblical quotation. It was a more rigorous treatment than most of the commercially inspired textbooks of the day, and is best compared with H. R. Mill's *Realm of nature*,[118] which first appeared in 1892.

The reasons which led to the educational decline of physiography in Britain, however, applied also in America. The decline was delayed for a time by W. M. Davis's insistence on the dual nature of geography, which he saw as a science of the relations between inorganic controls and organic responses, though this ceased to be a viable definition long before Davis's death.[119]

Meanwhile, physiography as a research discipline in America came to have a more distinctive meaning even than as a synonym for genetic geomorphology. Beginning with a series of essays edited by J. W. Powell in 1896,[120] it became increasingly applied to regional geomorphic studies, both monographs[121] and synthetic works,[122] and in this sense is still in use. While influential textbooks were written on systematic physiography early in the century,[123] these dealt purely with geomorphology, and the latter term alone is now used for works of this kind. Carl Sauer in particular objected to the use of physiography (a term properly applied since Kant and Humboldt to systematic rather than chorological knowledge) in regional contexts, but even more he objected to the causal framework which had so

[118] H. R. Mill, *The realm of nature: an outline of physiography* (London: John Murray, 1892).

[119] W. M. Davis, 'The progress of geography in the schools', *Yearbook of the National Society for the Scientific Study of Education*, 1 (2) (1902), 7–49; W. M. Davis, 'An inductive study of the content of geography', *Bulletin of the American Geographical Society*, 38 (1906), 67–84; J. Leighly, 'What has happened to physical geography?', *Annals of the Association of American Geographers*, 45 (1955), 309–318, reference on pp. 310–311.

[120] J. W. Powell, ed., *The physiography of the United States* (New York: American Book Company, 1896).

[121] C. Abbe, 'A general report on the physiography of Maryland', *Special Publications of the Maryland Weather Services*, 1 (2) (1899), 41–216; I. Bowman, 'Physiography of the Central Andes', unpublished Ph.D. thesis, Yale University (1909).

[122] I. Bowman, *Forest physiography of the United States and physiography of soils in relation to forestry* (New York: Wiley, 1911); N. M. Fenneman, 'Physiographic divisions of the United States', *Annals of the Association of American Geographers*, 6 (1917), 19–98; M. N. Fenneman, *Physiography of western Unites States* (New York: McGraw-Hill, 1931); N. M. Fenneman, *Physiography of eastern United States* (New York: McGraw-Hill, 1938); F. B. Loomis, *Physiography of the United States* (New York: Doubleday Doran, 1937); W. W. Atwood, *The physiographic provinces of America* (New York: Harper, Row, 1940), and other works.

[123] R. S. Tarr and L. Martin, *College physiography* (New York, Macmillan, 1914); R. D. Salisbury, *Physiography* (London: John Murray, 1907).

attracted Huxley and later Davis, and which Sauer recognized as inadequate for the human sciences.[124] By the mid-1930s physiography had almost disappeared as a component of American geography,[125] and geomorphology was largely carried out by geologists. Methodological debates at this time on the nature of physiography were largely of a technical geomorphic nature, and had little to do with its place among the sciences or its educational contribution.[126]

PHYSIOGRAPHY AND THE SCIENCES

We have seen that one of the primary causes of the decline of physiography was that its scope was too ambitious, attempting to include too much of other sciences. Some earlier writers had made this catholicity a virtue: Woodward, for example, argued that physiography 'endeavours to knit together the sum and substance of all that is known of the physical history of the universe – it is, in fact, a Cosmogony – though its chief aim is to develop the intimate connexion between all sciences, and to illustrate the Unity of Creation.'[127] Even Humboldt had attempted less.

Later workers were concerned to establish the field of physiography in more practical terms. While acknowledging that it covered the 'substance, form, arrangement and changes of all the real things of Nature in their

[124] C. O. Sauer, 'The morphology of landscape', *University of California Publications in Geography*, 2 (2) (1925), 19–53, reprinted in J. Leighly, ed., *Land and life* (Berkeley: University of California Press, 1963), 315–350; reference on p. 347.

[125] It is not mentioned by C. C. Colby, 'Changing currents of geographic thought in America', *Annals of the Association of American Geographers*, 26 (1936), 1–37.

[126] W. H. Hobbs, 'The frontier of physiography', *Science, N.Y.*, n.s. 18 (1903), 538–540; W. S. Glock, 'Dual nature of physiography', *Science, N.Y.*, n.s. 72 (1930), 3–5; D. W. Johnson, 'Physiography and the dynamic cycle', *Science, N.Y.*, n.s. 75 (1932), 636–640. But compare also T. C. Chamberlin, 'The relations of geology to physiography in our educational system', *National Geographical Magazine*, 5 (1894), 154–160; T. C. Chamberlin, 'The methods of the earth-sciences', *Congress of Science and Arts (Universal Exposition, St Louis, 1904)*, (1906), 477–487; W. M. Davis, 'The relations of the earth-sciences in view of their progress in the nineteenth century', *ibid.*, 488–503; A. Penck, 'The relations of physiography to other sciences', *ibid.*, 607–626; A. Penck, 'Geography among the earth sciences', *Proceedings of the American Philosophical Society*, 66 (1927), 621–644.

[127] H. B. Woodward, 'Reviews' (1879) (of S. J. B. Skertchly, *The physical system of the universe: an outline of physiography* [London: Daldy, Isbister, 1878]), *Geological Magazine*, (2) 6 (1879), 82–84. Skertchly (1874/1875) had published an early version of his book in a series of articles before Huxley's book appeared: 'The mechanism of the globe', *Fenland meteorological circular and weather report*, 1, 25–27, 33–35, 41–43, 51–52, 59–61, 67–68, 75–77; 1, 84–85.

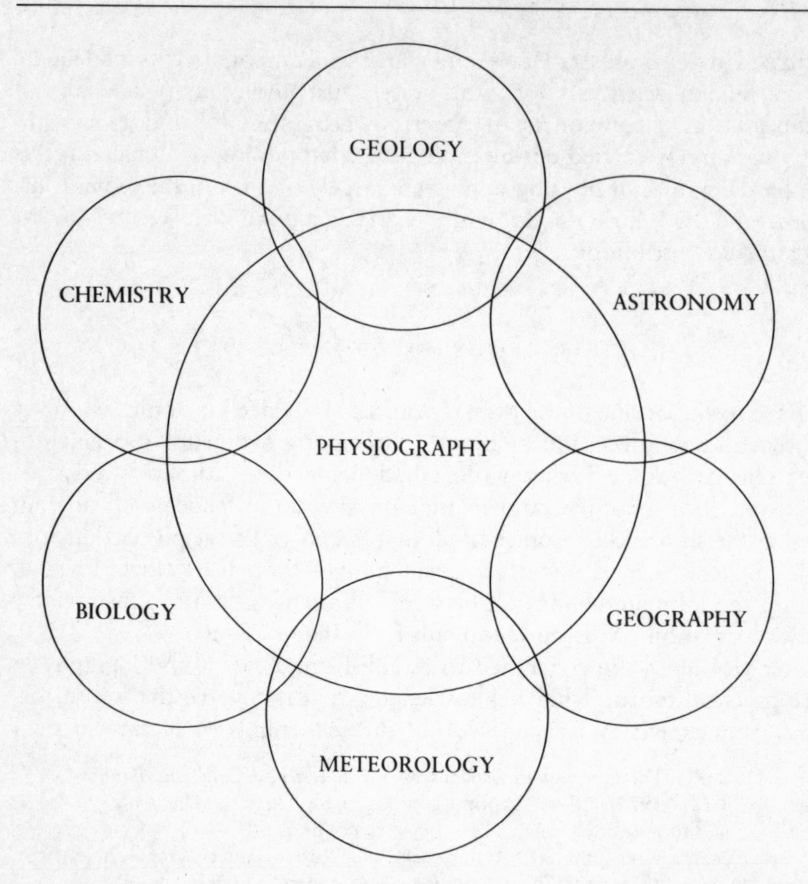

Figure 5 Physiography and the natural sciences according to H. R. Mill.
Source: H. R. Mill, *The realm of nature: an outline of physiography* (London: John Murray, 1892), p. 12.

relations to each other', H. R. Mill saw the function of the subject as a bridge, 'at once an introduction to all the sciences and a summing up of their result'.[128] Figure 5 gives his conception of the overlap between the separate systematic fields and that of physiography. Lobeck[129] much later also saw physiography as a field overlapping the systematic sciences, both natural and human, with geography occupying in each case the area of overlap (see

[128] H. R. Mill, *op. cit.* (note 118), pp. 3, 12, 13.
[129] A. K. Lobeck, *Geomorphology: an introduction to the study of landscapes* (New York: McGraw-Hill, 1939), p. 3.

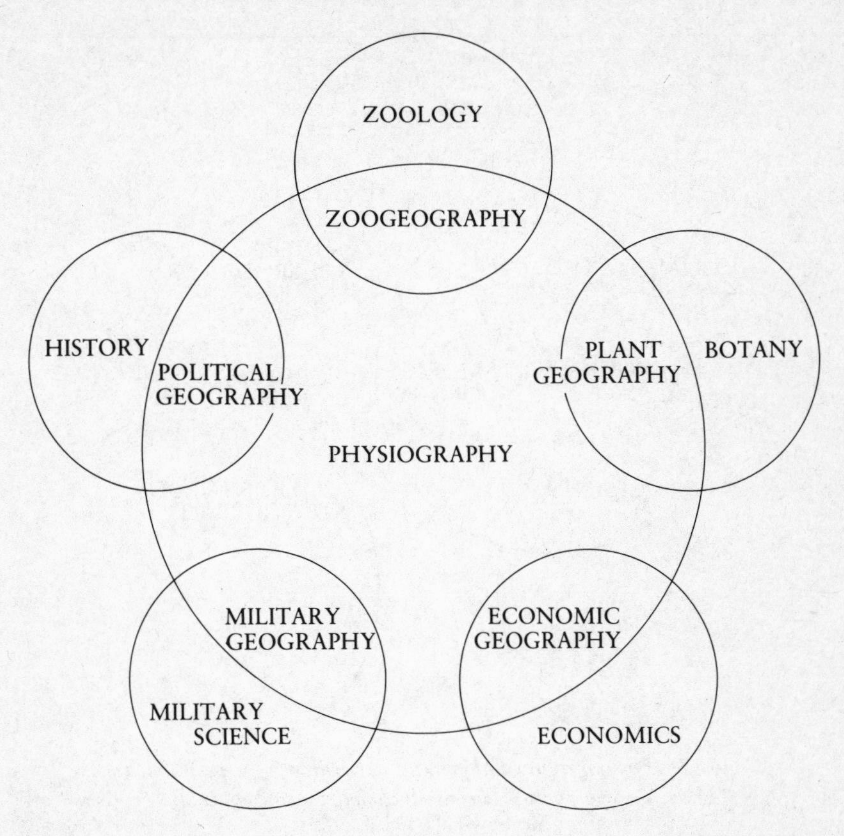

Figure 6 Physiography and other systematic fields according to A. K. Lobeck.
Source: A. K. Lobeck, *Geomorphology: an introduction to the study of landscapes* (New York: McGraw-Hill, 1939), p. 2.

figure 6): for him, physiography was purely the 'study of the physical environment', and man played no part in it.

There was especial concern about the relationships between physiography and geology, and later between physiography and geography. For Mackinder, 'the geologist looks at the present that he may interpret the past; the geographer looks at the past that he may interpret the present';[130] for Salisbury, 'Physiography may be said to be, on the one hand, a special phase of geography, namely, physical geography, and, on the other, a special

[130] H. J. Mackinder, *op. cit.* (note 95), p. 146.

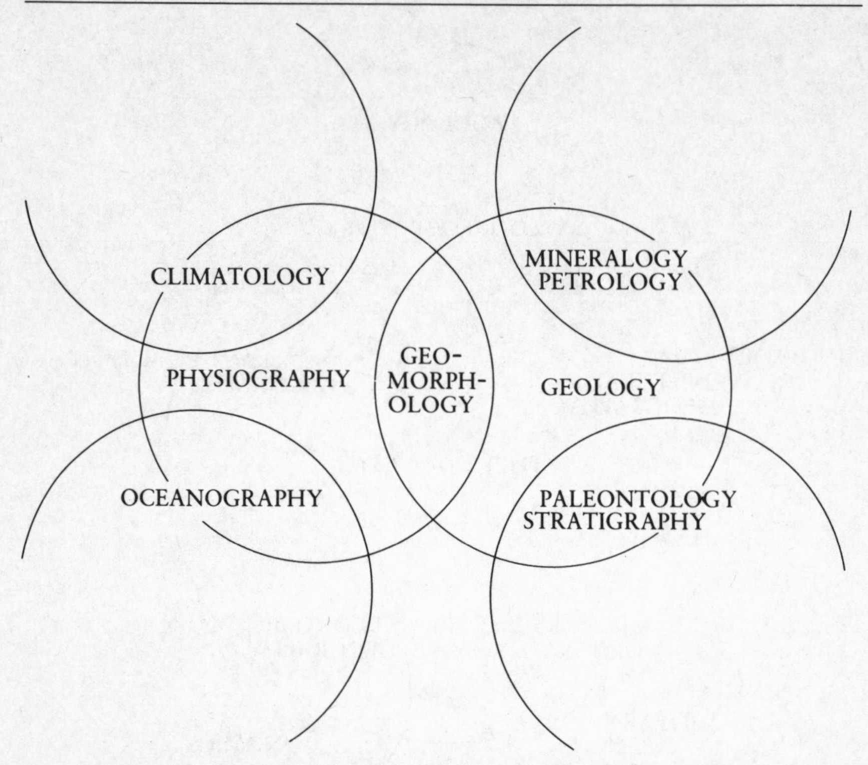

Figure 7 Physiography and geology according to A. K. Lobeck.
Source: A. K. Lobeck, *Geomorphology: an introduction to the study of landscapes* (New York: McGraw-Hill, 1939), p. 2.

chapter of geology, namely, the present.'[131] Lobeck (see figure 7) saw physiography and geology as distinct fields, overlapping in what was for him the historical science of geomorphology, with physiography also having strong links with climatology and oceanography.[132] In this sense physiography was more concerned with processes than with either historical development or spatial distributions, but such a definition still left unresolved the unity of the field so outlined. Fenneman (see figure 8) saw physiography as the area of overlap between geography and geology, but as we have seen he was among those who redefined physiography as regional geomorphology in North America.[133]

[131] R. D. Salisbury, *op. cit.* (note 123), p. 4.
[132] A. K. Lobeck, *op. cit.* (note 129), p. 2.
[133] N. M. Fenneman, 'The circumference of geography', *Annals of the Association of American Geographers*, 9 (1919), 3–11; reference on p. 4.

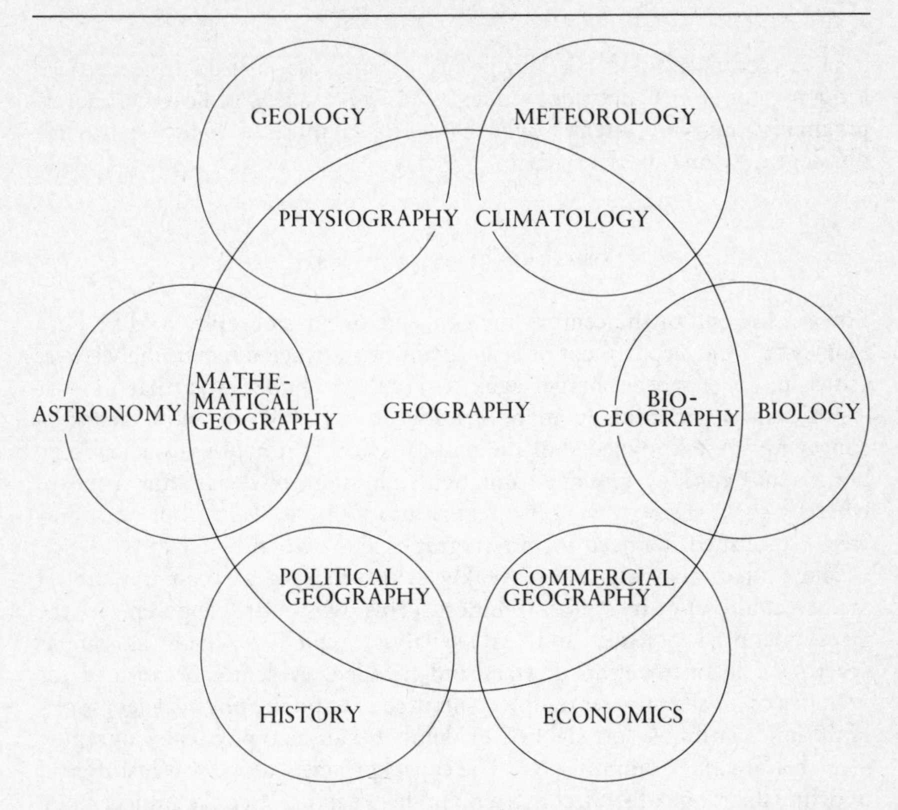

Figure 8 Geography and other systematic fields according to N. M. Fenneman.
Source: N. M. Fenneman, 'The circumference of geography', *Annals of the Association of American Geographers*, 9 (1919), p. 4.

Clearly attempts to find a place among the sciences for a field so ambitious outlined could hardly be successful; but this is not to say that physiography, in Huxley's original sense, did not continue to be of value as a framework for research as well as teaching. Philip Lake's textbook on *Physical geography*[134] was essentially an up-dated physiography in the range it covered, and it remained in print for over half a century. It represented teaching in the subject at Cambridge University, where physiography was still a field of undergraduate specialization up to 1969. The concept proved especially attractive to workers on coastal landforms, where the influences of the sea, of weather and climate, and of vegetation are all clearly seen. Steers in particular maintained a physiographical rather

[134] P. Lake, *Physical geography* (Cambridge: Cambridge University Press, 1915).

than a geomorphological viewpoint,[135] and his attitude has proved attractive to many Cambridge students.[136] This usage was, however, purely pragmatic, and no attempt was made to establish a logical place for physiography among the sciences.

OBJECTION TO PHYSIOGRAPHY

Toward the end of the century the concepts of physiography held by both Huxley and the Department of Science and Art attracted increasingly severe criticism. The main charge against Huxley's book was that it was insufficiently scientific. In many schools it was still the only means of conveying some knowledge of the natural sciences, for physics, chemistry, botany and zoology remained untaught. Physiography was thus a means whereby these sciences were 'smuggled into the schools',[137] but once they were established the need for physiography diminished.

The subject also was often weakly structured; as a result the subject matter could appear almost limitless. This was more apparent in the Department of Science and Art syllabus, which reached 'as far as spectroscopic observation of stars and nebulae, evidently because of the presence of an astronomer on the committee . . . and the physical geography of the lands is almost lost sight of, evidently because no physical geographer is on the directing committee.'[138] The causal principle alone was insufficient to define the range of subject matter, and hence in practice the subject could become impossible to teach, except according to the narrow rubric of the regulations. The lack of criteria of significance also led to the undue prominence given to particular spectacular phenomena, such as volcanoes and glaciers, which in a human context were of limited importance.[139] The regulations also undoubtedly accounted for the inclusion of specific

[135] J. A. Steers, 'The coast as a field for physiographical research', *Transactions of the Institute of British Geographers*, 22 (1957), 1–13; J. A. Steers, 'Physiography: some reflections and trends', *Geography*, 45 (1960), 1–15.

[136] O. H. K. Spate, *The compass of geography* (Canberra: Australian National University Press, 1953), p. 11.

[137] Kropotkin, *op. cit.* (note 51), p. 350.

[138] W. M. Davis, 'Physical geography in the high school', *School Review*, 8 (1900), 388–404, 449–456, reprinted in D. W. Johnson, ed., *Geographical essays by William Morris Davis* (Boston: Ginn, 1909), 129–145; reference on p. 134. It was this paper which provoked Gregory's defence of the South Kensington syllabus: *op. cit.* (note 16). Also W. M. Davis, 'The physical geography of the lands', *Popular Scientific Monthly*, 57 (1900), 157–170, reprinted in Johnson, *op. cit.*, pp. 70–86; reference on p. 76.

[139] H. J. Mackinder, *op. cit.* (note 95), p. 148.

chapters, notably on magnetism and electricity, often at illogical places in the text, which was characteristic of many of the textbooks of physiography of the time.

H. E. Armstrong, a student of Huxley's who had not been impressed by his teaching and who became one of his most severe critics, failed even to find a principle of organization in the *Physiography* – 'a study delightfully set out but none the less a book of mere fact, without any attempt to display the method of discovery – which is the background of our modern progress'.[140] Earlier he had described it as 'a type of the book to be avoided until method has been fully mastered'.[141] In another address, Armstrong was even more outspoken on 'that rank and pretentious hybrid Physiography . . . a shallow fraud as a means of teaching. Huxley never did a worse day's work than when he put his lectures in the form of a textbook; as lectures and when delivered by him they were doubtless admirable but as a book they are doing infinite injury to rational teaching. You cannot study Nature except by scientific methods.'[142] It is fair to add that Armstrong's experience as a student of Huxley's was apparently unhappy – he found Huxley a marvellous exponent, therefore, a 'bad teacher'[143] – and that he was himself concerned to establish a new philosophy of education in the so-called Heuristic Method. Yet this was 30 years after Huxley's book was published, and the situation in the schools after the Education Act of 1902 was very different from that after the Act of 1870. Armstrong's criticisms were in fact directed more to the twentieth century than to the 1870s.

The answer to the criticism that physiography was insufficiently scientific was to narrow the field, first by excluding man,[144] then by confining it to the study of landforms. In this, physiography became involved in a double rivalry. Many geologists, and notably T. McKenny Hughes, believed that they were perfectly capable of teaching physical geography themselves, and that separate recognition of the latter was unnecessary; others argued that

[140] H. E. Armstrong, 'Huxley's message in education', *Nature, Lond.*, 115 (1925), 743–747; reference on p. 744.

[141] H. E. Armstrong, 'Address to the British Association 1902', in *The teaching of scientific method and other papers on education* (London: Macmillan, 1903), reference on p. 86.

[142] H. E. Armstrong, 'Domestic science', *London Technical Education Gazette*, reprinted in *The teaching of scientific method*, *op. cit.* (note 141); reference on p. 408.

[143] H. E. Armstrong, 'Our need to honour Huxley's will', reprinted in W. H. Brock, ed., *H. E. Armstrong and the teaching of science 1880–1930* (Cambridge: Cambridge University Press, 1973), reference on p. 57; H. E. Armstrong *Pre-Kensington history of the Royal College of Science and the university problem* (London: Royal College of Science, 1921), pp. 4–5. Compare Beatrice Webb's portrait of Huxley following her meeting with him on 6 May 1887 (note 3).

[144] To which Kropotkin objected: *op. cit.* (note 51), p. 350.

geology should be extended to incorporate such teaching in the schools as well as in the universities.[145] Many geographers, conversely, interested mainly in human topics, were content to abandon physical geography entirely. Mackinder came increasingly to think in this way; others, such as Lyde, dismissed geomorphology in contemptuous terms as 'mere morbid futility'.[146] Fundamentally, however, it was the rise of Davisian geomorphology which usurped the field hitherto occupied by physiography, and which made the study of landforms a science in its own right, distinct from both geography and geology.

CONCLUSION

Physiography thus held a dominant position in British education for a quarter of a century before being displaced, partly as a result of its own changing nature under the influence of Norman Lockyer and the Department of Science and Art, partly because of the development of natural science subjects in their own right in the schools. In university geography it never held such an important position, partly because of the influence exerted by geology, partly because of the independent development of the 'New Geography' and its rather different concerns. And the new science of geomorphology, although still called physiography in the United States, was supplied with a new unifying principle, the cycle of erosion, a new technique, the historical analysis of landforms, and a new field, of regional analysis, while at the same time abandoning climate, oceanography, biogeography and the study of human geography to other disciplines.

Thus analysed, physiography in Huxley's sense was a development peculiarly suited to the expansion of popular education in the decades of rapid industrialization, population growth and social awareness in the half century following *The origin of species*. To later observers its philosophy appears simplistic. Sauer perceptively summarized its role: 'It was the bright hour of dawn in scientific monism, with Huxley officiating at the observation of the lands. Physiography served in such a canonical role in elementary scientific education until a later age of machinery sent it into

[145] J. Geikie, 'Geography and geology', *Fragments of earth lore: sketches and addresses geological and geographical* (Edinburgh: J. Bartholomew, 1893), pp. 1–13, reference on pp. 4–5; T. McKenny Hughes in discussion in Freshfield, *op. cit.* (note 94).

[146] S. W. Wooldridge, 'On taking the ge- out of geography', *Geography*, 34 (1949), 9–18; reference on p. 13. L. W. Lyde was appointed at University College London in 1903 but his appointment excluded physical geography, which was left to the geologists.

discard in favour of "general science".'[147]

It may in fact be argued that 1877 was something of a watershed in the study of the earth. Huxley's *Physiography* marks the end of three centuries of the study of landforms in Britain in Gordon Davies's perceptive work, *The earth in decay*;[148] it marks, too, the point when the scientific study of landforms passed from professional geologists to university geographers. But above all, in its time, and for 30 years, Huxley's book characterized one dramatic episode in Darwin's impact on the sciences, and as such it had a lasting impact on the structure and organization of knowledge.

APPENDIX

TEXTBOOKS OF PHYSIOGRAPHY

Textbooks are notoriously ephemeral publications and are often ignored both by libraries and by bibliographers. The holdings of the Cambridge University Library form the basis of this list. The most difficult books to trace are those which refer to physiography only in a sub-title; otherwise this list is probably reasonably complete. Items marked with an asterisk have not been studied by me.

Anon. *Recapitulation of physical geography for Remove: notes on the various sections of physiography and physical geography, drawn up for Eton science classes* (Eton: Williams & Sons, 1879)*

Anon. *Elementary physiography; or, earth lore. Adapted to the latest requirements of the Department of Science and Art* (London: Gill, 1894)

Ansted, D. T. *Elements of physiography* (London: W. H. Allen, 1877)

Carey, A. L., Bryant, F. L., Clendenin, W. W. and Morrey, W. T. *Physiography for high schools* (Boston: D. C. Heath, 1912)

Cartwright, T. '*Section one*' *physiography* (London: Thomas Nelson, 1897)

Cartwright, T. '*Section two*' *physiography* (London: Thomas Nelson, 1901)*

Cowham, J. H. *The school journey: a means of teaching geography, physiography, and elementary science* (London: Westminster School Book Depot, 1900)

Davies, A. M. *First stage physiography, for the elementary examination of the Science and Art Department* (London W. B. Clive, 1897)

Davies, A. M. *Physiography* (London: City of London Book Depot, 1897)

[147] C. O. Sauer, *op. cit.* (note 124), reference on p. 347.

[148] G. L. Davies, *The earth in decay: a history of British geomorphology 1578–1878* (London: Macdonald), n.d., pp. 353, 356.

Dickie, H. *Elements of physiography, adapted to the requirements of the Science and Art Department* (London: William Collins, 1893)

Douglas, J. *A sketch of the first principles of physiography* (London: Chapman & Hall, 1889)

Evans, J. *Physiography* (London: Institution of Civil Engineers, 1885)*

Findlater, A. *Elementary physiography, adapted to the syllabus of the South Kensington Science Department* (London: W. and R. Chambers, 1881)

Findlater, A. *Physiography: advanced course adapted to the syllabus of the South Kensington Science Department* (London: W. and R. Chambers, 1883). New editions revised by D. Forsyth 1892, 1894

Findlater, A. *Physiography: elementary course. Adapted to the syllabus of the South Kensington Science and Art Department* (Edinburgh: W. and R. Chambers, 1884). New editions revised by D. Forsyth 1891 and 1896*

Furneany, W. S. *Notes on elementary physiography, for the use of students in science classes* (London: Thomas Murby, n.d.)*

Gardner, T. W. *Elementary physiography* (London, 1904)*

Gee, W. *Short studies in nature knowledge: an introduction to the science of physiography* (London: Macmillan, 1895)

Gregory, R. A. *Elementary physiography: being a description of the laws and wonders of nature* (London: J. Hughes, 1892). 8th edition 1895; new edition 1896

Gregory, R. A. *Honours physiography* (London: J. Hughes, 1893)*

Gregory, R. A. and Christie, J. C. *Advanced physiography* (London: J. Hughes, 1894)

Gregory, R. A. and Simmons, A. T. *Experimental science: physiography (section I): an elementary course of physics and chemistry adapted to the syllabus of the Science and Art Department* (London: Macmillan, 1899)

Gregory, R. A. and Wells, H. G. *Honours physiography* (London: J. Hughes, 1893)

Harrison, J. A. *First steps in earth-knowledge: being an introduction to physiography (section I)* (London: Blackie, 1899)*

Harrison, W. A. *A manual of physiography* (London: Bemrose, 1879)*

Harrison, W. J. *Guide to examinations in physiography, and answers to questions* (London: Blackie, 1891)*

Harrison, W. J. and Wakefield, H. R. *Earth knowledge: a textbook of elementary physiography* (London: Blackie, 1887)*

Herbertson, A. J. *Outlines of physiography: an introduction to the study of the earth* (London: Edward Arnold, 1901). 3rd edition

Herberton, A. J. *A physiographical introduction to geography* (Oxford: Clarendon Press, 1910). 2nd edition 1912

Herbertson, F. D. *A first physiography (Elementary Geography, vol. I)* (Oxford: Clarendon Press, n.d., *c.* 1909)

Hull, E. *A text-book of physiography or physical geography, being an introduction to the study of the physical phenomena of our globe* (London: C. W. Deacon, 1888)

Hull, E. *Physiography: an introduction to the study of nature* (London: C. W. Deacon, 1891)*

Huxley, T. H. *Physiography: an introduction to the study of nature* (London: Macmillan, 1877)

Huxley, T. H. and Gregory, R. A. *Physiography: an introduction to the study of nature*. Rev. edition (London: Macmillan, 1904)

Lawson, W. *Outlines of physiography*. New edition (Edinburgh: Oliver & Boyd, 1887)

Lockyer, J. N. *Outlines of physiography: the movements of the earth* (London: Macmillan, 1887)

Mackay, A. *Physiography or physical geography with special reference to the instructions recently issued by the Science and Art Department* (Edinburgh: William Blackwood, 1877)

Mackay, A. *Physiography and physical geography* (Edinburgh: William Blackwood, 1883)

Martin, H. C. *Notes on elementary physiography compiled to meet the requirements of students preparing for elementary physiography* (Manchester: J. Heywood, 1891)

Mawer, W. *Physiography: an elementary text-book* (London: Marshall & Co., 1888)*

Mawer, W. *Elementary text-book of physiography* 3rd edition (London: A. Brown, 1900)

Mill, H. R. *The realm of nature: an outline of physiography* (London: John Murray, University Extension Manuals, 1892). 2nd edition 1913, 3rd edition 1924

Mills, J. *Lessons on elementary physiographical astronomy: designed to meet the requirements of students preparing for the elementary stage of physiography in the Science and Art Department examinations* (London: Chapman & Hall, 1889)

Mills, J. *Advanced physiography (physiographic astronomy) designed to meet the requirements of students preparing for the elementary and advanced stages of physiography in the Science and Art Department examinations, and as an introduction to physical astronomy* (London: Chapman & Hall, 1890)

Mills, J. *Solutions to the questions set at the May examinations of the Science and Art Department, 1881 to 1886: physiography* (London: Chapman & Hall, 1890)

Morgan, A. *Elementary physiography treated experimentally* (London: Longmans, Green, 1901)*

Morgan, A. *Advanced physiography* (London: Longmans, Green, 1898)*

Nancarrow, J. H. *Advanced physiography* (London: Ralph's Science Notebooks, 1897)*

Perry, W. J. *The local examination physiography* (London: Relfe Brothers, 1904)

Prince, J. J. *The elements of physiography: for the use of science classes, and elementary and middle-class schools* (Manchester: J. Heywood, 1877). Later edition 1885

Prince, J. J. *The elements of physiography, for the use of science classes, elementary and middle-class schools, and pupil teachers. Part I – elementary stage.* 7th edition (London: J. Heywood, 1889)

Prince, J. J. *The elements of physiography: for the use of science classes, elementary and middle-class schools, and pupil teachers. Part II – advanced stage.* 7th edition (London: J. Heywood, 1889)

Salisbury, R. D. *Physiography* (London: John Murray, 1907). Later edition 1919

Salisbury, R. D. *Physiography for schools* (London: John Murray, 1909)

Salisbury, R. D. *Elementary physiography* (New York: Henry Holt, 1911)*

Shaler, N. S. *Outliners of the earth's history: a popular study in physiography* (London: William Heinemann, 1898)

Simmons, A. T. *Physiography for beginners* (London: Macmillan, 1806). Second edition 1897

Simmons, A. T. *Physiography for advanced students* (London: Macmillan, 1897)

Skertchly, S. B. J. *The physical system of the universe: an outline of physiography* (London: Daldy, Isbister, 1878)

Spencer, J. *Physiography: a class book for the elementary stage of the Science and Art Department* (London: Percival, 1891)

Stewart, R. W. *First stage physiography* (London, 1905)*

Tarr, R. S. and Martin, L. *College physiography* (New York: Macmillan, 1914)

Thom, G. *Outlines of elementary physiography, with illustrations and questions from the science examinations papers, South Kensington* (Edinburgh: J. Thin, 1881)

Thornton, J. *Elementary physiography: an introduction to the study of nature* (London: Longmans, Green, 1888). 9th edition 1896, 10th edition 1899

Thornton, J. *Advanced physiography* (London: Longmans, Green, 1890). 4th edition 1896, 6th edition 1901

Thornton, J. *Elementary practical physiography (section I): a course of lessons and experiments in elementary science* (London: Longmans, Green, 1897). New edition 1906

Thornton, J. *Elementary practical physiography (section II): a course of lessons and experiments in elementary science for the Queen's Scholarship Examination* (London: Longmans, Green, 1900)

Van Baren, J. *De Vormen der Aardkorst: inleiding tot de Studie der Physiographie* (Groningen: J. B. Wolters, 1907)

Wilkinson, P. *Experimental physiography: section I* (London: Simkin, Marshall, Hamilton, Kent, 1903)

Williams, J. *Physiography* (London: Stewart's educational Series, 1882)*

Young, E. *Elementary physiography* (London: Lyon, 1900)*

10

'Grandeur in This View of Life'

It was in Sydney, where he had landed from H.M.S. *Beagle* for a fortnight's exploration and collecting, that on 12 January 1836 Charles Darwin lay down on a grassy bank and wondered at the profound differences between the life forms of Australia compared with those of all the other continents. As he did so, he allowed himself the scandalous speculation that perhaps there had been two creators at work, not just one – one for Australia, and another for everywhere else.[1] It was a solution to one of the most basic problems of biogeography that Darwin's own lifework was to demonstrate to be completely untenable, but which was not to be revealed for another two decades. In this chapter I wish to consider how Darwin's experience on board the *Beagle* helped to shape the quite different solution he was to find, and to show how the lessons in scientific method which Darwin learned on board the ship have immediate relevance still. I will first consider some aspects of Darwin's experience at sea; I will then discuss his reaction to nature during the voyage of the *Beagle*; and finally I will draw attention to the way he carried out his scientific work, with particular reference to his coral-reef theory.

DARWIN AT SEA

For Darwin, of course, the voyage of the *Beagle* was by far the most important event in his life, and determined his whole career.[2] The voyage itself lasted nearly five years. He was only 22 when it began and 27 when it ended, though the difference it made in him was not just to be measured in years, or even, as his father exclaimed when it ended, in the changed shape of his head.[3] During the voyage Darwin spent 500 days, or one-third of the

[1] N. Barlow, ed., *Charles Darwin's diary of the voyage of H.M.S. Beagle* (Cambridge: Cambridge University Press), p. 383.

[2] N. Barlow, ed., *The autobiography of Charles Darwin 1809–1882* (London: William Collins, 1958), p. 76.

[3] *Ibid.*, p. 79.

total time, at sea; and nearly 95 per cent of the entire expedition was in the southern hemisphere.

It is perhaps paradoxical to emphasize Darwin's connection with the sea. He was a landsman all his life, from his boyhood in rural Shropshire to his old age in rural Kent. He frankly said at the outset that his idea of the inside of a ship was like that of most people of the inside of a man – 'a large cavity containing air, water and food mingled in hopeless confusion'.[4] He was never happy on the ocean. He had been almost sick at anchor on 5 December 1831, three weeks before the ship sailed, and was dreadfully ill as soon as she passed the breakwater into the open Channel.[5] His misery, he says, was excessive. He was too ill even to go on deck to see Madeira, their first sight of land, but lay in his hammock nibbling raisins and trying to keep down sago and hot wine.[6] Thereafter it was constant – he was 'unspeakably miserable' across the Atlantic;[7] ill round Cape Horn;[8] suffered 'misery and vexation of spirit' on the coast of Chile;[9] had 'plenty of sea sickness' in New Zealand;[10] and suffered more at the Cape of Good Hope than at the beginning.[11] No wonder he burst out in a letter to his sister Susan on the final leg of the voyage – 'I loathe, I abhor the sea, and all ships which sail on it.'[12]

He found himself also in cramped quarters, living with naval officers who had little idea what he was about. He messed with the captain, FitzRoy, a moody, difficult, occasionally even somewhat insane man who years later committed suicide, and with whom conversational disagreement was almost tantamount to mutiny.[13] 'Damned beastly devilment', Lt Wickham called Darwin's collections,[14] and said he brought more dirt on board than any ten other men: 'if I were skipper', he told poor Darwin, 'I would soon have you and all your damned mess out of the place.'[15] Below decks they called him

[4] N. Barlow, ed., *op. cit.* (note 1), entry for 23 November 1831, p. 9.

[5] *Ibid.*, p. 13.

[6] *Ibid.*, p. 19.

[7] N. Barlow, ed., *Charles Darwin and the voyage of the Beagle* (New York: Philosophical Library, 1946), p. 59. For an interesting commentary on the structure of the voyage, see J. Tallmadge, 'From chronicle to quest: the shaping of Darwin's "Voyage of the Beagle"', *Victorian Studies*, 23 (1980), 325–345.

[8] N. Barlow, ed., *op. cit.* (note 7), p. 81.

[9] *Ibid.*, p. 113.

[10] *Ibid.*, p. 128.

[11] *Ibid.*, p. 140. See also F. Darwin, ed., *The life and letters of Charles Darwin, including an autobiographical chapter* (London: John Murray, 3 volumes, 1887), volume 1, p. 224.

[12] N. Barlow, ed., *op. cit.* (note 7), p. 145: letter of 4 August 1832.

[13] N. Barlow, ed., *op. cit.* (note 2), pp. 73–76.

[14] N. Barlow, ed., *op. cit.* (note 7), p. 103.

[15] F. Darwin., *op. cit.* (note 11), volume 1, p. 223.

'Flycatcher'.[16] It must have taken courage and resolution, especially at the outset, to persist in his work.

But there was another side to the voyage. First, he could do science, at sea as well as on shore. His first trawl was on 10 January 1832, after leaving the Cape Verde Islands, and he was amazed at the beauty, delicacy and colour of the animals in his homemade net. It reawakened an interest in marine life which he had formed when, as a student of 16 at Edinburgh University, he explored the tide pools of the Firth of Forth, and during the voyage he made a special study of zoophytes and other marine invertebrates. Second, he had the opportunity to think and write on board,[17] and within a year he was venturing to hope that he could do 'some original work in natural history'.

Third, he had the example of the sailors to live up to. The voyage taught him the 'characteristic qualities of the great number of sailors' – 'good humoured patience, unselfishness, the habit of acting for himself, and of making the best of everything, or contentment'.[18] Experience and practical competence were what mattered at sea. As Huxley later said of sailors on the *Rattlesnake*, they were 'emphatically men trained to face realities and to have a wholesome contempt for more talkers. Any one of them was worth a wilderness of phrase-crammed undergraduates.'[19] For Darwin, as for Huxley, knowledge came not from books written by desk-bound scholars, but from direct confrontation with the earth itself, and it was the *Beagle* voyage which led him to discover this.

DARWIN AND NATURE

How did Darwin react to the vast complexity of the world he explored? He went out as a natural historian (Whewell did not invent the word scientist until some years after the voyage was over), with his hammer, barometer, clinometer, butterfly net and vasculum: his aim was to define, identify, measure and enumerate the great diversity of nature. Some even then saw such an aim as shattering the sense and unity of the organic world with all its intricate connections, wrenching components out of context and depriving them of their intrinsic meaning. But for Darwin the perception of nature was also a deeply emotional and personal experience.

We see it first at St Iago in the Cape Verde Islands – black volcanic rocks

[16] *Ibid.*, volume 1, p. 221.
[17] N. Barlow, ed., *op. cit.* (note 7), p. 59.
[18] N. Barlow, ed., *op. cit.* (note 1), p. 430.
[19] L. Huxley, ed., *Life and letters of Thomas Henry Huxley* (London: Macmillan, 2 volumes, 1900), volume 2, p. 280.

largely destitute of vegetation: 'There was a grandeur in such scenery, and to me the unspeakable pleasure of walking under a tropical sun on a wild and desert island.'[20] Elsewhere he 'first saw the glory of tropical vegetation. Tamarinds, Bananas and Palms were flourishing at my feet. I returned to the shore, treading on volcanic rocks, hearing the notes of unknown birds, seeing new insects fluttering about still newer flowers. It has been for me a glorious day, like giving to a blind man eyes – he is overwhelmed with what he sees and cannot justly comprehend it.'[21] When he reached Brazil he speaks of 'the sublime solitude of the forest' and the unspeakable pleasure it causes him.[22] 'The mind is a chaos of delight, out of which a world of future and more quiet pleasure will arise.'[23] At the end of the voyage he recalled some of the most moving landscapes as 'temples filled with the varied productions of the God of Nature. No one can stand unmoved in these solitudes, without feeling that there is more in man than the mere breath of his body.'[24]

The intensity of feeling was matched by the sadness of his final diary entry as the ship called once more at Brazil before making for England.

> In the last walk I took, I stopped again and again to gaze on such beauties, and tried to fix for ever in my mind, an impression which at the time I knew must sooner or later fade away. The forms of the Orange tree, the Cocoa nut, the Palms, the Mango, the Banana, will remain clear and separate, but the thousand beauties which unite them all into one perfect scene must perish: yet they will leave, like a tale heard in childhood, a picture full of indistinct but most beautiful figures.

– and then he abruptly breaks the spell: 'In the afternoon, weighed anchor, and stood out to sea.'[25]

My purpose is to note the profound paradox of Darwin's attitude to nature and to science, and the complementarity of his modes of approach to landscape, landforms and life.[26] On the one hand, Darwin's reaction was emotional and integrative: he grasped, like Humboldt, the harmony and wholeness of the Cosmos. His was also a largely visual apprehension: he grappled to describe it. In a sense, Darwin's vision was epitomised in the

[20] N. Barlow, ed., *op. cit.* (note 1), p. 26.
[21] *Ibid.*, pp. 24–25: entry for 16 January 1832.
[22] N. Barlow, ed., *op. cit.* (note 7), p. 70.
[23] N. Barlow, ed., *op. cit.* (note 1), p. 32.
[24] *Ibid.*, p. 427: September 1836.
[25] *Ibid.*, pp. 416–417.
[26] F. Darwin, ed., *op. cit.* (note 11), volume 2, p. 315.

Romantic notion of sublimity, and it is reflected in his vocabulary: he finds stillness, desolation, solitude, nobility, gloom, wild luxuriance, grandeur; it excited feelings of wonder, awe, admiration, devotion, profundity. *Place* was both the cause and the focus for these perceptions.[27]

But on the other hand there was his analytical, instrumental, scientific approach, measuring and comparing, dissecting objects out from their background for closer investigation. The visually apprehended landscape became a conceptualized structure not limited to what could be directly observed. His perspective thus expanded from the here and now, bounded by the horizon and the heavens, to a vast panorama in space and time: a vision on a continental, indeed a global scale. It was here that instead of harmony and integration Darwin saw disequilibrium, struggle, change, constant diversification, the driving force of evolution. The *place* observed in the real world became transformed into the concept of the *niche*: the physical landscape and its emotional content became simultaneously a mental landscape intellectually perceived. Yet Darwin *saw* it still. Throughout all his works we are constantly invited to *see* with him, first with the eye the landscape he observes, then with the mind the insights he develops.

DARWIN AND SCIENTIFIC METHOD

This duality of vision had important consequences for how Darwin did science, and indeed what he saw science to be. We can resolve the problem into the question of the status of facts – facts as infallible, the building blocks of science, inductively assembled to generate descriptive laws; or facts as statements given meaning and sense only by their theoretical connotations.

Darwin, of course, pretended he believed in the former view, but in reality acted in accordance with the latter. A famous passage in his *Autobiography*, written in the last few years of his life, tells how he would have us believe in the genesis of *The origin of species*: 'After my return to England, it appeared to me that by following the example of Lyell in Geology, and by collecting all facts which bore in any way on the variation of animals and plants . . . some light might perhaps be thrown on the whole subject. . . . I worked on true Baconian principles, and without any theory collected facts on a wholesale scale.'[27] As a result, 'My mind seems to have become a kind of machine for grinding general laws out of large collections of facts.'[28]

[27] J. Paradis, 'Darwin and landscape', *Annals of the New York Academy of Sciences*, 360 (1981), 85–109.

[28] N. Barlow, ed., *op. cit.* (note 2), pp. 119 and 139.

Of course it was not like that at all. Darwin was impelled towards evolutionary ideas on the *Beagle* voyage, started his notebooks in 1837, read Malthus in 1838, wrote his abstract in 1842 and his sketch in 1844 – everything he did for over twenty years before 1859 was conditioned by his belief in evolution and its mechanism. That is what made his 'facts' significant, why he collected them, and why he knew what to collect. Darwin's invocation of Lyell is quite misleading, for with Lyell too ideas had primacy and facts were marshalled to support them. Indeed, for Adam Sedgwick this was both Lyell's characteristic and his tragedy: poor Lyell, he wrote to David Livingstone in 1865, 'he never could look steadily in the face of nature, except through the spectacles of an hypothesis.'[29] No one knew this better than Darwin himself. FitzRoy (of all people) had given him volume 1 of *Principles of geology* before the *Beagle* set sail; volume 2 from Henslow reached him in Montevideo; and volume 3 caught up with him at Valparaiso. As Darwin wrote to Leonard Horner in 1844, 'The great merit of the *Principles* was that it altered the whole tone of one's mind, and therefore that, when seeing a thing never seen by Lyell, one yet saw it partially through his eyes.'[30]

No finer example of the primacy of theory over fact can be given than in the accounts of Darwin's excursions to North Wales. The first was with Sedgwick himself in 1831. 'Neither of us saw a trace of the wonderful glacial phenomena all around us; we did not notice the plainly scored rocks, the perched boulders, the lateral and terminal moraines. Yet these phenomena are so conspicuous that . . . a house burned down by fire did not tell its story more plainly than did this valley.'[31] Agassiz brought his theory of continental glaciation to Britain in 1840, and when Darwin again visited Wales in 1842 it was to see a landscape transformed.[32] No wonder Francis Darwin tells us that his father often remarked that no one could be a good observer unless he was an active theorist;[33] indeed, he made this quite explicit in a letter to Lyell shortly after *The origin* was published.[34] To nineteenth-century philosophers of science – the Whewells, the Herschels –

[29] *Ibid*., p. 139.

[30] J. W. Clark and T. McK. Hughes, eds, *The life and letters of the Reverend Adam Sedgwick* (Cambridge: Cambridge University Press, 2 volumes, 1890), volume 2, p. 412.

[31] F. Darwin and A. C. Seward, eds, *More letters of Charles Darwin: a record of his work in a series of hitherto unpublished letters* (London: John Murray, 2 volumes, 1903), volume 2, p. 117.

[32] N. Barlow, ed., *op. cit*. (note 2), p. 70.

[33] P. H. Barrett, 'The Sedgwick–Darwin tour of North Wales', *Proceedings of the American Philosophical Society*, 118 (1974), 146–164.

[34] F. Darwin, ed., *op. cit*. (note 11), volume 1, p. 149.

this was a rather scandalous state of affairs. Let us see what it meant in Darwin's own scientific work during the *Beagle* years.

The process of development of Darwin's theory of coral reefs is well known.[35] He had been much impressed with the evidence of recent uplift throughout Patagonia and Chile, and the great earthquake of 20 February 1835, which destroyed Concepción, demonstrated the mechanism. FitzRoy subsequently found putrid fish and molluscs 3 m above high-water mark.[36] Later Darwin climbed to a height of 4,000 m in the Andes, where he also found marine fossils, and he speculated that the continuing uplift of the mountains must be balanced by some equivalent subsidence elsewhere. He found this in the ocean basins. 'The whole theory was thought out on the coast of South America before I had seen a true coral reef. I had therefore only to verify and extend my views by a careful examination of living reefs.'[37] From South America he sailed through the Tuamotus, seeing atolls from the deck of the ship, and then from the mountains of Tahiti looked across to the intermediate stage of a reef-encircled island at Moorea. He wrote the theory out in full in November 1835, on passage from Tahiti to New Zealand. By the time he made his only visit to an atoll, at Cocos-Keeling in April 1836, there was little left to do. Nevertheless, he was 'glad we have visited these Islands; such formations surely rank high amongst the wonderful objects of this world. It is not a wonder which at first sight strikes the eye of the body, but rather after reflection the eye of reason.'[38]

The theory was thus a mental construct of the sequential development of forms through time: its argument was deductive, was not derived from observation, and was not susceptible to direct test. The movement of subsidence, if coral growth was to keep pace with it, would have to be too slow to be observable. Such evidence for subsidence as there was (in, for example, the erosion of shorelines of coral islands) was more readily explicable by other means, such as storm activity: it is notable that the owner of Cocos-Keeling, Clunies Ross, dismissed the whole theory out of

[35] D. R. Stoddart, '*Coral Islands* by Charles Darwin, with an introduction, map and remarks', *Atoll Research Bulletin*, 88 (1962), 1–20.

[36] N. Barlow, ed., *op. cit.* (note 7), p. 119.

[37] N. Barlow, ed., *op. cit.* (note 2), pp. 98–99.

[38] N. Barlow, ed., *op. cit.* (note 1), pp. 399–400: entry for 12 April 1836. See also H. E. Gruber and V. Gruber, 'The eye of reason: Darwin's development during the *Beagle* voyage', *Isis*, 53 (1962), 186–200, and D. R. Stoddart, 'Darwin and the seeing eye', *Museums and Art Galleries Board of the Northern Territory [Darwin] Charles Darwin Annual Lecture 1984* (in press).

hand.[39] And the unequivocal evidence generally indicated recent uplift rather than subsidence. Indeed, support for the theory was largely conceptual rather than observational. It had a plausible logic, but it could hardly have been seriously considered had this not been so. It had the great advantage that it made sense of geodynamic patterns in the Pacific basin: indeed Darwin made this the theme of his first paper on the subject in 1838.[40] But above all it supplied an essential link in a wider scientific controversy.

To understand why this was so, it is necessary to consider Darwin's visit to the Galapagos Islands in October 1835. These are active volcanic islands, and Darwin correctly inferred that in the recent geological past they had probably not existed above sea level at all. Nevertheless they had an extraordinary endemic biota. One hundred of 175 plant species he collected were new; 25 of the 26 land birds, including the dozen or so finches, were endemic. There were also the tortoises and the iguanas. Darwin realized that many were peculiar not just to the archipelago but even to individual islands. No wonder he commented that 'The natural history of these islands is eminently curious and well deserves attention. . . . The archipelago is a little world within itself. . . . Here both in space and time we seem to be brought somewhat nearer to that great fact – that mystery of mysteries – the first appearance of new beings on this earth.'[41] 'If there is the slightest foundation for these remarks the Zoology of Archipelagoes will be well worth examining: for such facts would undermine the stability of species.'[42] Notice his style: the intimacy (we are invited to join him); the visual impact as we are invited to see the landscape with him; the grounding in the concrete phenomena of lava, birds, reptiles; the conceptualization of space and time; and the appeal to an almost mystic vision of the mystery of

[39] J. C. Ross, 'Review of the theory of coral-formation set forth by Ch. Darwin in his book entitled: Researches in Geology and Natural History', *Natuurkundig Tijdschrift voor Nederlandsch-Indie*, 8 (1855), 1–43.

[40] C. R. Darwin, 'On certain areas of elevation and subsidence in the Pacific and Indian Oceans, as deduced from the study of coral formations', *Proceedings of the Geological Society of London*, 2 (1838), 552–554.

[41] C. R. Darwin, *The voyage of the Beagle*, edited by L. Engel (Garden City, New York: Doubleday, 1962), pp. 378–379.

[42] N. Barlow, ed., *op. cit.* (note 7), pp. 246–247. The passage is from Darwin's Ornithological Notebooks, subsequently published by N. Barlow, ed., 'Darwin's ornithological notes', *Bulletin of the British Museum (Natural History), Historical Series*, 2 (7) (1963), 203–278, reference on p. 262. F. J. Sulloway, 'Darwin's conversion: the Beagle voyage and its aftermath', *Journal of the History of Biology*, 15 (1982), 325–396, has argued that these notes were written during the homeward sector of the Beagle voyage, between the Cape of Good Hope and Ascension Island, 18 June–19 July 1836.

mysteries, Herschel's phrase which appeared again in the first paragraph of *The origin of species* in 1859.

Now the fundamental question about the Galapagos plants and animals was how they had got there in the first place. By separate creation of each individual entity on each of the islands? By land bridges from the continents? Or, as Darwin believed, by overwater dispersal? The coral-reef theory provided a series of stepping stones of former high islands each now marked by an atoll, thus giving powerful impetus to the dispersal theory. In this argument geology and biology closely interact in the explanation of present distribution patterns, just as they do in the formation of the coral reefs themselves.[43]

IMPLICATIONS

The direction of my argument is clear. For Darwin, theory took precedence over fact; indeed fact only had significance in terms of theory. The basis for theory was initially insight and intuition, not induction: the procedure was the reverse of what the philosophers of science and indeed Darwin himself said it should be. The routine science came later, in building up the edifice of confirmation and qualification.

The method succeeded spectacularly in the case of coral reefs: Darwin 'reflected with high satisfaction' at the end of his life 'on solving the problem of coral islands'.[44] It succeeded too in the long road to *The origin of species*. But it was, equally, profoundly dangerous, even for a Darwin. There is no better example than Darwin's paper in 1839 on the 'parallel roads' of Glen Roy, explained by Darwin as identical to the raised marine beaches he had seen at Coquimbo in Chile but which in reality formed in ice-dammed glacial lakes. Plausible though it was, the paper was 'a great failure' and Darwin came to be 'ashamed of it'.[45]

This danger points to what I believe to be one of Darwin's most fundamental and lasting contributions. Since his theories were intuitional, we accept or reject them not primarily in terms of truth or falsity, right or wrong, but in terms of their *usefulness*. The coral-reef theory could not be proved to be either right or wrong but it had enormous explanatory value.

[43] D. R. Stoddart, 'Darwin, Lyell, and the geological significance of coral reefs', *British Journal for the History of Science*, 9 (1976), 199–218.

[44] N. Barlow, ed., *op. cit.* (note 2), p. 80.

[45] M. J. S. Rudwick, 'Darwin and Glen Roy: a "great failure" in scientific method?', *Studies in the History and Philosophy of Science*, 5 (1974), 97–185.

The current debate over whether Darwin was 'right' or 'wrong' over evolution similarly misses the point: these terms have no definable real meaning set against the achievement of *The origin*. I like to think that in this view of theories as tools for understanding problems rather than as shorthand factual generalizations, Darwin brought into science the common-sense empiricism of the sailors with whom he lived for nearly five years: that they gave him an operational rather than simply a logical view of life.

This is, of course, not a comfortable interpretation for any of us. We all have filing cabinets filled with facts. We have batteries of techniques for manipulating them. But, if we are honest, we also know that there are very few occasions when we had an idea, when something truly new occurred to us. Perhaps we ought to think more deeply in education and research about the importance of ideas and the triviality of facts: too many books, too much xeroxing, too many examinations, too little thought, too little contact I fear – and that too remote – with the great world of nature. Is it coincidence that all the great field naturalists of the eighteenth and nineteenth centuries engaged in their life's work, often without formal training, while they were very young? Darwin and Hooker were both 22, Huxley was 21, Wallace and Georg Forster were only 15. FitzRoy himself as captain of the *Beagle* was only 26, and Darwin thought him younger. None had their observation stereotyped by the process of education; indeed, fieldwork was their education. We need to rediscover Darwin's vision, both the profundity and intensity of his response to the world of nature, and the daring of his speculation.

Let us return to Darwin himself. First, Darwin and the ocean world. As he sailed home towards the Channel at the end of voyage, he asked himself, 'What are the glories of the illimitable ocean?' His answer was emotional: 'A tedious waste, a desert of water . . . [but also] a moonlight night, with the clear heavens, the dark glittering sea, the white sails filled by the soft air of a gently blowing trade wind; a dead calm, the heaving surface polished like a mirror, and all quite still except the occasional flapping of the sails . . . a squall, with its rising arch and coming fury, or the heavy gale and mountainous waves'.[46] Grandeur, then, in this view of life, as Darwin phrased it so memorably in the famous last paragraph of *The origin of species*:

> Thus from the war of nature, from the famine and death, the most exalted object which we are capable of conceiving, namely, the

[46] N. Barlow, ed., *op. cit.* (note 1), p. 426.

> production of the higher animals, directly follows. There is grandeur in this view of life, with its several powers, having been originally breathed into a few forms or into one; and that, whilst this planet has gone on according to the fixed law of gravity, from so simple a beginning endless forms most beautiful and most wonderful have been, and are being evolved.[47]

Whether he was right or whether he was wrong in this or that particular does not, it seems to me, matter. His achievement was not to provide the answers, but to show us, during the *Beagle* voyage and after, how to seek them. This is surely what science is all about, on whatever level we practise it, and it is something we all need to demonstrate anew.

[47] C. R. Darwin, *On the origin of species by means of natural selection, or the preservation of favoured races in the struggle for life* (London: John Murray, 1859), p. 490.

11

Organism and Ecosystem as Geographical Models

INTRODUCTION

Geography and ecology are concerned with the distribution, organization and morphology of phenomena on the surface of the earth,[1] and both disciplines have developed similar concepts and techniques to handle similar problems.[2] Overt geographical interest in ecological techniques, however, has been largely confined to a small group of biogeographers, whose influence on the rest of the subject has been marginal, and to a group of American sociologists who sought for a time to restate the aims of human geography in terms of human ecology.

The influence of biological concepts in geography, however, has been both deeper and more pervasive than explicit reference might suggest. Thus, in spite of a traditional insistence on the importance of areal differentiation as a methodological framework for geography, derived by Hartshorne from the work of von Richthofen and Hettner, much geographical work in the past hundred years has taken its inspiration directly from Darwin and the biological revolution which he began. Elsewhere I have indicated some of the main strands in the evolutionary impact on geographical thought since 1859: particularly the emphasis on changing form through time, expressed in organic analogies of ageing; the popularity of natural selection and environmental models, particularly in early American human geography; and, latterly, the application of Haeckel's concept of ecology and Tansley's of the ecosystem, which may be traced back to the third chapter of *The origin of species.*[3]

[1] L. B. Slobodkin, 'Energy in animal ecology', *Advances in Ecological Research*, 1 (1962), 69–101.

[2] W. Bunge, 'Patterns of location', *Michigan Inter-University Community of Mathematical Geographers Discussion Paper* 3 (1964), 1–36.

[3] See chapter 8.

This chapter treats the biological impact on a methodological level. In Kuhn's[4] terminology, we are concerned with the evolution of geographic *paradigms* of largely biological inspiration: paradigms, or interrelated networks of concepts on a sufficiently general level, which serve to define, at least for a time, the nature of geographical goals and the conventional frameworks within which these are pursued. The paradigms examined here are those of *organism* and *ecosystem*: they are not the only ones which may have been chosen, nor necessarily the most influential, but both concern fundamental issues in the methodology of geography. The organic paradigm in particular includes several examples of smaller-scale paradigms which can be defined with greater precision: Davis's cyclic and Clements's successional concepts among them. A distinction is thus made between such generalized conceptual models, for which the term paradigm is usefully employed, and the puzzle-solving procedures (again to use Kuhn's terminology), within the paradigm framework, with which most geographic work is concerned. Ecology itself, especially in recent years, has become more concerned with empirical studies of a puzzle-solving nature, as in the problem of the definition and statistical treatment of plant and animal communities:[5] and many of these techniques are of direct geographic interest at a comparable level of enquiry.[6] Puzzle-solving procedures of this sort, leading to the development of explanatory or predictive models for particular problems, depend on the acceptance of common standards of procedure and of common scientific aims, but they only become of methodological importance when they require the modification of the prevailing paradigm itself.

Organism and ecosystem are of interest as alternative approaches to a central theme in geographical enquiry: that of the relationship of man and environment in area. Thus Hettner states the classical position when he writes that 'both nature and man are intrinsic to the particular character of areas, and indeed in such intimate union that they cannot be separated from each other.'[7] It is also in this relationship that geography has faced two of its

[4] T. S. Kuhn, *The structure of scientific revolutions* (Chicago: University of Chicago Press, 1962). The reader will recall from chapter 1 that I would now take a less enthusiastic view of Kuhn's analysis.

[5] P. Grieg-Smith, *Quantitative plant ecology* (London: Butterworth, 4th edition, 1983); K. A. Kershaw, *Quantitative and dynamic ecology* (London: Edward Arnold, 1954); C. B. Williams, *Patterns in the balance of nature: and related problems in quantitative ecology* (London: Academic Press, 1964).

[6] D. W. Harvey, 'Models of the evolution of spatial patterns in human geography', in *Models in geography*, edited by R. J. Chorley and P. Haggett (London: Methuen, 1965), 549–608.

[7] A. Hettner, 'Das Wesen und die Methoden der Geographie', *Geographische Zeitschrift*, 11 (1905), 545–564, 615–629, 671–686; reference on p. 554.

most difficult methodological problems, of the dualism between man and environment, and of that between human and physical geography.[8] Both organic and ecosystem concepts have been used to overcome these problems, and to provide coherent frameworks for the organization of geographical data. This essay deals with the nature of these borrowing from the biological sciences, the way in which geographic interpretations have changed with developments in the biological sciences themselves, and the potential paradigmatic value of biological models in geographic methodology.

THE ORGANIC ANALOGY

Organic analogies have been used as 'explanations' of the real world since classical times. In the biological sciences in particular there has been much controversy between mechanists, vitalists and organicists over the reducibility of biological concepts and the scope and significance of organic holistic models.[9] Clements's work in plant ecology may be taken as an example of the use of organic analogy in ecology itself. Clements, following Warming, developed the ideas of climax and succession in vegetation, but went beyond the empirical observation of such changes to assert that 'the developmental study of vegetation necessarily rests upon the assumption that the unit or climax formation is an organic entity. As an organism the formation arises, grows, matures, and dies. . . . Each climax formation is able to reproduce itself.'[10] The plant community is 'a complex organism, or superorganism, with characteristic development and structure. . . . It is more than the sum of its individual parts . . . it is indeed an organism of a new order.'[11]

As with W. M. Davis's scheme of the geographical cycle,[12] to which Clements's successional model bears close resemblance, Clements and his

[8] R. Hartshorne, *Perspective on the nature of geography* (London: John Murray, 1959); see pp. 65–80.

[9] J. Loeb, *Untersuchungen zur physiologischen Morphologie der Thiere*, volume 2: *Organbildung und Wachsthum* (Wurzburg: G. Hertz, 1892); E. Nagel, 'Mechanistic explanation and organismic biology', *Philosophy and Phenomenonological Research*, 11 (1951), 327–338; L. von Bertalanffy, *Modern theories of development: an introduction to theoretical biology* (London: Oxford University Press, 1933; reissued New York: Harper, 1962). See also D. C. Phillips, 'Organicism in the late nineteenth and early twentieth centuries', *Journal of the History of Ideas*, 31 (1970), 413–432.

[10] F. E. Clements, *Plant succession and indicators* (Washington: H. W. Wilson, 1928), p. 3.

[11] F. E. Clements and V. Shelford, *Bio-ecology* (New York: J. Wiley, 1939), p. 24.

[12] W. M. Davis, 'The geographical cycle', *Geographical Journal*, 14 (1899), 481–504.

followers regarded the organismic idea as a revealing conceptual framework for understanding the real world: 'this concept is the "open sesame" to a whole new vista of scientific thought, a veritable *magna carta* for future progress.'[13] European ecologists never wholeheartedly accepted Clements's insistence on succession or his metaphysical interpretation of community: while succession may be adequately demonstrated on a small scale (as in marshland, on sand dunes or in pieces of dung), organismic views of community have little practical relevance at such levels; while at larger scales field workers concentrated on empirical descriptive studies in which organismic concepts had no place.[14] Similarly in pedology, the earlier organismic views of Marbut, Shaler and Whitney have been rejected by such later workers as Jenny[15] and particularly Nikiforoff.[16]

For the early ecologists, the value of organic analogies lay in their ability to bring large quantities of discrete and often apparently unrelated data into meaningful relationship, to emphasize the organization and functional relationships existing in nature, which, once recognized, served as at least a partial explanation of their complexity. Organic analogies were particularly successful when the problems were most complex and the analytical techniques too undeveloped to produce substantive results. Hence, with the rise of descriptive ecology, with its diverse body of sophisticated descriptive and analytical techniques, vague analogy no longer served as a valid explanation of natural complexity. To take one example, while Clements emphasized the organic nature of the great vegetational units of the world, such as the selva and the coniferous forests, more detailed studies, as in East Asia, have shown that these 'do not represent the discrete units of vegetation, but are no more than the vague patterns joined together by gradual transition.'[17] Figure 9 shows the overlapping distribution of 24 conifer species along a thermal gradient in Japan, and suggest that in this case the vegetation units have no discrete existence. Similar cases may be cited from animal ecology.[18]

[13] F. E. Clements and V. Shelford, *op. cit.* (note 11), p. 24.

[14] R. H. Whittaker, 'A consideration of climax theory: the climax as a population and pattern', *Ecological Monographs*, 23 (1953), 41–78; C. H. Muller, 'Science and philosophy of the community concept', *American Scientist*, 46 (1958), 294–308; E. W. Selleck, 'The climax concept', *Botanical Review*, 26 (1960), 534–545.

[15] H. Jenny, *Factors of soil formation: a system of quantitative pedology* (New York: McGraw-Hill, 1941).

[16] C. C. Nikiforoff, 'Reappraisal of the soil', *Science*, 129 (1959), 1886–196.

[17] T. Kira, H. Ogawa and K. Yoda, 'Some unsolved problems in tropical forest ecology', *Proceedings, 9th Pacific Science Congress*, 4 (1962), 124–134; reference on p. 125.

[18] E. J. Clark, 'Studies in the ecology of British grasshoppers', *Transactions of the Royal Entomological Society*, 99 (1946), 173–222.

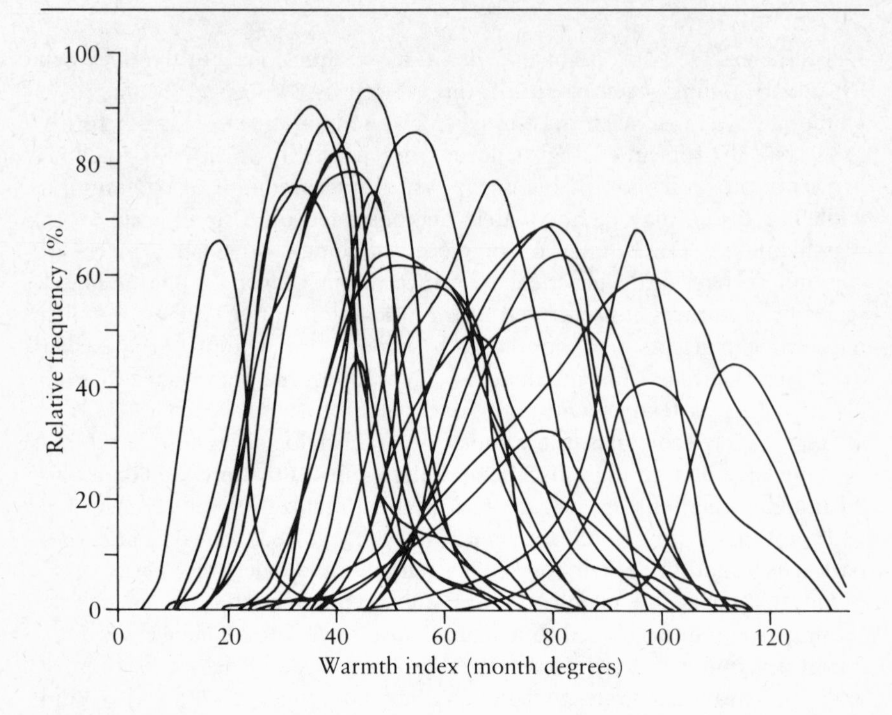

Figure 9 The distribution of 24 conifer species along a thermal gradient in the Central District, Japan.

Source: T. Kira, H. Ogawa and K. Yoda, 'Some unsolved problems in tropical forest ecology', *Proceedings, 9th Pacific Science Congress*, 4 (1962), 124–134.

THE ORGANIC ANALOGY IN GEOGRAPHY

This brief discussion demonstrates the appeal of organic analogy in ecological problems similar to those faced by geography. In geography itself, organic analogies are of considerable antiquity, but received fresh impetus from the Darwinian revolution, and maintained popularity until the 1930s. Carl Ritter's inspiration was in no sense biological, but for him and his followers the organic model not only provided a satisfying explanation of the relations of man and nature, but provided a religious and moral justification also. The earth for him was not 'a mere dead, inorganic planet, but an organism, a living work from the hand of a living God',[19] in which both animate and inanimate components form 'in a higher and

[19] C. Ritter, *The comparative geography of Palestine and the Sinaitic Peninsula*, translated by W. L. Gage (Edinburgh: William Blackwood, 4 volumes, 1866), reference in volume 2, p. 4.

comprehensive sense a cosmical life . . . one great organism'.[20] Ritter's pupil, Arnold Guyot, considered that:

> Few subjects seem more worthy to occupy thoughtful minds than the contemplation of the great harmonies of nature and history. The spectacle of the good and the beautiful in nature reflecting everywhere the idea of the Creator, calms and refreshes the soul. . . . Every being, every individual, necessarily forms a part of a greater organism than itself, out of which one cannot conceive its existence, and in which it has a special part to act. . . . All is order, all is harmony in the universe.[21]

Man himself, in Guyot's optimistic teleology, forms 'the bright consummate flower of this admirable organization'.[22] 'The continents are made for human societies, as the body is made for the soul. The conclusion is inescapable that the entire globe is a grand organism, every feature of which is the outgrowth of a definite plan of the all-wise Creator for the education of the human family, and the manifestation of his own glory.'[23]

More specifically, the organismic analogy has operated on three distinct levels in geographical work: those of the earth, its regions and its states; and on each level its use long predates Darwinian evolutionary theory.[24] Organic theories of the state and of the earth go back to classical and medieval times; they were revived by such philosophers as Hobbes; and were thoroughly worked out by Heinrich Ahrens in his *Organische Staatslehre* in 1850.[25] Much of this earlier work was abstract and

[20] C. Ritter, *Comparative geography*, translated by W. L. Gage (Edinburgh: William Blackwood, 1865), reference on p. 1.

[21] A. Guyot, *The earth and man: lectures on comparative physical geography* (London: R. Bentley, 1850), references on pp. 6 and 85. For analyses of 'wholeness' in geography, see A. Hettner, 'Der Begriff der Ganzheit in der Geographie', *Geographische Zeitschrift*, 40 (1934), 141–144; W. Volz, 'Ganzheit, Rhythmus und Harmonie in der Geographie', *Die Erde*, 3 (1951), 97–116; and D. Bartels, 'Der Harmoniebegriff in der Geographie', *Die Erde*, 100 (1969), 124–137.

[22] A. Guyot, *op. cit.* (note 21), p. 297.

[23] A. Guyot, quoted by C. R. Dryer, 'A century of geographic education in the United States', *Annals of the Association of American Geographers*, 14 (1924), 117–149; reference on p. 131.

[24] W. E. Ritter and E. W. Bailey, 'The organismal conception: its place in science and its bearing on philosophy', *University of California Publications in Zoology*, 31 (1928), 307–358; S. Günther, 'Der Erdkörper als Organismus: ein Beitrag zur Geschichte der Irrlehren in der physikalischen Geographie', *Naturwissenschaftliche Wochenschrift, Deutsche Gesellschaft für Volkstümliche Naturkunde in Berlin*, 1 (1902), 385–391.

[25] H. Ahrens, *Die organische Staatslehre auf philosophisch-anthropologischer Grundlage* (Wien: C. Gerold, 1850).

metaphysically teleological, as in Ritter's conception of terrestrial unity and Humboldt's cosmological philosophy, but from at least 1750 the analogy with the natural sciences of chemistry, mechanics and biology was being pursued more closely. Johann Bluntschli[26] imbued states with the attributes of human organisms, even to details of personality and sex, while a fundamental precept of Comte's positivist philosophy was that sociology could only be understood in biological terms.[27] Only after Darwin, however, were these somewhat metaphysical and often inchoate ideas given full expression, and social thought from about 1870 to 1900 was dominated by Darwinian thinking. In England and America Herbert Spencer and in France René Worms helped popularize organic analogies in the social sciences, and these retained vitality in geography long after they had been abandoned in other branches of human studies.

Among geographers it is to Butte[28] and especially to Ritter that the idea of the earth as a functioning organism may best be traced. Similar views were expressed by Alexander von Humboldt and half a century later by Vidal de la Blache, who acknowledged his indebtedness to Ritter in a much quoted aphorism: 'La Terre est un tout, dont les parties sont coordonnées.'[29] Vidal did not of course restrict the analogy to the earth: in his own substantive work, and in that of his pupils, the theme of harmonious interrelationships reappears at the state and local–regional level.[30] Organic concepts of terrestrial unity are found in the work of Brunhes[31] and Lucien Febvre,[32] and for a time in the United States also: 'The earth itself is an organism,'

26 J. Bluntschli, *Lehre vom modernen Stat*, volume 1: *Allgemeine Statslehre. Fünfte umgearbeiftete Auflage der ersten Bandes der allgemeine Statsrechts* (Stuttgart: J. G. Cotta, 1873).

27 F. W. Coker, 'Organizational theories of the state: nineteenth century interpretations of the state as organism or person', *Columbia University Studies in History, Economics, etc.*, 38 (2) (1910), 1–209.

28 W. Butte, *Die Statistik als Wissenschaft* (Landshut: J. Thomman, 1808).

29 P. Vidal de la Blache, 'Le principe de la géographie générale', *Annales de Géographie*, 5 (1896), 129–142, reference on p. 129; also 'Des caractères distinctifs de la géographie', *Annales de Géographie*, 22 (1913), 289–299, reference on p. 289.

30 P. Vidal de la Blache, *Tableau de la géographie de la France*, volume 1 of E. Lavisse, *Histoire de la France illustrée* (Paris: Hachette, 1911); *La France de l'Est (Lorraine-Alsace)* (Paris: A. Colin, 1917).

31 J. Brunhes, *Human geography: an attempt at a positive classification: principles and examples*, edited by I. Bowman and R. E. Dodge (Chicago: Rand-McNally, 1920), see pp. 26–27; *Human geography*, abridged by M. J.-B. Delamarre and P. Deffontaines (London: Harrap, 1952), see pp. 26–28.

32 L. Febvre, *A geographical introduction to history* (London: Kegan Paul, Trench, Trübner & Co., 1925), p. 137.

wrote Dryer in a more extreme statement, 'and geography is its anatomy, physiology and psychology.'[33]

The idea of the terrestrial organism served as a unifying theme in an increasingly particularistic discipline: as such, it was a synthetic rather than an analytical concept.[34] At the regional level, however, the organic analogy attracted more attention as an analytical tool, particularly in England. Herbertson raised the matter in his discussion of the 'major natural regions' in 1905, using the term 'macro-organism' for the 'complex entity' of physical and organic elements of the earth's surface: 'the soil itself the flesh, the vegetation its epidermal covering with its animal parasites, and the water the circulating life-blood automatically stirred daily and seasonally by the great solar heat. . . . If we regard the Earth as an individual, and these geographical regions, district, localities, as representing organs, tissues and cells, we perhaps get nearest to a useful comparison.'[35] The concept of the natural region was for Herbertson a tool for assimilating multitudes of diverse facts: within this macro-organism, men are the nerve-cells, in some cases forming amorphous groups, in others organized and specialized in function. In his last paper Herbertson goes on to make the organism analogy more specific: the natural region is comparable to the botanist's plant, the zoologist's animal, or the chemist's mass of matter: natural regions are 'definite associations of inorganic and living matter with definite structures and functions, with as real a form and possessing as regular and orderly changes as those of a plant or an animal', and like plants and animals they can be hierarchically ranked into species, genera, orders and classes.[36]

Similar thinking pervades later English work on regional methodology. Unstead, while building up a synthetic regional scheme from smaller units to larger ones, nevertheless regarded individual regions as living organisms, with distinctive morphology and physiography. He developed the analogy to include evolution, in the sense of increasing complexity, and pathology, in the case of conditions such as floods and droughts inimical to man.

[33] C. R. Dryer, 'Genetic geography: the development of the geographic sense and concept', *Annals of the Association of American Geographers*, 10 (1920), 3–16; reference on p. 13.

[34] J. Batalha-Reis, 'On the definition of geography as a science, and on the conception and description of the earth as an organism', *Report of the 6th International Geographical Congress, London, 1895* (1896), 753–766; also J. S. Duncan, 'The superorganic in American cultural geography', *Annals of the Association of American Geographers*, 70 (1980), 181–198.

[35] A. J. Herbertson, 'The higher units: a geographical essay', *Scientia*, 14 (1913), 203–213, reference on p. 205; reprinted in *Geography*, 50 (1965), 332–342.

[36] A. J. Herbertson, 'Natural regions', *Geographical Teacher*, 7 (1913–1914), 158–163; see pp. 158–159, 161.

While compelled to admit that regions cannot be said to die, as do biological organisms, Unstead neatly sidestepped the issue by comparing continuity of existence in the region with that of the 'germ-plasm of organisms through the successive generations',[37] thus endowing regions at one time with the properties of individuals, at other times with those of populations. Stevens, while criticizing the idea of a 'natural' region, allows himself to sink even deeper into the organism analogy.

> Unity within diversity, active functioning, progressive evolution are all to be associated with an organized whole, and it is this organism we have to seek. . . . The organizing agent is the human community: community and region are but aspects of the same organism. . . . The unity of the natural geographical region is achieved, maintained and developed, by organization, by cohesion, and this cohesion is attained and extended by intercourse, at first within and later beyond the region, provided the organic development is an indigenous growth.[38]

The most extreme statement of the organismic concept in England came from the palaeontologist Swinnerton, who speaks of the relationship between man and land as analogous to that between oyster and shell. In anthropomorphic language (fighting a losing battle, adaptive response, adaptive radiation, secreting bricks and mortar), he asserts that 'those geographers who think of the world as an organism are making no fantastic comparison. Man in his relation to his environment is still a child of nature, for his reactions closely resemble those of other organisms . . . even at the limits of cultivation the cultural landscape reflects the activities of a living, throbbing organism.'[39] Similar, if more moderate, views were expressed in America by Finch and James, and in Germany by Gradmann,[40] Hans Bluntschli[41] and others.[42]

Organic theories of the state go back at least to Plato, and formed the

[37] J. F. Unstead, 'Geographical regions illustrated by reference to the Iberian Peninsula', *Scottish Geographical Magazine*, 42 (1926), 159–170; see p. 168.

[38] A. Stevens, 'The natural geographical region', *Scottish Geographical Magazine*, 55 (1939), 305–317; see pp. 308–310.

[39] H. H. Swinnerton, 'The biological approach to the study of the cultural region', *Geography*, 23 (1938), 83–89; see pp. 88–89.

[40] R. Gradmann, 'Das harmonischer Landschaftsbilde', *Zeitschrift der Gesellschaft für Erdkunde zu Berlin*, 59 (1924), 129–147.

[41] H. Bluntschli, 'Die Amazonasniederung als harmonischer Organismus', *Geographische Zeitschrift*, 27 (1921), 49–67.

[42] W. Volz, 'Der Begriff der Rhythmus in der Geographie', *Mitteilungen der Gesellschaft für Erdkunde zu Leipzig*, 1923–1925 (1926), 8–41.

basis of Hobbes's *Leviathan* in pre-Darwinian times. In geographical terms, however, the concept owes its importance to Friedrich Ratzel, whose whole work is coloured by Darwinian and Spencerian evolutionary thinking[43] In his *Politische Geographie*, Ratzel states that 'the earth is for us an organism, not only because it is a union of the living Volk with a rigid soil, but because the union is strengthened by the effect of one upon the other to the extent that the two can no longer be visualized as separated from each other':[44] Ratzel's first chapter is entitled 'Der Staat als bodenständiger Organismus'. This mystical conception of the indivisibility of people and land thus goes far beyond the simple Spencerian analogy of lines of communication and arteries, seats of government and the brain, and so on, developed in *Principles of sociology*. The state organism depends on the fundamental properties of organization and interdependence; it then assumes properties of growth and competition, and in so doing goes beyond the organismic conceptions of earth and region. In a brief but well-known paper in 1896 Ratzel developed his seven laws of the growth of states, and went on to outline the subsequently notorious idea of *Lebensraum*: 'Just as the struggle for existence in the plant and animal world always centres about a matter of space, so the conflicts of nations are in great part only struggles for territory.'[45] While there is undoubtedly a danger that selective quotation of this sort may do violence to Ratzel's undoubtedly scholarly position, as both Broek[46] and Wanklyn[47] argue, it is clear that the organic analogy for Ratzel not only provided a simple and powerful model in analytical political geography, but provided an apparently scientific justification for political behaviour. Ratzel's political geography was welcomed by others, including Vidal de la Blache,[48] but Miss Semple, in her exposition of Ratzel's work, omitted the cruder Spencerian analogies as already abandoned even in

[43] J. Steinmetzler, 'Die Anthropogeographie Friedrich Ratzels und ihre ideengeschichtlichen Würzeln', *Bonner geographische Abhandlungen*, 19 (1956), 1–151; H. Wanklyn, *Friedrich Ratzel: a biographical memoir and bibliography* (Cambridge: Cambridge University Press, 1961).

[44] F. Ratzel, *Politische Geographie* (München: R. Oldenbourg, 1897), p. 4, quoted by J. Mattern, *Geopolitik: doctrine of national self-sufficiency and empire* (Baltimore: Johns Hopkins Press, 1942), p. 96.

[45] F. Ratzel, 'Studies in political areas', *American Journal of Sociology*, 3 (1897–1898), 279–313, 449–463; see p. 458.

[46] J. O. M. Broek, 'Friedrich Ratzel in retrospect' (mimeographed paper, 1954); abstract in *Annals of the Association of American Geographers*, 44 (1954), 207.

[47] H. Wanklyn, *op. cit.* (note 43).

[48] P. Vidal de la Blache, 'La géographie politique à propos des écrits de M. Frédéric Ratzel', *Annales de Géographie*, 8 (1898), 97–111; see pp. 108–109.

sociology.[49] Ratzel's views served as a source for the *Geopolitik* developed in Europe between the wars in the writings of Rudolf Kjellen and Karl Haushofer. For Kjellen, the state was 'a biological manifestation or form of life', 'deeply rooted in historic and factual realities', endowed not only with morality but also with 'organic lusts'.[50] Modern political geography is at pains to emphasize its empirical nature and freedom from the organismic preconceptions which stem back to Ratzel.[51]

THE ORGANIC ANALOGY: COMPONENTS AND CRITICISM

Organic analogies have thus been used by geographers at a variety of levels, and it is useful to enumerate briefly the organic properties said to be possessed by geographic areas. The fundamental criterion used by geographers is the possession of organization of constituent components into a functionally related, mutually interdependent complex, which in spite of a continuous flux of matter or energy through it remains in an equilibrium condition. This complex in equilibrium is said to possess the properties of adaptation, cohesion, reaction and recreation.[52] Many writers stress, with Fleure, that 'the whole is usually more than the sum of its parts; as a whole it has functions and relations which may not be functions or relations attached to any of its parts,'[53] and Anuchin argues that:

> We are fully justified in regarding as a certain unity such fundamentally different forms of matter, as dead forms, living forms, social forms, which together make up a certain whole of the material world. The whole make-up of all these forms is of course not a simple sum of

[49] E. C. Semple, *Influences of geographic environment: on the basis of Ratzel's system of anthropo-geography* (New York: H. Holt & Co., 1911), p. v.

[50] R. Kjellen, *Der Staat als Lebensform* (translation by J. Sandmeier of *Staten som Lifsform* [Stockholm, 1916]) (Berlin-Grunewald: K. Vowinckel, 4th edition, 1924), p. 203. Commentary by H. W. Weigert, *Generals and geographers: the twilight of geopolitics* (London: Oxford University Press, 1942), pp. 106–107; R. H. Fifield and G. E. Pearcy, *Geopolitics in principle and practice* (Boston: Ginn, 1944), p. 11; and C. Troll, 'Geographic science in Germany during the period 1933–1945: a critique and justification', *Annals of the Association of American Geographers*, 39 (1949), 99–137.

[51] H. W. Weigert, *Principles of political geography* (New York: Appleton-Century-Crofts, 1957).

[52] C. Vallaux, 'La surface terrestre assimilée à un organisme', in *Les sciences géographiques* (Paris: F. Alcan, 1925), chapter 2, pp. 28–57.

[53] H. J. Fleure, *An introduction to geography* (London: Ernest Benn, 1929), p. 13.

> these forms but more than the sum in view of the process of interaction within the whole.[54]

One may here distinguish the influence of biological vitalism,[55] and of the holistic philosophy of Bergson,[56] Smuts[57] and Whitehead.[58]

Secondly, and on a more empirical level, there is the emphasis on morphology, stemming from Sauer's early work on landscape. With recent quantitative studies of pattern in landscape, there has been renewed interest in purely morphological studies, taking its inspiration from the work of D'Arcy Wentworth Thompson.[59] Sauer's work, however, led directly to a third organic property, that of development in the sense of change through time.[60] The work of the Berkeley school has been fundamentally of the historical type, tracing the geographic influence of culture groups on the landscape through time. Such work finds close parallel in Derwent Whittlesey's concept of sequent occupance, explicitly derived from the ecological concept of succession.

> Human occupance of area, like other biotic phenomena, carried within itself the seed of its own transformation. . . . Because of its obedience to rule analogous to that governing human organisms, the study of human occupance of area rests on secure foundations. . . . The view of geography as a succession of stages of human occupance establishes the genetics of each stage in terms of its predecessor.[61]

[54] V. A. Anuchin, *Teoreticheskiye problemy geografii* (Moscow: Gosudarstvenoye Izdatel-'stvo Geograficheskoy Literatury, 1960), p. 222; translated as *Theoretical problems of geography* (Columbus: Ohio State University Press, 1977).

[55] H. A. E. Driesch, *The science and philosophy of the organism: the Gifford lectures delivered before the University of Aberdeen in 1907 and in 1908* (London: Adam & Charles Black, 2 volumes, 1908–1909).

[56] H. Bergson, *L'Évolution creatrice* (Paris: F. Alcan, 1907); translated by A. Mitchell as *Creative evolution* (London: Macmillan, 1911).

[57] J. C. Smuts, *Holism and evolution* (New York: Macmillan, 1926).

[58] A. N. Whitehead, *Science and the modern world* (New York: Macmillan, 1925).

[59] D'Arcy Wentworth Thompson, *On growth and form* (Cambridge: Cambridge University Press, 1917), and also W. R. Tobler, 'D'Arcy Thompson and the analysis of growth and form', *Papers of the Michigan Academy of Science, Arts and Letters*, 48 (1963), 385–390.

[60] C. O. Sauer, 'The morphology of landscape', *University of California Publications in Geography*, 2 (1925), 19–54; see pp. 30, 41.

[61] D. Whittlesey, 'Sequent occupance', *Annals of the Association of American Geographers*, 19 (1929), 162–165. The idea has been explored more fully by M. W. Mikesell, 'The rise and decline of "sequent occupance": a chapter in the history of American geography', in D. Lowenthal and M. J. Bowden, eds, *Geographies of the mind: essays in historical geosophy* (New York: Oxford University Press, 1975), 149–169.

Similar ideas are of course found in historical geography and in the frontier hypothesis of F. J. Turner (though here bound up with other Darwinian ideas), as well as in contemporary plant ecology: much of what passed for 'dynamic ecology' at this time was in fact successional or historical interpretation.

Organic analogies have thus been promoted by a wide range of geographers, to whom the subject owes its present status as an independent science. One must therefore ask why such widely held views produced so little in the way of substantive results or new lines of investigation. The debate here closely parallels that between vitalists and mechanists in biology, and similar themes were also raised in philosophy. The theme of organic unity is superficially satisfying, and has the appearance of profound insight, but as Beck states of Driesch's vitalistic views in biology, it poses no questions, and hence obtains no answers.[62] The vitalists' hypothesis cannot be tested by observation, for vitalist definitions, e.g. of entelechy, are metaphysical rather than operational in nature. The quality of 'explanation' is on a different level from that normally accepted in the physical and biological sciences. One of Vidal's own pupils, Camille Vallaux, while at first impressed by the aesthetic unity and harmony of the idea of the terrestrial whole, came to see that no fresh insights could be obtained from it, and that all that remained was a certain poetic attractiveness:

> Qu'on la mette, si l'on veut, au Panthéon scientifique, avec tous les honneurs funèbres; mais qu'on néglige pas de sceller sa challe; qu'elle ne soute point du tombeau.[63]

The vitalist position in biology was fundamentally to observe life, to fail to understand it, and to call up essentially undemonstrable, and hence unprovable, causes – the *élan vital* of Bergson and the *entelechy* of Driesch – to fill the gap. Such procedures preclude the objective formulation and testing of hypotheses, because they lie outside hypothesis;[64] and the same objection holds for organismic beliefs in geography. The use of such terms as quasi-organism or super-organism simply defers the problem of definition.[65]

[62] W. S. Beck, *Modern science and the nature of life* (New York, Harcourt, Brace, 1957).

[63] C. Vallaux, *op. cit.* (note 52), p. 49.

[64] E. Nagel, *op. cit.* (note 9); E. Caspari, 'On the conceptual basis of the biological sciences', in R. G. Colodny, ed., *Frontiers of science and philosophy* (London: Allen & Unwin, 1964), 131–145.

[65] A. Remane, 'Die Gemeinschaft als Lebensform in der Natur', *Kieler Blätter*, 2 (1939), 43–61; 'Ordnungsformen der lebenden Natur', *Studium Generale*, 3 (1950), 404–410.

The major objection to organic interpretations in geography, however, is that they give no assistance in actual investigation: they are synthetic, not analytical conceptions. Thus, as geographic methodology of this type became increasingly metaphysical, so it became increasingly divorced from geographical writing. With a few exceptions, such as the regional monographs of the French school, regional geography became a stereotyped catalogue of categories of regional data, while the metaphysicians talked inspiringly of an organic unity which existed only in their methodological addresses.

The vitalist position in biology on which the extreme organismic view rests has been met and resolved, largely by recent work on molecular biology, and is no longer a live issue.[66] Even those biologists who will not admit to the more extreme mechanistic positions, such as von Bertalanffy in *Problems of life*,[67] take the view that an organism or whole is comprehended if not only its parts are known but also the relationships between them: whether this is a practical proposition in many branches of study at the moment does not affect the argument. Transferring this idea to geography removes the necessity for any metaphysical interpretation of organic unity. Further, as Friederichs[68] points out, most of the qualities by which the organism is defined (such as steady-state conditions, self-regulation and development) are not restricted to living organisms but are also possessed by non-living systems. Finally, the concept of the organism, and of organic unity, is largely, if not entirely, idiographic.[69] As such it can make little or no contribution to an increasingly nomothetic science: though this is not to deny its importance in the development of biological thought.[70] It is of interest that it is precisely where the idiographic conception of geography has been maintained longest, as at Oxford, that the idea of organic unity of regions has had its most favourable reception. With these criticisms, the organismic idea is reduced to a metaphor of dubious value, which hinges on gross formal and functional comparisons between living matter and complexly interrelated facts in areas. Organization and interrelation cannot be denied, but there were few geographers, at

[66] W. S. Beck, *op. cit.* (note 62).

[67] L. von Bertanlanffy, *Problems of life* (London: Watts, 1952).

[68] K. Friederichs, 'A definition of ecology and some thoughts about basic concepts', *Ecology*, 39 (1958), 154–159.

[69] W. R. Siddall, 'Idiographic and nomothetic geography: the application of some ideas in the philosophy and history of science to geographic methodology' (University of Washington, Ph.D. thesis, 1959).

[70] L. E. R. Picken, *The organization of cells and other organisms* (Oxford: Clarendon Press, 1960).

least before 1939, to follow Crowe in his emphatic denial that organization implied any kind of organism.[71] After 1939, with the eclipse of vitalism in biology and philosophy, and its disgrace in *Geopolitik*, the idea of organism has dropped out of geographical writing, except in occasional reverent mention of the work of Herbertson and Vidal de la Blache.

HUMAN ECOLOGY AND THE URBAN SOCIOLOGISTS

With the development of ecology as a formal branch of study, several workers attempted to restate the problem of the relationship between man and land less in terms of organic analogy or Davisian cause–effect determinism, than of the complexity of the interrelationships between the two and its ecological interpretation. This movement was led particularly by the Chicago group of urban sociologists, who coined the term *human ecology* as a scientific substitute for a discredited human geography.[72] Robert Ezra Park's programmatic statements of the scope of human ecology deal with the web of life, the balance of nature, concepts of competition, dominance and succession, biological economics and symbiosis – all concepts derived from plant and animal ecology.[73] For Park, human ecology investigated the processes involved in biotic balance, in which man interacts with nature through culture and technology, as demonstrated in urban life.[74] McKenzie expressed similar ideas, with a more economic

[71] P. R. Crowe, 'On progress in geography', *Scottish Geographical Magazine*, 54 (1938), 1–19.

[72] R. E. Park and E. W. Burgess, *Introduction to the sciences of sociology* (Chicago: University of Chicago Press, 1921).

[73] R. E. Park, 'Human ecology', *American Journal of Sociology*, 43 (1934), 1–15; 'Succession as an ecological concept', *American Sociological Review*, 1 (1936), 171–179. Park regarded ecology, however, as a nomothetic, geography as an idiographic subject. It is of great interest, in terms of intellectual linkages and the development of ideas, that Park carried out his doctoral research under Wilhelm Windelband, who had signalled this distinction in his rectoral address at Strasbourg in 1894 ('Geschichte und Naturwissenschaft', in W. Windelband, *Präludien* [Tübingen: J. Mohr, 1907], volume 2, pp. 355–379). At Heidelberg Hettner was simultaneously developing his neo-Kantian view of geography; he was closely associated with Windelband, and indeed was one of Park's examiners. The nature of Park's thought, and the relationship between human ecology and human geography, is beautifully analysed by J. N. Entrikin, 'Robert Park's human ecology and human geography', *Annals of the Association of American Geographers*, 70 (1980), 43–58.

[74] R. E. Park, 'The city: suggestions for the investigation of human behaviour in the city environment', *American Journal of Sociology*, 20 (1915), 577–612; R. E. Park, E. W. Burgess and R. D. McKenzie, *The city* (Chicago: University of Chicago Press, 1925); F. C. Evans, 'Ecology and urban areal research', *Scientific Monthly*, 73 (1951), 37–38; A. H. Hawley, 'The approach of human ecology to urban areal research', *Scientific Monthly*, 73 (1951), 48–59.

bias.[75] Human ecology developed a wide body of methodological literature,[76] some of which – dealing for example with problems of distance[77] or position[78] – was purely geographical in nature. Cities had of course already been interpreted by geographers in organic terms,[79] but on a Spencerian analogical level.

The potential of human ecology as stated in purely biological terms by Park and his school, proved attractive to geographers. H. H. Barrows, in his presidential address to the Association of American Geographers in 1923, stated that:

> geography is the science of *human ecology*. . . . Geography will aim to make clear the relationships existing between natural environments and the distribution and activities of man. Geographers will, I think, be wise to view this problem in general from the standpoint of man's adjustment to environment, rather from that of environmental influence. . . . The centre of geography is the study of human ecology in specific areas. This notion holds out to regional geography a distinctive field, an organising concept throughout, and the opportunity to develop a unique group of underlying principles.[80]

Barrow's address aroused little support among geographers,[81] at least

[75] R. D. McKenzie, 'The ecological approach to the study of human community', *American Journal of Sociology*, 30 (1924), 287–301; 'The scope of human ecology', *Publications of the American Sociological Society*, 20 (1926), 141–154; 'Demography, human geography, and human ecology', in L. L. Bernard, *The fields and methods of sociology* (New York: R. Long & R. R. Smith, 1934), 52–66.

[76] C. C. Adams, 'The relation of general ecology to human ecology', *Ecology*, 16 (1935), 316–335; A. E. Emerson, 'Ecology, evolution, and society', *American Naturalist*, 77 (1943), 97–118; R. Gerard, 'Organisation, society, and science', *Scientific Monthly*, 50 (1940), 340–350, 403–412, 530–535; A. H. Hawley, 'Ecology and human ecology', *Social Forces*, 22 (1944), 398–405; F. F. Darling, 'The ecological approach to the social sciences', *American Scientist*, 39 (1951), 244–254.

[77] R. D. McKenzie, 'Spatial distance', *Sociology and Social Research*, 13 (1929), 536–545.

[78] R. E. Park, 'The concept of position in sociology', *Publications of the American Sociological Society*, 20 (1925), 1–14.

[79] For example, H. J. Mackinder, *Britain and the British seas* (Oxford: Clarendon Press, 2nd edition, 1907), p. 257 (on London).

[80] H. H. Barrows, 'Geography as human ecology', *Annals of the Association of American Geographers*, 13 (1923), 1–14; see pp. 3, 9. For commentary, R. J. Chorley, 'Geography as human ecology', in R. J. Chorley, ed., *Directions in geography* (London: Methuen, 1973), 155–169.

[81] C. R. Dryer, *op. cit.* (note 33), p. 16; G. F. Carter, 'Ecology – geography – ethnobotany', *Scientific Monthly*, 70 (1950), 73–80. For commentary, G. Fuchs, 'Das Konzept der Okologie in der amerikanischen Geographie', *Erdkunde*, 21 (1967), 81–93.

methodologically, and only the Berkeley school, following Carl Sauer, has consistently sought to interpret the relationship between man and land in ecological terms, particularly in frontier areas of the Americas. Studies from Berkeley which make explicit mention of human ecology as an aim include Aschmann's account of the history of settlement in Baja California,[82] Gordon's on the Sinú country of Colombia,[83] Innis on Jamaica,[84] and Talbot on Masailand.[85] Methodologically, however, these human ecology studies by geographers are indistinguishable from similar work by members of the Chicago school (where Sauer himself studied): compare, for example, Aschmann's work on Baja California with that on the Hawaiian Islands by Park's student Lind.[86] Both show concern with historical (successional) development within a spatial setting, weaving geography and narrative together.

Meanwhile, the sociologists themselves gradually moved away from the specifically biological position of Park. J. W. Bews, himself a botanist, followed Park closely,[87] but beginning with Alihan's *Social ecology* (1938), sociologists turned to community as their field of study and expressed dissatisfaction with concepts borrowed from the natural sciences,[88] and with some exceptions, including a textbook by White and Renner,[89] the field delimited by Barrows and Park was abandoned by both geographers and human ecologists alike.[90] For the geographer, the study of the relationship between man and environment became too specialized in the

[82] H. Aschmann, 'The central desert of Baja California: demography and ecology', *Ibero-Americana*, 42 (1959), 1–282.

[83] B. LeR. Gordon, 'Human geography and ecology in the Sinú country of Colombia', *Ibero-Americana*, 39 (1957), 1–117.

[84] D. Q. Innis, 'Human ecology in Jamaica with a detailed study of peasant agriculture in the Mollison District of northern Manchester' (University of California at Berkeley, Ph.D. thesis, 1959, 248 pp.).

[85] L. M. Talbot, 'The ecology of western Masailand, East Africa' (University of California at Berkeley, Ph.D. thesis, 1963).

[86] A. W. Lind, *An island community: ecological succession in Hawaii* (Chicago: University of Chicago Press, 1938).

[87] J. W. Bews, *Human ecology* (London: Oxford University Press, 1935).

[88] W. E. Gettys, 'Human ecology and social theory', *Social Forces*, 18 (1940), 469–476; A. H. Hawley: Social ecology: a theory of community structure (New York: Ronald Press, 1950).

[89] C. L. White and G. T. Renner, *Geography: an introduction to human ecology* (New York: D. Appleton-Century, 1936).

[90] M. A. Alihan, *Social ecology: a critical analysis* (New York: Columbia University Press, 1938), pp. 1–18; L. F. Schnore, 'Geography and human ecology', *Economic Geography*, 37 (1961), 207–217.

physiological sense, and though its importance was stressed by Fleure and Biasutti, there have been few substantive studies by geographers.[91] Geographers are generally unaware of the work which has been done on environmental influences on, for example, fertility and reproduction,[92] though much relevant literature has been assembled by Bresler.[93] Both the geographer and the sociologist found that ecological concepts applied to human affairs only on a general and even philosophical level.[94] At the same time, the ecologists themselves were moving away from the classical statements of Clements, Wheeler and others,[95] in the search for more useful analytical ideas which finally resulted in the concept of the ecosystem. Some attempts have been made in recent years to revive a 'human ecology' based on the descriptive interpretation of man–environment interrelationships in particular unique localities,[96] but such treatments lack methodological rigour, lean heavily on a largely discarded regional approach, appear to be mainly pedagogical in aim and ignore developments in ecology itself over the last 30 years.

THE ECOSYSTEM AS A MODEL OF REALITY

The term *ecosystem* was formally proposed by Tansley, the plant ecologist,

[91] C. Monge, *Acclimatisation in the Andes* (New York: American Geographical Society, 1948); D. H. K. Lea, *Climate and economic development in the tropics* (New York: Harper Brothers, 1957). Compare J. R. Martin, *Influence of tropical climates in producing the acute endemic diseases of Europeans* (London: J. Churchill, 1855); G. E. Burch and N. P. DePasquale, *Hot climates, man, and his heart* (Springfield: C. C. Thomas, 1962); M. T. Newman, 'Adaptation of man to cold climates', *Evolution*, 9 (1955), 101–105.

[92] T. D. Glover and D. H. Young, 'Temperature and the production of spermatozoa', *Fertility and Sterility*, 14 (1963), 441–450; D. Grahn and J. Kratchman, 'Variation in neonatal death rate and birth weight in the United States and possible relations to environmental radiation, geology and altitude', *American Journal of Human Genetics*, 15 (1963), 329–352.

[93] J. B. Bresler, ed., *Human ecology: selected essays* (Reading, Mass.: Addison-Wesley, 1966).

[94] J. A. Quinn, 'The nature of human ecology: re-examination and re-definition', *Social Forces*, 18 (1939), 161–168; W. Sprout and M. Sprout, *The ecological perspective on human affairs, with special reference to international politics* (Princeton: Princeton University Press, 1965).

[95] W. M. Wheeler, 'The ant colony as an organism', *Journal of Morphology*, 22 (1911), 307–325; W. M. Wheeler, *Emergent evolution and the development of societies* (London: Kegan Paul, Trench, Trubner & Co., 1927); J. F. V. Phillips, 'Succession, development, the climax and the complex organism: an analysis of concepts', *Journal of Ecology*, 22 (1934), 554–571, 23 (1935), 210–246, 488–508.

[96] S. R. Eyre and G. R. J. Jones, eds, *Geography as human ecology: methodology by example* (London: Methuen, 1966).

in 1935, as a general term for both the biome ('the whole complex of organisms – both animals and plants – naturally living together as a sociological unit')[97] and its habitat. 'All the parts of such an ecosystem – organic and inorganic, biome and habitat – may be regarded as interacting factors which, in a mature ecosystem, are in approximate equilibrium: it is through their interactions that the whole system is maintained.'[98] Tansley's concept effectively broadens the scope of ecology itself, which is no longer purely biological in content.[99] In so doing it gives formal expression to a variety of concepts covering habitat and biome which date back at least to Forbes's use of *microcosm* in the ecosystem sense.[100] Friederichs[101] termed this new ecology *Wissenschaft von der Natur*; Thienemann[102] distinguished it from autecology and biocenology (synecology) as *holography*; Clements[103] terms it *bioecology*; and Markus[104] used the term *Naturkomplex*. Friederichs himself notes that the boundary between this extended ecology and geography 'has never been sharp, for ecology deals sometimes with landscapes . . . like geography. But the latter does not go into detail so far.'[105] In his rationale, however, Friederichs's picture of geography leans heavily on an outdated Romanticism drawn from Goethe and Alexander von Humboldt; his picture of 'the cosmos of life . . . the great, all-life-embracing holocene – the unity of nature',[106] accords more with Guyot and Ritter than with the subsequent development of Tansley's concept.

Fosberg[107] has developed Transley's definition as follows:

> An ecosystem is a functioning interacting system composed of one or more living organisms and their effective environment, both physical

[97] A. G. Tansley, *Introduction to plant ecology* (London: Allen & Unwin, 1946), p. 206.

[98] *Ibid.*, p. 207.

[99] For a more restrictive view, see J. R. Bray, 'Notes toward an ecologic theory', *Ecology*, 39 (1958), 770–776; and C. McMillan, 'The status of plant ecology and plant geography', *Ecology*, 37 (1956), 600–602.

[100] S. A. Forbes, 'The lake as a microcosm', *Bulletin of the Illinois Natural History Society*, 15 (1925), 537–550.

[101] K. Friederichs, *Okologie als Wissenschaft von der Natur, oder biologische Raumforschung* (Leipzig: J. A. Barth, 1937).

[102] A. Thienemann, 'Vom Wesen der Okologie', *Biologia Generalis*, 15 (1942), 312–331.

[103] F. E. Clements and V. E. Shelford, *op. cit.* (note 11).

[104] E. Markus, 'Naturkomplexe', *Protokoly obschestva Estestviospytatelei pri imperatorskom Yur'evskom Universitete* (Yur'ev, Dorpat, Jurjew or Tartu), 32 (1929), 79–94.

[105] K. Friederichs, *op. cit.* (note 68), p. 154.

[106] *Ibid.*, p. 158.

[107] F. R. Fosberg, 'The island ecosystem', in F. R. Fosberg, ed., *Man's place in the island ecosystem* (Honolulu: Bishop Museum Press, 1963), 1–6; reference on p. 2.

and biological. . . . The description of an ecosystem may include its spatial relations; inventories of its physical features, its habitats and ecological niches, its organisms, and its basic reserves of matter and energy; the nature of its income (or input) of matter and energy; and the behaviour or trend of its entropy level.

Properties of ecosystems have been recently outlined by Sjörs, Whittaker, Odum, Duvigneaud and McIntosh.[108] Evans [109] insists on the categorical nature of the term ecosystem, which includes a hierarchy of systems at different levels of complexity and extent. The whole terrestrial ecosystem has been termed the *ecosphere*, derived from ecosystem and biosphere, by Cole.[110]

A parallel development has been that of the Russian school of ecologists, using the terms geocenosis for the physical habitat and biocenosis for the biome, the two uniting to form the *geobiocenosis*.[111] Sukachev's work has had considerable influence on Soviet 'landscape science',[112] and its use in geography has been discussed by Morgan and Moss.[113] The term geobiocenosis, however, is an unwieldy one, and fails to communicate the single most important characteristic of the ecosystem: that it is a *system* and not simply a random aggregation of discrete phenomena. Lindeman inclusively defined the ecosystem in his classic paper of 1942 as 'any system

[108] H. Sjörs, 'Remarks on ecosystems', *Svensk Botanisk Tidskrift*, 49 (1955), 155–169; R. H. Whittaker, 'Ecosystems', *McGraw-Hill Encyclopaedia of Science and Technology*, 4 (1960), 404–408; E. P. Odum, *Fundamentals of Ecology* (Philadelphia: Saunders, 1953); E. P. Odum, 'Relationship between structure and function in the ecosystem', *Japanese Journal of Ecology*, 12 (1962), 108–118; P. Duvigneaud: *Ecosystèmes et biosphere: l'écologie, science moderne de la synthèse* (Brussels: Centre d'Écologie générale, 2 volumes, 1962); R. P. McIntosh, 'Ecosystems, evolution and relational patterns of living organisms', *American Scientist*, 51 (1963), 246–267.

[109] F. C. Evans, 'Ecosystem as the basic unit in ecology', *Science*, 123 (1956), 1127–1128.

[110] L. Cole, 'The ecosphere', *Scientific American*, 198 (4) (1958), 83–92.

[111] Y. N. Sukachev, 'On the principles of genetic classification in biocenology', *Zhurnal Obschei Biologii*, 5 (1944), 213–227 [in Russian]; 'Biogeozonose', *Bol'sheia Sovetskaia Entsiklopediia*, 5 (1950), 180–181 [in Russian]; 'On the principles of genetic classification in biocenology', *Ecology*, 39 (1958), 364–367; see also J. Blydenstein, 'The Russian school of phytocenology', *Ecology*, 42 (1961), 575–577.

[112] A. I. Perel'man, 'Geochemical principles of landscape classification', *Soviet Geography*, 2 (3) (1961), 63–73; Yu. K. Yefremov, 'The concept of landscape and landscapes of different orders', *Soviet Geography*, 2 (10) (1961), 32–43; V. I. Prokayev, 'The facies as the basic and smallest unit in landscape science', *Soviet Geography*, 3 (6) (1962), 21–29.

[113] W. B. Morgan and R. P. Moss, 'Geography and ecology: the concept of the community and its relationship to environment', *Annals of the Association of American Geographers*, 55 (1965), 339–350.

composed of physical–chemical–biological processes within a space–time unit of any magnitude',[114] a definition which clearly includes the operational range of geography.

The ecosystem concept has four main properties which recommend it in geographical investigation.[115] First, it is *monistic*: it brings together environment, man and the plant and animal worlds within a single framework, within which the interaction between the components can be analysed. Hettner's methodology, of course, emphasizes this ideal of unity, and some synthesis was achieved in the regional monographs of the French school, but the unity here was aesthetic rather than functional, and correspondingly difficult to define. Ecosystem analysis disposes of geographic dualism, for the emphasis is not on any particular relationship, but on the functioning and nature of the system as a whole.

Secondly, ecosystems are *structured* in a more or less orderly, rational and comprehensible way. The essential fact here, for geography, is that once structures are recognized they may be investigated and studied, in sharp contrast to the transcendental properties of the earth and its regions as organisms or organic wholes. Much geographical work in the past has been concerned with the framework of systems, and the current concern with the geometry of landforms, settlement patterns and communication networks may be interpreted on this level. As an example of a structural investigation in biology, reference may be made to the work of Hiatt and Strasburg[116] on the food web and feeding habits of over 200 species of fish in coral reefs of the Marshall Islands in the Pacific. Observation showed that the fish could be classified into five trophic groups, which were related in a rather complex manner, forming the structure shown in figure 10: this includes all levels from plankton and algae to sharks and other carnivores. Among structural studies at a fairly low level of complexity made by geographers, we may cite Holdridge's work on tropical American vegetation types. Holdridge seeks to delimit ecosystem boundaries on the basis of descriptive climatic parameters with presumed physiographical significance. Figure 11 shows the climatic limits of ecosystem units which define a mosaic of terrestrial ecosystems,[117]

[114] R. L. Lindeman, 'The trophic–dynamic aspect of ecology', *Ecology*, 23 (1942), 399–418; and commentary by R. E. Cook, 'Raymond Lindeman and the trophic–dynamic concept in ecology', *Science*, 198 (1977), 22–26.

[115] D. R. Stoddart, 'Geography and the ecological approach: the ecosystem as a geographic principle and method', *Geography*, 50 (1965), 242–251.

[116] R. W. Hiatt and D. W. Strasburg, 'Ecological relationships of the fish fauna on coral reefs of the Marshall Islands', *Ecological Monographs*, 30 (1960), 65–127.

[117] J. Tosi, 'Climatic control of the terrestrial ecosystem: a report on the Holdridge model', *Economic Geography*, 40 (1964), 173–181.

and area studies have been published on this basis for a number of Latin American states.[118]

Third, ecosystems *function*: they involve continuous through-put of matter and energy.[119] To take a geographical example, the system involves not only the framework of the communication net, but also the goods and people flowing through it. Once the framework has been defined, it may be possible to quantify the interactions and interchanges between component parts, and at least in simple ecosystems the whole complex may be quantitatively defined. Odum and Odum[120] in a pioneering study, again on

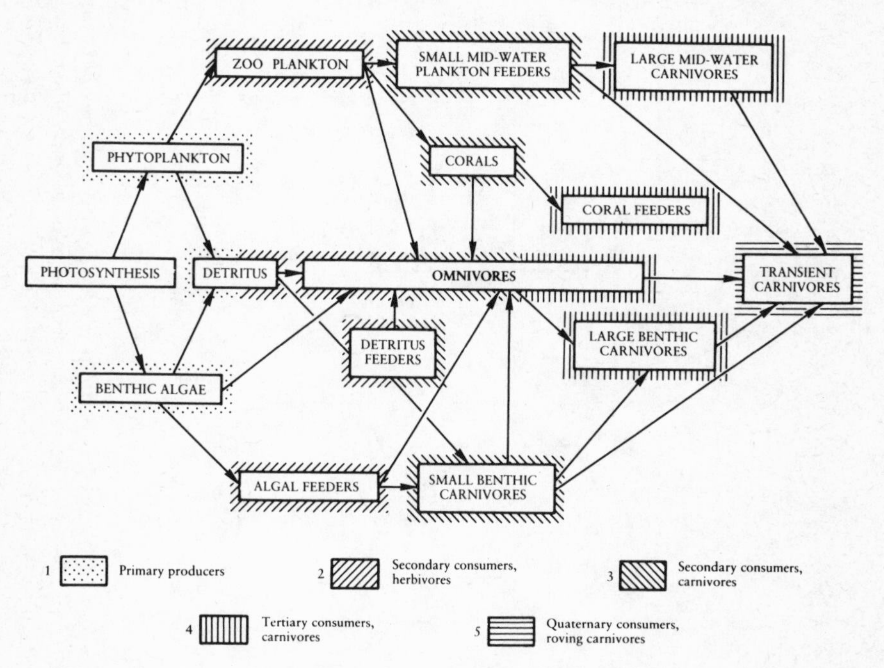

Figure 10 Food web of coral reefs in the Marshall Islands, showing the trophic structure in a qualitative manner.
Source: R. W. Hiatt and D. W. Strasburg, 'Ecological relationships of the fish fauna on coral reefs of the Marshall Islands', *Ecological Monographs*, 30 (1960), 65–127.

[118] For example, J. Tosi, 'Zones de vida natural en el Peru: Memoria explicativa sobre el Mapa ecologico del Peru', *Instituto Interamericano de Ciencias Agricolas Boletin Technico*, 5 (1960), 1–271.

[119] E. P. Odum, *Ecology* (New York: Holt, Rinehart & Winston, 1963), p. 10.

[120] H. T. Odum and E. P. Odum, 'Trophic structure and productivity of a windward coral reef community on Eniwetok Atoll', *Ecological Monographs*, 25 (1955), 291–320.

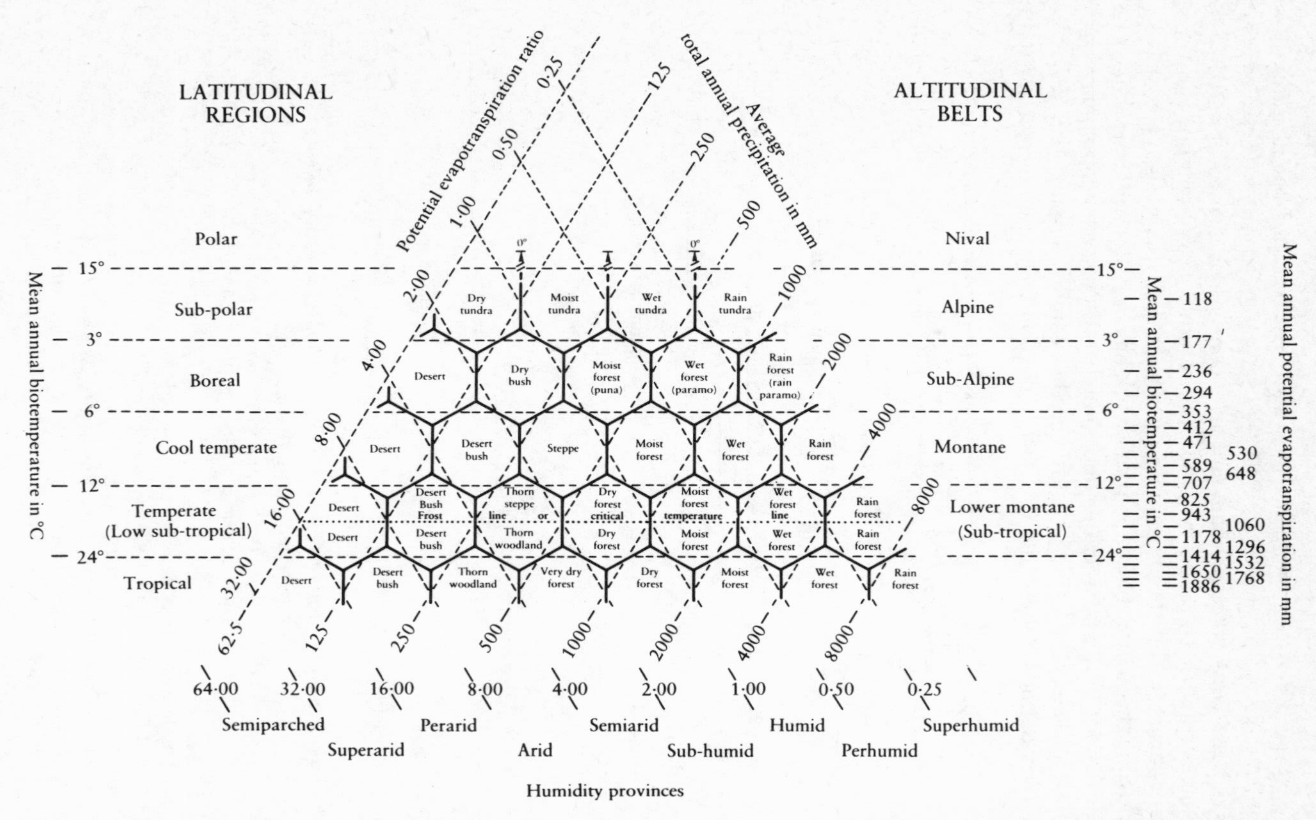

Figure 11 Mosaic diagram for the classification of natural life zones or ecosystems in terms of world plant formations, based on the method of Holdridge. The mean annual biotemperature is given by (mean monthly temperatures 0°C)/12.
Source: After J. Tosi, 'Zones de vida natural en el Peru: Memoria explicativa sobre el Mapa ecologico del Peru', *Instituto Interamericano de Ciencias Agricolas Boletin Technico*, 5 (1960), 1–271.

a Marshall Island coral reef, attempted to quantify the major trophic stages in the coral reef community – the primary producers, the herbivores and the carnivores. Figure 12A shows a biomass pyramid for a measured quadrat near the seaward edge of a reef; figure 12B is a mean biomass pyramid generalized from quadrats across a whole reef flat. While the details of the intepretation, particularly the trophic status of the corals, is open to question, the Odums, in this and in other studies,[121] have certainly demonstrated the possibility of quantifying the gross structural characteristics of small ecosystems. Equally remarkable is Teal's study of a salt marsh

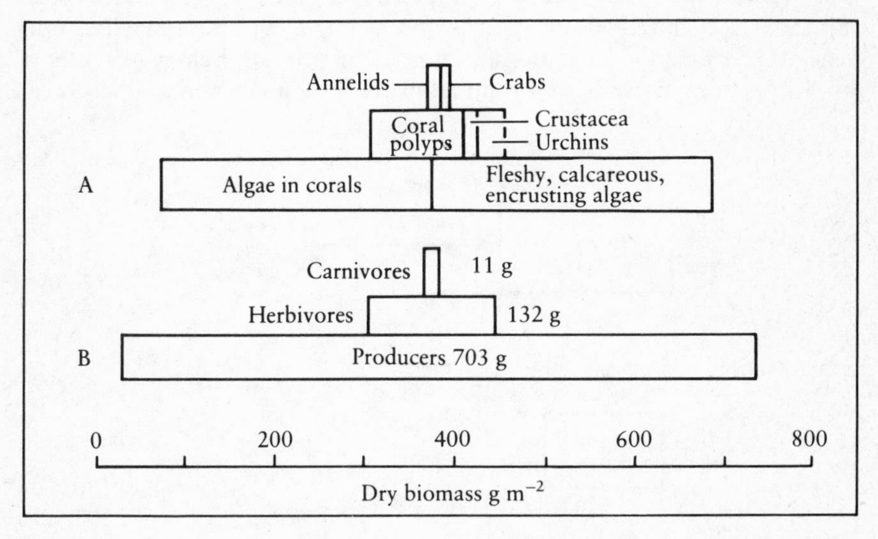

Figure 12 Biomass pyramids showing the dry weights of living materials in quadrats on the reef of Enewetak Atoll in the Marshall Islands. A: a quadrat on the reef edge; B: average biomass for the reef. The gross trophic structure is here shown in a quantitative way.

Source: H. T. Odum and E. P. Odum, 'Trophic structure and productivity of a windward coral reef community on Eniwetok Atoll', *Ecological Monographs*, 25 (1955), 291–320.

[121] H. T. Odum, 'Trophic structure and productivity of Silver Springs, Florida', *Ecological Monographs*, 27 (1957), 55–112; H. T. Odum and R. C. Pinkerton, 'Time's speed regulator, the optimum efficiency for maximum output in physical systems', *American Scientist*, 43 (1955), 331–343; E. P. Odum and A. E. Smalley, 'Comparison of population energy flow of a herbivorous and a deposit-feeding invertebrate in a salt marsh ecosystem', *Proceedings of the National Academy of Science*, 45 (1959), 617–622. For the fullest statement see H. T. Odum, *Systems ecology: an introduction* (New York: Wiley, 1983).

ecosystem in Georgia.[122] Teal constructed a food web for the salt marsh, and then measured standing crop, production and respiration for each of its components. Figure 13 shows in diagrammatic form the energy flow through this ecosystem, and the part played by each component of the web, with an energy input (light) of 600,000 kilocalories per square metre per year.

Fourthly, the ecosystem is a type of general system, and possesses the attributes of general systems. In general system terms, the ecosystem is an open system tending towards a steady state under the laws of open-system thermodynamics. Many of the properties of such systems have been implicitly recognized in the past – for example, the idea of climax in vegetation, of maturity in soils and of grade in geomorphology – but most of these conceptions have been, in effect, the application of classical

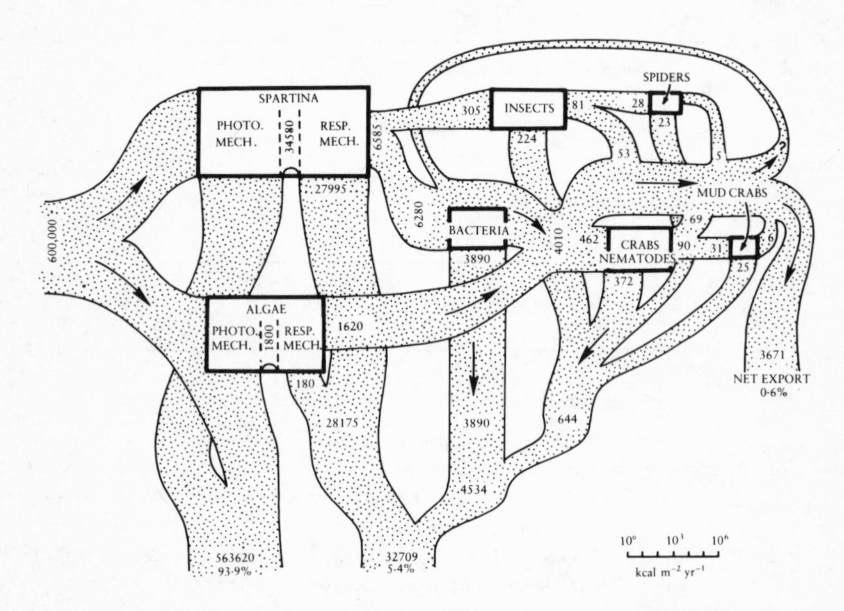

Figure 13 Energy flow diagram for a salt marsh in Georgia. The numbers are in units of kcal m^{-2} yr^{-1}.

Source: After J. M. Teal, Jr, 'Energy flow in the salt marsh ecosystem of Georgia', *Ecology*, 43 (1962), 614–624.

[122] J. M. Teal, Jr, 'Energy flow in the salt marsh ecosystem of Georgia', *Ecology*, 43 (1962), 614–624.

thermodynamic ideas to closed systems situations. With the development of open-system thermodynamics,[123] many of these older ideas are being reinterpreted in a dynamic rather than a static manner. Whittaker[124] has thus revised Clements's view on succession and climax; Jenny[125] and more recently Nikiforoff[126] and Auerbach[127] have done the same for soils; and Chorley[128] and others[129] have reinterpreted landforms in open-system terms.

Ecosystems in a steady state possess the property of self-regulation (action and reaction), and this is similar in principle to a wide range of mechanisms such as homeostasis in living organisms, feedback principles in cybernetics and servomechanisms in systems engineering.[130] Systems such as ecosystems, moreover, may be conceived on different levels of complexity, and it is the task of the geographer to search out aspects of reality which are significant at the levels at which the system is conceived. Systems, in fact, possess many of the structural properties of theoretical models, and a first approximation to system structure may be reached in a model-building manner, by selection, simplification and ordering of data at a series of levels.[131] Thus systems may be constructed at the framework level (e.g. settlement hierarchies or transportation nets) or as simple cybernetic systems (e.g. the mechanism of supply and demand), or at the more complex level of social systems and living organisms. Often in the case of highly complex systems, the system must be conceived at a very much lower level

[123] K. G. Denbigh, *The thermodynamics of the steady state* (London: Methuen, 1951); I. Prigogine, *Introduction to thermodynamics of irreversible processes* (Springfield: C. C. Thomas, 1955).

[124] R. H. Whittaker, *op. cit.* (note 14).

[125] H. Jenny, *op. cit.* (note 15).

[126] C. C. Nikiforoff, *op. cit.* (note 16).

[127] S. I. Auerbach, 'The soil ecosystem and radioactive waste disposal to the ground', *Ecology*, 39 (1958), 522–529.

[128] R. J Chorley, 'Geomorphology and general systems theory', *United States Geological Survey Professional Paper*, 500–B (1962), 1–10.

[129] J. T. Hack, 'Interpretation of erosional topography in humid temperate regions', *American Journal of Science*, 258–A (1960), 80–97; A. D. Howard, 'Geomorphological systems – equilibrium and dynamics', *American Journal of Science*, 263 (1965), 302–312; A. N. Strahler, 'Systems theory in physical geography', *Physical Geography*, 1 (1980), 1–27.

[130] N. Wiener, 'Cybernetics, or control and communication in the animal and the machine', *Actualités scientifiques et industrielles*, 1053 (1953), 1–194; G. E. Hutchinson, 'Circular control systems in ecology', *Annals of the New York Academy of Sciences*, 50 (1948), 221–246.

[131] R. J. Chorley, 'Geography and analogue theory', *Annals of the Association of American Geographers*, 54 (1964), 127–137.

of complexity, in the hope of gaining insight into problems where the data are too involved or the techniques inadequate for complete understanding, or indeed where the problem itself has been insufficiently defined. In geography, for example, the study of such highly complex systems as human groups has often been carried out at the level of 'clockwork' systems, such as simple deterministic, cause–effect relationships. The potential value of a system clearly depends on the correct selection of components at the initial structuring stage, and this normally presupposes considerable experience with the problems or data involved.[132]

APPLICABILITY

The ecosystem concept is in origin a biological idea, and most of its applications, including those already quoted, have been from the non-human world. Some attempts have been made, however, to describe fairly complex ecosystems in which man may play some part. Fosberg,[133] for example, after many years' work on coral atolls, attempted a general qualitative description of the coral atoll ecosystem, in terms of the media involved, the inflow of energy, primary productivity and successive elaboration, transformation and decomposition of its animal and plant community, excretion and accumulation of matter, and total turnover of matter and energy. Fosberg subsequently convened a symposium[134] to discuss the role of man in the isolated ecosystems of islands, in which the discussion ranged from man's own appraisal of his ecological status to more detailed consideration of the effects of over-population on island life. Islands in fact provide small laboratories for the testing and analysis of relatively simple and well-defined ecosystem structures. Sachet[135] has described the effects of the introduction of pigs on the ecology of Clipperton Island: vegetation was severely checked by crabs, until men introduced pigs, which ate crabs and allowed the vegetation to grow again. The pigs have recently been killed, and the ecological readjustments are awaited. In a similar situation, I have shown how coral islands in the Caribbean, when

[132] K. Boulding, 'General systems theory – the skeleton of science', *General Systems*, 1 (1956), 11–17.

[133] F. R. Fosbery, 'Qualitative description of the coral atoll ecosystem', *Atoll Research Bulletin*, 81 (1961), 1–11, and *Proceedings 9th Pacific Science Congress*, 4 (1962), 161–167.

[134] F. R. Fosberg, ed., *op. cit.* (note 107).

[135] M.-H. Sachet, 'History of change in the biota of Clipperton Island', in J. L. Gressitt, ed., *Pacific Basin biogeography* (Honolulu: Bishop Museum Press, 1963), 525–534.

covered with natural vegetation, are in equilibrium with major storms, and are even built up during hurricanes, but when man replaces the natural vegetation by coconut plantations the storms begin to cause catastrophic erosion.[136] A classic study of an island ecosystem involving man is that of Thompson[137] on the interaction of man, plants and animals in Fiji, and other works on island ecology of relevance here include those by Lind[138] on Hawaii and Harris[139] on the Leeward Islands.

Most ecosystems involving man are more complex than the salt-marsh and coral-reef systems already described, and attempts to describe ecosystems at such complex levels are likely to be difficult until experience is gained with relatively simple or restricted systems. Fosberg's focus on islands is one way out of this problem; another, which has received considerable attention recently, is to concentrate on primitive human and sub-human groups, in the hope of obtaining insight into the structure and function of more complex organization. Schaller's extraordinary study[140] of the mountain gorilla, *Gorilla gorilla berengei*, its territoriality, population structure, ecology and behaviour, for example, and DeVore's of the baboon[141] demonstrate the intriguing possibilities of primate geography.[142] Among geographers, Sauer has been pre-eminent in the study of the ecology of man in the Pleistocene, a subject currently being actively developed in relation to the South African Pleistocene[143] and the Australian aborigine;[144]

[136] D. R. Stoddart, 'Storm conditions and vegetation in equilibrium of reef islands', *Proceedings 9th Conference on Coastal Engineering, Lisbon 1964* (1964), 893–906.

[137] L. Thompson, 'The relations of man, animals and plants in an island community (Fiji)', *American Anthropologist*, 51 (1949), 253–267.

[138] A. W. Lind, *op. cit.* (note 86).

[139] D. R. Harris, 'The invasion of oceanic islands by alien plants: an example from the Leeward Islands, West Indies', *Transactions of the Institute of British Geographers*, 31 (1962), 67–82.

[140] G. B. Schaller, *The mountain gorilla: ecology and behaviour* (Chicago: University of Chicago Press, 1963).

[141] S. L. Washburn and I. DeVore, 'Social behaviour of baboons and early man', in S. L. Washburn, ed., *Social life of early man* (Chicago: Aldine Press, 1962), 91–105; I. DeVore and S. L. Washburn, 'Baboon ecology and human evolution', in F. C. Howell and F. Bourlière, eds, *African ecology and human evolution* (London: Methuen, 1964), 335–367.

[142] I. DeVore, ed., *Primate behaviour: field studies of monkeys and apes* (New York: Holt, Rinehart & Winston, 1965).

[143] J. D. Clark, 'Human ecology during Pleistocene and later times in Africa south of the Sahara', *Current Anthropology*, 1 (1960), 307–324; R. B. Lee, 'The population ecology of man in the early Upper Pleistocene of southern Africa', *Proceedings of the Prehistoric Society*, n.s. 29 (1963), 235–257.

[144] J. B. Birdsell, 'Some environmental and cultural factors influencing the structuring of Australian aboriginal population', *American Naturalist*, 87 (1953), 171–207.

while Daryll Forde,[145] in a classic volume, studied the ecology of some two dozen modern primitive peoples. Most of these studies, however, have been conducted on traditional lines, and not within an explicit system framework: with some of the simpler groups it should be possible to delineate ecosystems with as much precision as in the non-human world.

The power of ecosystem analysis to pose new problems in geography, and hence to seek new answers, is demonstrated by Clifford Geertz's discussion of shifting cultivation and wet-rice cultivation in Indonesia.[146] Geertz points out that most discussions of shifting cultivation emphasize its negative characteristics,[147] but that it is more profitably viewed in its system characteristics in relation to the tropical forest it replaces. Both are highly diverse systems, in which matter and energy cycle rapidly among the vegetation components and the topmost soil layer: the soil itself plays little part in this energy flow, and may often be impoverished. Burning is seen as a means of channelling the nutrients locked up in the vegetation into certain selected crop plants: the general ecological efficiency is lowered, but the yield to man increased. In a well-developed shifting cultivation system both structure and function are comparable to those in the tropical forest, but the equilibrium is more delicately poised. By contrast, in wet-rice cultivation, the ecosystem structure is quite different, the productivity is high and the system equilibrium more stable. The analysis is given in qualitative terms, but points the way to several lines of quantitative investigation, with clear import for land-use planning and rural reform programmes.

Apart from Geertz's work, there have been few specific system-building studies in geography. The ecologist Dice, after working on natural communities, has produced a survey of ecosystem properties which may serve as a programme for human ecosystem research:[148] to the normal ecological feedback mechanisms of starvation, predation, disease, migration and competition, he adds[149] public opinion, punishment and rewards, wealth, taxation, supply and demand, cooperation and the democratic process. Dice suggest that human ecosystems may be conceived at

[145] C. D. Forde, *Habitat, economy and society: a geographical introduction to ethnography* (London: Methuen, 1934).

[146] C. Geertz, *Agricultural involution: the process of agricultural change in Indonesia* (Berkeley: University of California Press, 1963).

[147] For example, P. Gourou, *The tropical world* (London: Longmans, 1953).

[148] L. R. Dice, *Natural communities* (Ann Arbor: University of Michigan Press, 1952); *Man's nature and nature's man: the ecology of human communities* (Ann Arbor: University of Michigan Press, 1955).

[149] L. R. Dice (1955), *op. cit.* (note 148), pp. 82–119.

successively larger-scale intervals: tribe, homestead, village, town, city, national and international levels.[150] Systems theory is also being used in many branches of land-use planning, for example the study of water resources;[151] Chorley[152] has carried systems analysis into geomorphology; Brookfield[153] has briefly noted the potentiality of ecosystem studies; and Ackerman,[154] in a major paper, has pointed to systems analysis as geography's great research frontier. A beginning has been made in substantive geographic work with the McGill University Savanna Research Project, explicitly organized in ecosystem terms: a first report has appeared on the Rupununi savannas of British Guiana.[155] The ecosystem as here interpreted, however, is largely restricted to the climate–vegetation–soils complex,[156] and is thus narrower than the view taken in this paper.

The potential practical value of ecosystem studies in geography may be demonstrated particularly in agricultural and land-use studies, and may be illustrated by a number of comprehensive recent symposia: on the biological productivity of Britain,[157] on the exploitation of natural animal populations,[158] on the ecology of grazing,[159] and on the ecology of

[150] *Ibid.*, pp. 251–266. See also A. Wolman, 'The metabolism of the city', *Scientific American*, 213 (1965), 179–190; I. Douglas, 'The city as an ecosystem', *Progress in Physical Geography*, 5 (1981), 315–367; K. Newcombe, 'Nutrient flow in a major urban settlement: Hong Kong', *Human Ecology*, 5 (1977), 179–208; K. Newcombe, J. D. Kalma and A. R. Aston, 'The metabolism of a city: the case of Hong Kong', *Ambio*, 7 (1979), 3–15; and S. Boyden, S. Millar, K. Newcombe and B. O'Neill, *The ecology of a city and its people: the case of Hong Kong* (Canberra: Australian National University Press, 1981).

[151] R. N. McKean, *Efficiency in government through systems analysis, with emphasis on water resources development* (New York: Wiley, 1958).

[152] R. J. Chorley, *op. cit.* (note 128).

[153] H. C. Brookfield, 'Questions on the human frontiers of geography', *Economic Geography*, 40 (1964), 283–303.

[154] E. A. Ackerman, 'Where is a research frontier?', *Annals of the Association of American Geographers*, 53 (1963), 429–440.

[155] M. J. Eden, 'The savanna ecosystem, northern Rupununi, British Guiana', *McGill University Savanna Research Series*, 1 (1964), 1–216.

[156] K. C. Edwards, 'The importance of biogeography', *Geography*' 49 (1964), 85–97; S. R. Eyre, 'Determinism and the ecological approach in geography', *Geography*, 49 (1964), 369–376; I. R. Simmons, 'Ecology and land use', *Transactions and Papers of the Institute of British Geographers*, 38 (1966), 59–72.

[157] W. B. Yapp and D. J. Watson, eds, 'The biological productivity of Britain', *Symposia of the Institute of Biology*, 7 (1958), 1–128.

[158] E. D. Le Cren and M. W. Holdgate, eds, 'The exploitation of natural animal populations', *Symposia of the British Ecological Society*, 2 (1962), 1–399.

[159] D. J. Crisp, ed., 'Grazing in terrestrial and marine environments', *Symposia of the British Ecological Society*, 4 (1964), 1–322.

industrial air and water pollution, pesticides and waste-land reclamation.[160] Ryther[161] has discussed potential productivity in the oceans; Ovington[162] has summarized a great deal of work on woodland productivity throughout the world; and Newbould[163] has reviewed developments in production ecology. Some of the literature on production ecology is complex because of the diverse meanings attributed to terms such as productivity and efficiency,[164] but Macfadyen[165] has proposed the use of L. Dudley Stamp's[166] Standard Nutritional Unit (SNU), equivalent to 10^6cal./yr as a measure of productivity. Table 4, from Macfadyen,[167] gives sample net annual production for some terrestrial ecosystems managed by man, in SNU/hectare. Summary tables of gross and net primary productivity for a variety of natural and man-managed ecosystems are given by Odum and

Table 4 Net annual production of some man-managed terrestrial ecosystems

	SNU/hectare
Maize, unfertilized	8.2
Maize, fertilized	24.4
British improved grazing	1.0
British farming, national average	7.0
British farming, best	13.6
Europe, wheat maximum	8.3
Sugar beet	60.5
Sugar cane, mean	65.5
Sugar cane, maximum	254.0
Cassava	280.0

[160] G. T. Goodman, R. W. Edwards and J. M. Lambert, eds, 'Ecology and the industrial society', *Symposia of the British Ecological Society*, 5 (1965), 1–395.

[161] J. H. Ryther, 'Potential productivity of the sea', *Science, N.Y.*, 130 (1959), 602–608.

[162] J. D. Ovington, 'Quantitative ecology and the woodland ecosystem concept', *Advances in Ecological Research*, 1 (1962), 103–192.

[163] P. J. Newbould, 'Production ecology', *Science Progress*, 51 (1963), 91–104.

[164] L. B. Slobodkin, *op. cit.* (note 1).

[165] A. Macfadyen, 'Energy flow in ecosystems and its exploitation by grazing', in D. J. Crisp, ed., *op. cit.* (note 159), 3–20.

[166] L. D. Stamp, 'The measurement of land resources', *Geographical Review*, 48 (1958), 1–15.

[167] A. Macfadyen, *op. cit.* (note 165), p. 10.

Odum[168] and by Westlake.[169] It is interesting to note the extraordinarily high productivity of coral reefs by comparison with almost all other ecosystems. Measurements of gross primary productivity of Pacific reefs range from 1,800–11,690 g C/m^2/yr.[170] At Kauai, Hawaiian Islands, where a fringing reef gave a measured productivity of 2,900 g C/m^2/yr, an adjacent sugar-cane plantation, managed over the years to give maximum production, averaged only 1,775 g C/m^2/yr. It is concluded that 'even under the optimum conditions and techniques employed by modern agriculturists, cultivated terrestrial areas may not be as productive as certain shallow-water areas in the sea in which an abundance of benthic algae is exposed to high insolation and a continuous replenishment of nutrients.'[171] Similarly, Ovington,[172] working in the East Anglian Breckland, found that pine plantations produced more than twice the average annual production of adjacent farmland, and in ecological terms are undoubtedly the more efficient means of fixing solar energy. Such comparative data are clearly of great importance in rational land-use planning, particularly in overpopulated areas. At this empirical level geographers may make a considerable contribution to the understanding of terrestrial ecosystems:[173] active exploration into the value of the ecosystem concept, especially in land-use studies, is being carried out by Simmons.[174]

PROBLEMS

It may be objected that the study of ecosystems in geography is either (a) not new, or (b) not geography. In a sense, it is true that study of systems is implicit in most geographic work: in economics system-building goes back to Smith and Ricardo, and in human and physical geography elements of

[168] H. T. Odum and E. P. Odum, 'Principles and concepts pertaining to energy in ecological systems', in E. P. Odum, *op. cit.* (1953) (note 108), chapter 3, 43–87.

[169] D. F. Westlake, 'Comparisons of plant productivity', *Biological Reviews*, 38 (1963), 385–425.

[170] For a review, see P. Helfrich and S. J. Townsley, 'The influence of the sea', in F. R. Fosberg, ed., *op. cit.* (note 107), 39–53.

[171] *Ibid.*, p. 50.

[172] J. D. Ovington, 'Dry-matter production by *Pinus sylvestris* L.', *Annals of Botany*, 21 (1957), 287–314.

[173] P. J. Newbould, 'Production ecology and the International Biological Programme,' *Geography*, 49 (1964), 98–104; L. D. Stamp, *Our developing world* (London: Faber & Faber, 1960).

[174] I. R. Simmons, *op. cit.* (note 156).

systems are even older. Mackinder himself[175] went so far as to state the content of geography in closed-system terms. Discussion on the geographic relevance of the ecosystem concept, however, has been tentative and vague. McMillan[176] has argued the case for excluding geography from the ecological field, and while Rowe[177] believes that all ecosystems are necessarily geographical, he bases his interpretations on Hartshorne's arguments of the uniqueness of geographical phenomena, and thus, from a nomothetic point of view, vitiates his argument.[178]

The charge that ecosystem study is 'not geography' lies in the fact, presumably, that the ecosystem definition does not explicitly define the earth's surface as a field of operation. 'Ecology is the study of environmental relationships; geography is the study of space relationships,' states Davies,[179] but he goes on to add that 'what is not clear is where the one stops and the other starts.' Troll,[180] among geographers, has used the ecosystem concept in his definition of *Landschaftselemente*, which he subsequently modified to *ökotop* (ecotope), an unfortunate usage since the term has already been pre-empted by Tansley. Perel'man[181] has classified landscapes in terms of system-processes, and similar procedures are implicit in Holdridge's classification of tropical ecosystems. The study of space relationships, however, if it is to be more than mere nominal-scale classification of areas, must involve system-building: the study of the ecosystem requires the explicit elucidation of the structure and functions of a community and its environment, with the ultimate aim of the quantification of the links between components.

POTENTIALITIES

The ecosystem is a type of general system, defined as a 'set of objects

[175] H. J. Mackinder, 'The content of philosophical geography', *Proceedings of the [12th] International Geographical Congress, Cambridge 1928* (Cambridge: Cambridge University Press, 1930), 305–312; see p. 310.

[176] C. McMillan, *op. cit.* (note 99).

[177] J. S. Rowe, 'The level-of-integration concept and ecology', *Ecology*, 42 (1961), 420–427.

[178] W. R. Siddall, *op. cit.* (note 69).

[179] J. L. Davies, 'Aim and method in zoogeography', *Geographical Review*, 51 (1961), 412–417.

[180] C. Troll, 'Die geographische Landschaft und ihre Erforschung', *Studium Generale*, 3 (1950), 163–181.

[181] A. I. Perel'man, *op. cit.* (note 112).

together with relationships between the objects and between their attributes'.[182] Partaking in general system theory, the ecosystem is potentially capable of precise mathematical structuring within a theoretical framework, a very different matter from the tentative and incomplete descriptions of highly complex relationships which too often pass for geographical 'synthesis'. The limits of the ecosystem may be set at any desirable areal extent, and so flexible is the concept that it may be employed at any level from that of acorns[183] or pieces of dung[184] to that of the universe itself, and is currently being used in the study of artificial ecosystems within space capsules and interplanetary rockets.[185] Within any areal framework the ecosystem concept will give point to enquiry, and thus highlight both form and function in a spatial setting. Simplistic ideas of causation and development, or of geographic dualism, are in this context clearly irrelevant: ecosystem analysis gives geographers a tool with which to work.

The value of systems analysis lies not only in its emphasis on organization, structure and functional dynamics: through its general system properties, it brings geography back into the realm of the natural sciences, and allows us to participate in the scientific revolutions of this century from which the Kantian exceptionalist position excluded us.[186]

A most significant implication of the ecosystem approach in geography is that systems may be interpreted in terms of cybernetics, information and communication theory, and related mathematical techniques. Quantitative interpretation of ecosystem dynamics dates back to Lindeman's paper[187] on the energy relationships and trophic structure of the ecosystem. This approach has been developed theoretically,[188] and by the use of simulation techniques. Odum[189] attempted to construct a simple electric analogue for

[182] A. D. Hall and R. E. Fagen, 'Definition of system', *General Systems*, 1 (1956), 18–28.

[183] P. W. Winston, 'The acorn microsere with special reference to arthropods', *Ecology*, 37 (1956), 120–132.

[184] C. O. Mohr, 'Cattle droppings as ecological units', *Ecological Monographs*, 13 (1943), 275–298.

[185] E. B. Konecci, 'Space ecological systems', in K. E. Schaefer, ed., *Bioastronautics* (New York: Macmillan, 1964), 274–304.

[186] F. K. Schaefer, 'Exceptionalism in geography: a methodological examination', *Annals of the Association of American Geographers*, 43 (1953), 226–249.

[187] R. L. Lindeman, *op. cit.* (note 114).

[188] H. T. Odum and E. P. Odum, *op. cit.* (note 168); H. T. Odum and R. C. Pinkerton, *op. cit.* (note 121); H. T. Odum and E. P. Odum, *Energy basis for man and nature* (New York: McGraw-Hill, 1976).

[189] H. T. Odum, 'Ecological potential and analogue circuits for the ecosystem', *American Scientist*, 48 (1960), 1–8.

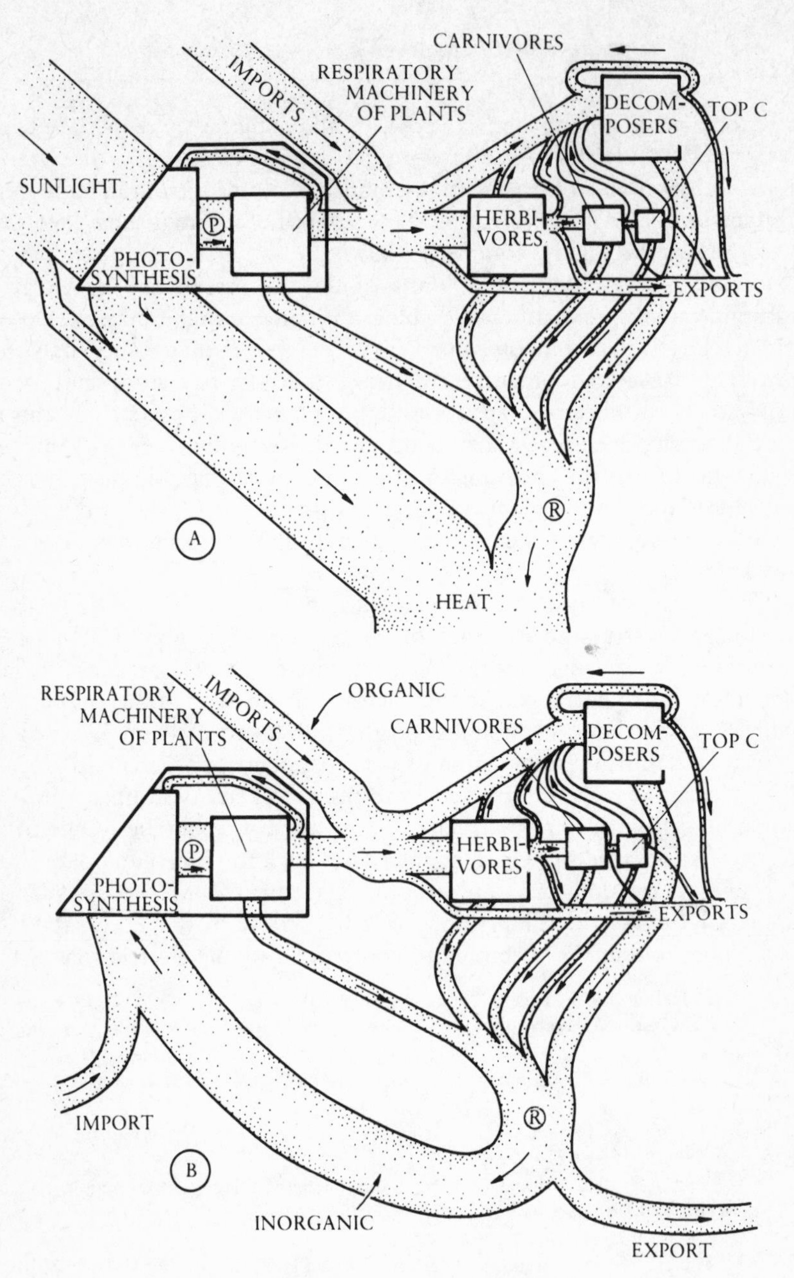

Figure 14 Odum's conception of (A) energy flow and (B) materials flow (the 'ecomix cycle') in a simple terrestrial ecosystem.

Source: After H. T. Odum, 'Ecological potential and analogue circuits for the ecosystem', *American Scientist*, 48 (1960), 1–8.

the ecosystem, based on energy flow (see figure 1A) and materials (see figure 14B) in a simple ecosystem. Following Ohm's Law (A=cv, where A is the flow of electric current, v the voltage and c the conductivity), Odum derives an equation for force and flux in the ecosystem, of the form

$$J_e = c_e x_e$$

where J_e is the ecoflux (e.g. flow of food through a food chain circuit), c_e is the ecological conductivity of the food chain, and x_e the thermodynamic force or ecoforce. Figure 15 shows Odum's analogue circuit of the ecosystem, with energy input derived from batteries, food flows simulated by currents, and energy dissipation by amperage and voltage changers. Similarly, Olson has constructed more complex analogue models for the

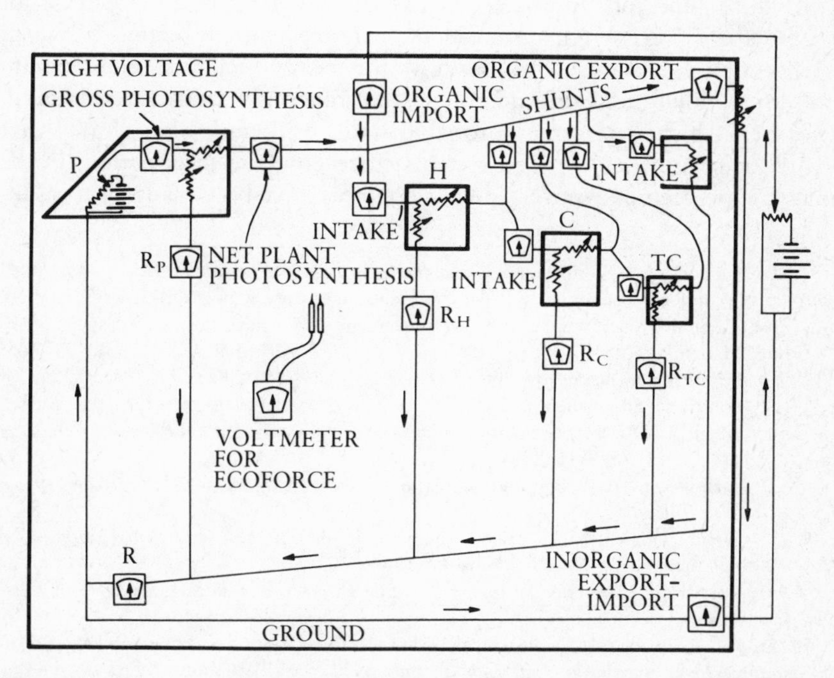

Figure 15 Electric analogue circuit for a steady-state ecosystem modelled on the ecosystem materials flow in figure 14B.

Source: After H. T. Odum, 'Ecological potential and analogue circuits for the ecosystem', *American Scientist*, 48 (1960), 1–8.

simulation of small ecosystems, incorporating both positive and negative feedback links.[190]

The formulation of mathematical models for simple ecosystems also suggests the possibility of digital computer simulation for the derivation of steady-state ecosystem properties.[191] Garfinkel and Sack[192] have used an ecosystem model based on Volterra's mass-action law (that the rate at which two species interact is proportional to the product of their populations)[193] to derive ecosystem properties for a six-species system which includes grass, bushes, trees, a small and a large herbivore, and a carnivore. The results are clearly only applicable to such simple systems, but they suggest further lines of enquiry in the simulation of more complex systems, and the techniques used may be relevant in, for example, the use of gravity models in population geography.[194]

A third line of approach stems from Wiener's development of cybernetics.[195] Ecosystems are ordered arrangements of matter, in which energy inputs carry out work. Remove the energy input and the structure will break down until the components are randomly arranged (maximum entropy), which is the most probable state. Brillouin[196] has shown that order, or negative entropy, in systems corresponds to information (in the information-theoretic sense).[197] First attempts have been made, as a result,

[190] R. B. Neel and J. S. Olson, *Use of analog computers for simulating the movement of isotopes in ecological systems* (Oak Ridge National Laboratory Report ORNL–3172, 1962); J. S. Olson, 'Analog computer models for movement of nuclides through ecosystems', in V. Schultz and A. W. Klement, Jr, eds, *Radioecology* (New York: Reinhold, 1963), 121–125; J. S. Olson, 'Gross and net production of terrestrial ecosystems', *British Ecological Society Jubilee Symposium*, Supplement to *Journal of Ecology*, 52 (1964), 99–118. For further discussion, see R. G. Wiegert, 'Simulation models of ecosystems', *Annual Reviews of Ecology and Systematics*, 6 (1975), 311–338.

[191] D. Garfinkel, 'Digital computer simulation of ecological system', *Nature*, 194 (1962), 856–857.

[192] D. Garfinkel and R. Sack, 'Digital computer simulation of an ecological system, based on a modified mass-action law', *Ecology*, 45 (1964), 502–507.

[193] V. Volterra, *Leçons sur la théorie mathématique de la lutte pour la vie* (Paris: Gauthier-Villars, 1931).

[194] G. A. P. Carrothers, 'An historical review of the gravity and potential concepts of human interaction', *Journal of American Institute of Planners*, 22 (1958), 94–102.

[195] N. Wiener, *op. cit.* (note 130).

[196] L. Brillouin, *Science and information theory* (New York: Academic Press, 2nd edition, 1962); *Scientific uncertainty and information* (New York: Academic Press, 1964).

[197] H. Quastler, 'A primer on information theory', in H. P. Yockey, R. L. Platzman and H. Quastler, eds, *Symposium on information theory in biology* (New York: Pergamon Press, 1958), 3–49.

to apply information theory to ecosystem analysis,[198] and to interpret ecosystems in terms of cybernetics.[199] Hairston[200] and MacArthur[201] have used these techniques in specific problems of ecological populations. Similar methods are being more extensively used in purely biological fields.[202]

Many ecosystem models proposed are conceptually simple, and may be criticized on the grounds that they ignore the complexities of the real world. Slobodkin[203] points out that in the ecosystem the structure of the system itself depends on continuing energy flow, whereas in analogue systems such as Odum's electric circuit the existence of the structure is independent of energy flow. More important from a theoretical point of view, and with special geographical significance, is the fact that in complex-feedback systems, such as ecosystems,[204] Prigogine's Theorem (that steady-state systems tend to a condition of minimum entropy production) is not necessarily applicable.[205] If this is indeed the case in ecosystems and geographical systems, static-structured analogue models will have little predictive power. This general argument underlies von Bertalanffy's suggested replacement of purely mechanistic models in biology with his 'organismic theory',[206] and Slobodkin's statement[207] that Odum's electric analogue model not only violates common sense, but fails to produce either substantive results or suggestions for further work.[208]

[198] R. Margalef, 'La teoria de la informacion en ecologia', *Memorias de la Real Academia de Ciencias y Artes de Barcelona*, 32 (1957), 373–449; 'Information theory in ecology', *General Systems*, 2 (1958), 36–71.

[199] B. C. Patten, 'An introduction to the cybernetics of the ecosystem: the trophic-dynamic aspect', *Ecology*, 40 (1959), 221–231.

[200] N. G. Hairston, 'Species abundance and community organization', *Ecology*, 40 (1959), 404–416.

[201] R. MacArthur, 'On the relative abundance of species', *American Naturalist*, 94 (1960), 25–36.

[202] J. W. L. Beament, ed., 'Models and analogues in biology', *Symposia of the Society for Experimental Biology*, 14 (1960), 1–255; F. S. Grodins, *Control theory and biological systems* (New York: Columbia University Press, 1963); H. Quastler, ed., *Information theory in biology* (Urbana: Illinois University Press, 1953); H. P. Yockey *et al.*, eds., *op. cit.* (note 197); F. H. George, 'Models in cybernetics', *Symposia of the Society for Experimental Biology*, 14 (1960), 169–191; F. H. George, *Cybernetics and biology* (Edinburgh: Oliver & Boyd, 1964).

[203] L. B. Slobodkin, 'Ecological energy relationships at the population level', *American Naturalist*, 94 (1960), 213–236.

[204] G. E. Hutchinson, *op. cit.* (note 130).

[205] C. Foster, A. Rappoport and E. Trucca, 'Some unsolved problems in the theory of non-isolated systems', *General Systems*, 3 (1957), 9–29.

[206] L. von Bertalanffy, *op. cit.* (note 67).

[207] L. B. Slobodkin, *op. cit.* (note 1), p. 82.

[208] See B. Greer-Wootten, 'General systems theory – a new backbone for the "formless"

On a general level, Maruyama's distinction between two types of mutual causal processes is especially illuminating in the geographical context.[209] Wiener's cybernetic model is a negative feedback system, typified by such machines as the Watt's Governor and ships' automatic steering mechanisms. Negative feedback systems in geography are numerous: they include the supply and demand process in equilibrium economics,[210] Malthus' model of population control,[211] Gilbert's principle of dynamic equilibrium in river erosion,[212] and the 'vicious circle' situation in tropical economic development,[213] all of which are, in effect, mechanisms for damping down fluctuations. In addition, Maruyama distinguishes 'deviation-amplifying mutual causal processes' as the basis for a 'second cybernetics'. Deviation-amplifying or positive feedback processes are common in geography, and in ecology itself have led to the great emphasis given to problems of growth, development and succession. Modern technological development is an example of deviation-amplification on a massive scale. The recent expansion of development economics[214] is based on positive feedback principles: if a factory is built or a new road constructed in a rural area, previously existing deviation-damping processes cease to operate, and the system structure is itself transformed. Problem analyses within the earlier frameworks become redundant. Maruyama's process lies at the root of geographical discontent with, for example, 'uniform plain' suppositions in location models;[215] and it is clear that many ecosystems or geographical systems need to have built into them either deterministic or stochastic 'growth' processes. Geographical needs here begin to converge with those of

---

discipline of geography?', *McGill University Graduate Seminar Group in Arctic Geography Discussion Paper* (1965), 1–18; 'The role of general systems theory in geographic research', *Department of Geography, York University, Toronto, Discussion Paper*, 3 (1972), 1–14.

[209] M. Maruyama, 'Morphogenesis and morphostasis', *Methodos*, 12 (1960), 251–296; 'The second cybernetics: deviation-amplifying mutual causal processes', *American Scientist*, 51 (1963), 164–179.

[210] A. Marshall, *Principles of economics* (London: Macmillan, 1890).

[211] D. V. Glass, ed., *Introduction to Malthus* (London: Watts, 1953).

[212] G. K. Gilbert, *Report on the geology of the Henry Mountains* (Washington: Government Printing Office, 1877).

[213] R. Nurkse, *Problems of capital formation in underdeveloped countries* (Oxford: Basil Blackwell, 1953).

[214] W. W. Rostow, *The stages of economic growth: a non-Communist manifesto* (Cambridge: Cambridge University Press, 1960); G. M. Meier, *Leading issues in development economics* (Oxford: Oxford University Press, 1965).

[215] J. D. Nystuen, 'Identification of some fundamental spatial concepts', *Papers of the Michigan Academy of Science, Arts and Letters*, 48 (1963), 373–384.

development economics in such techniques as linear and non-linear programming, rather than with most ecosystem models currently envisaged. Positive feedback requirements thus impose severe restraints on the value of mechanical or electrical analogues,[216] and suggest that computerized mathematical model-building may be more productive.[217]

While simulation models may be constructed within given theoretical frameworks, therefore, they are largely limited by the level of sophistication of the theoretical models available. Slobodkin[218] has suggested that the *categorization* of the theoretical models on the basis of either quantitative or formal properties will be of greater heuristic value than the construction of more and more detailed specific models: such categorization he terms a *meta-model*. The value of meta-models of this sort in geography has been discussed by Chorley,[219] and in the present state of techniques may be the most useful way in which concepts may be clarified and specific new approaches suggested.

GEOGRAPHY, THE ECOSYSTEM AND GENERAL SYSTEMS

This review of biological models in geographic methodology has demonstrated the pervasiveness of first, organic analogies, with intellectual origins far back in antiquity but deriving major impetus from vitalist biology, and second, and more recently, of the more formal framework of ecosystem analysis. While the ecosystem concept has proved useful in several branches of geographical work, it has become apparent that its influence is seminal rather than definitive, and lies not in the ecosystem concept as such, but in its general system properties. Recent quantitative and simulation studies of ecosystems by biologists have geographical value chiefly insofar as they suggest similar lines of work: the biological emphasis on the energetic and trophic structure of ecosystems, for example, is clearly of peripheral geographic significance, but the fundamental concept of systems in geography is central to the development of the subject as a nomothetic science.

Paradoxically, therefore, one is led beyond the concept of the ecosystem

[216] But see J. S. Olson, (1963, 1964) *op. cit.* (note 190).
[217] D. Garfinkel, *op. cit.* (note 191).
[218] L. B. Slobodkin, 'Meta-models in theoretical ecology', *Ecology*, 39 (1958). 550–551.
[219] R. J. Chorley, *op. cit.* (note 131).

to the recognition of the importance of system studies in geography.[220] Geography is clearly concerned with systems on a multitude of levels. A preliminary attempt to develop a science of 'geocybernetics' has been made in a little-known paper by Polonskiy,[221] and several cases of system-building in geography have been discussed in this chapter. The study of these *geosystems* may now replace that of ecosystems in geography, which remains concerned with precisely the same body of data, which, intuitively grasped in all its complexity, led to the use of simplistic organic analogies and subsequently to Tansley's fundamental concept. As Berry rightly concludes, 'Geography's integrating concepts and processes concern the world-wide ecosystem of which man is the dominant part.'[222]

[220] V. B. Socava, 'Das Systemparadigma in der Geographie', *Petermanns Geographische Mitteilungen*, 118 (1974), 161–166; E. A. Ackerman, *op. cit.* (note 154); L. von Bertalanffy, 'The theory of open systems in physics and biology', *Science, N.Y.*, 111 (1950), 23–29; L. von Bertalanffy, 'An outline of general system theory', *British Journal for the Philosophy of Science*, 1 (1951), 134–165; L. von Bertalanffy, 'General system theory', *General Systems*, 1 (1956), 1–10; L. von Bertalanffy, C. G. Hempel, R. E. Bass and H. Jones, 'General system theory: a new approach to unity of science', *Human Biology*, 23 (1951), 302–361; J. M. Blaut, 'Object and relationship', *Professional Geographer*, 14 (6) (1962), 1–7; K. Boulding, *op. cit.* (note 132).

[221] M. L. Polonskiy 'Geokibernetkia, predmet i metod' (mimeographed, Minsk, 1963).

[222] B. J. L. Berry, 'Approaches to regional analysis: a synthesis', *Annals of the Association of American Geographers*, 54 (1964), 2–11. See also Yu. G. Saushkin and A. M. Smirnov, 'Geosystems and geostructures', *Soviet Geography*, 11 (1968), 149–154; A. F. Plakhotnik, 'The subject and structure of geosystems theory', *Soviet Geography*, 14 (1974), 429–436; N. Beroutchachvili and J.-L. Mathieu, 'L'éthologie des géosystèmes', *Espace géographique*, 6 (1977), 73–84; K. V. Zvorykin and A. Yu. Reteyum, 'Systems research in the study of nature', *Voprosy geografii*, 104 (1977), 1–232; J. Demek, 'The landscape as a geosystem', *Geoforum* 9 (1978), 29–34.

12

Putting the Geography Back in the Bio-

Much of this book has been concerned with the nature of scientific change, in terms both of the history of ideas and of the social and institutional context in which change occurs. I have argued that Kuhn's concept of paradigm change is insufficiently subtle to subsume all the diverse aspects of these subjects within a single framework.

The previous chapter discussed the history and usefulness of two general ways of organizing our views of the natural world, more particularly as they were applied during the formative period of academic geography. In this final chapter I wish to examine more contemporary issues, specifically the nature of the field of study of biogeography. This is not, however, a formal contribution to the history of biogeography:[1] rather, it explores the contradictions explicit today in an extremely active field, in the light of the approaches developed in this book. I leave the reader to judge to what extent Kuhn's notion of the paradigm, Gramsci's concept of hegemony or other interpretative structures are the most appropriate tools for making sense of the wildly divergent views I shall describe.[2] It will be clear from my argument where my own sympathies lie.[3]

Biogeography is rapidly becoming one of the most popular branches of physical geography at undergraduate level in Britain; it has its own professional groups within both the Institute of British Geographers and the Association of American Geographers; and it now has specialist journals devoted to it. But these external marks of academic coherence hide a central

[1] P. Stott, 'History of biogeography', in J. A. Taylor, ed., *Themes in biogeography* (London: Croom Helm, 1983), 1–24.

[2] Much of the literature is frankly polemical. Nowhere is this more explicit than in Leon Croizat's denunciation of Ernst Mayr. See L. Croizat, 'Mayr vs. Croizat: Croizat vs. Mayr – an enquiry', *Tuatara*, 27 (1984), 49–66.

[3] Likewise the cases I cite are mainly from areas in which I have worked or on groups in which I am interested: they are illustrative, not comprehensive.

paradox: the intellectual thrust of modern biogeography derives not from geographers at all, but from theoretical directions given over the last twenty years either by ecologists, notably such students of G. Evelyn Hutchinson as Robert MacArthur, or by systematists working on museum collections, such as Gareth Nelson and Norman Platnick. Meanwhile geographers themselves have remained heavily committed to the historical elucidation of environmental change, especially in the late Pleistocene and Holocene; to the study of ecosystem structure and function; and to problems of conservation and management – all of which, though labelled as biogeographical by their proponents, are either temporal in focus or at least not overtly spatial in design.

This paradox is the more striking when it is recalled that 'classical' biogeography relied heavily on the analysis of spatial distributions. Alfred Russel Wallace in the East Indies,[4] H. W. Bates in South America,[5] J. D. Hooker in the Himalayas as well as in southern temperate and sub-Antarctic latitudes,[6] and of course Darwin himself, were all interested in spatial distributions and especially discontinuities, and their work led not only to classificatory and regional schemes[7] but also to wide inferences on tectonic history, climatic change and movements of sea level. Geographers in general failed to follow this line of advance because of the depth as well as the breadth of taxonomic expertise required, and it was gradually abandoned to the systematists.

Instead, they looked for linkages with other forms of enquiry in physical geography, and found them in the study of palaeoenvironments, notably in Britain the study of the changing habitats of the later Pleistocene, and this remains a major research focus.[8] But its central techniques – of pollen

[4] A. R. Wallace, *The distribution and abundance of animals* (London: Macmillan, 2 volumes, 1876); A. R. Wallace, *Island life; or, The phenomena and causes of insular faunas and floras, including a revision and attempted solution of the problem of geological climates* (London: Macmillan, 1892).

[5] H. W. Bates, *The naturalist on the River Amazons, a record of adventures, habits of animals, sketches of Brazilian and Indian life and aspects of nature under the Equator during eleven years of travel* (London: John Murray, 2 volumes, 1863); H. P. Moon, *Henry Walter Bates 1825–1892: explorer, scientist and Darwinian* (Leicester: Museums, Art Galleries and Records Service, 1976).

[6] J. D. Hooker, 'Introductory essay', in *Flora Novae-Zelandiae, Part I, Flowering plants* (London: Lovell Reeve, 1853), i–xxxix; J. D. Hooker, 'Introductory essay', in *Flora Tasmania, Part I, Dictoyledones* (London: Lovell Reeve, 1860), ix–cxxviii.

[7] G. G. Simpson, 'Too many lines; the limits of the Oriental and Australian zoogeographic regions', *Proceedings of the American Philosophical Society*, 121 (1977), 107–120.

[8] L. F. Curtis and I. G. Simmons, eds, 'Man's impact on past environments', *Transactions of the Institute of British Geographers*, N.S. 1 (1976), 257–384.

analysis, faunal reconstruction and stratigraphy – are not specifically geographical, and there is a clear overlap with the work of palaeobiologists. Geographers, it is true, often emphasize the role of man in palaeoenvironmental modification, but so too do other palaeoenvironmental scientists.

In an attempt to develop a new and more individual orientation, biogeographers have turned since the 1950s to the ecosystem concept and used it as a vehicle to develop principles and concepts for the subject.[9] But such studies, emphasizing structure and function, have practically abandoned classical concerns with the distribution, ranges and limits of particular plants and animals. When plate tectonics gave dramatic new insight into these questions in the early 1970s, the new interpretations came from zoologists and botanists,[10] and the geographers were strangely silent.

Not surprisingly, therefore, there remains considerable controversy over what biogeography ought to be: practitioners in search of a paradigm continue to mirror the divergent concerns of the subject over the last hundred years. From outside geography, Fosberg[11] has roundly told us that our prime task as biogeographers is to study 'organisms . . . as components of patterns. Spatial and temporal patterns, major processes, secular changes, areas rather than sites, total populations rather than individuals or local populations.' Yet Gersmehl's subsequent call[12] for an 'alternative biogeography', based on the Hubbard Brook mineral cycling studies, virtually abandons distribution in favour of specific ecosystem studies. He emphasizes their pedagogic utility, but what he proposes is hardly bio*geography*, except by radical redefinition of the language. Within geography, only Stott in recent years has held to the theme that the core of biogeography lies in the analysis and explanation of pattern and distribution.[13] Meanwhile, there continue also to be concerns with practical and management issues as a framework for the subject, and continuing suggestions that the philosophi-

[9] See chapter 11, and also D. Watts, *Principles of biogeography: an introduction to functional mechanisms of ecosystems* (New York: McGraw-Hill, 1971).

[10] C. B. Cox, 'Vertebrate palaeodistributional patterns and continental drift', *Journal of Biogeography*, 1 (1974), 75–94; A. Hallam, 'Changing patterns of provinciality and diversity of fossil animals in relation to plate tectonics', *Journal of Biogeography*, 1 (1974), 213–225; T. C. Whitmore, 'Plate tectonics and some aspects of Pacific plant geography', *New Phytologist*, 72 (1973), 1185–1190; R. F. Thorne, 'Plate tectonics and angiosperm distribution', *Notes from the Royal Botanic Gardens, Edinburgh*, 36 (1978), 297–315.

[11] F. R. Fosberg, 'Geography, ecology and biogeography', *Annals of the Association of American Geographers*, 66 (1976), 117–128; reference on p. 119.

[12] P. J. Gersmehl, 'An alternative biogeography', *Annals of the Association of American Geographers*, 66 (1976), 223–241.

[13] P. Stott, *Historical plant geography: an introduction* (London: Allen & Unwin, 1981).

cal ideas of man–environment relationships have potential significance in structuring a definition and a field of study.[14]

In this chapter I wish to examine the dominant research concerns of recent years, and to demonstrate the lack of and need for a new spatial emphasis in biogeography. My intention – following the earlier and broader concerns of both W. M. Davis and S. W. Wooldridge about taking the *ge-* out of *geography*[15] – is to put the *geography* back in the *bio-*.

ISLAND BIOGEOGRAPHY AND ITS IMPLICATIONS

Since 1967 there is no doubt that the major intellectual thrust of a spatially orientated biogeography has come from MacArthur and Wilson's *Theory of island biogeography*.[16] In this book they postulated that the number of species on an island is the result of a dynamic equilibrium between species arriving and colonizing (a function of distance from source) and species becoming extinct (a function of island area). The close relationship between area and species number has frequently been empirically confirmed, but there is nevertheless considerable uncertainty over its meaning, in the general as well as in the insular case. Connor and McCoy[17] found that there is no unique theoretical basis for the relationship, which could derive from the properties of area itself or from habitat diversity, or simply be an artefact of sampling; that a variety of best-fit curves can be fitted to it; and that no specific biological meaning can be assigned to its slope and intercept parameters. These arguments, developed in general terms, are applied particularly to the equilibrium theory of island biogeography by Gilbert.[18] He was concerned to show that the species–area relationship,

[14] A. R. Hill, 'Biogeography as a sub-field of geography', *Area*, 7 (1975), 156–161; G. B. Kolata, 'Human biogeography: similarities between man and beast', *Science, N.Y.*, 185 (1974), 134–135.

[15] My title is derived from S. W. Wooldridge, 'On taking the ge- out of geography', *Geography*, 34 (1949), 9–18, reference on p. 15; reprinted in S. W. Wooldridge, *The geographer as scientist: essays on the scope and nature of geography* (London: Nelson, 1956), 7–25. The phrase was originally used by W. M. Davis in 'A restrospect of geography', *Annals of the Association of American Geographers*, 22 (1932), 211–230, reference on p. 229, though this was not cited by Wooldridge.

[16] R. H. MacArthur and E. O. Wilson, *The theory of island biogeography* (Princeton: Princeton University Press, 1967); see also M. Williamson, *Island populations* (Oxford: Oxford University Press, 1981).

[17] E. F. Connor and E. D. McCoy, 'The statistics and biology of the species–area relationship', *American Naturalist*, 113 (1979), 791–833.

[18] F. S. Gilbert, 'The equilibrium theory of island biogeography: fact or fiction?, *Journal of Biogeography*, 7 (1980), 209–235.

while generally applicable, could not be determinate, and that the way in which area affected species numbers was unknown. Meaning could only be assigned to any specific relationship by consideration of individual instances. Nevertheless, attempts continue to be made to refine the equilibrium model, particularly by incorporating probabilistic terms.[19]

The paucity of empirical data in MacArthur and Wilson's original formulation – they relied heavily on floristic data from the islands of Kapingamarangi Atoll in the Carolines[20] and the Dry Tortugas in Florida – has encouraged a wide range of investigations which in general have yielded impressive empirical support for the observed regularities they sought to explain. Studies have been made of insect diversity on flowers[12] and tea plants;[22] in patches of nettles,[23] caves[24] and ponds;[25] in city parks[26] and laboratory 'islands';[27] of plants on reef islands[28] – and the list is far from exhaustive. Other studies have added refinement to knowledge of classic island areas such as the Galapagos, either through improved taxonomic

[19] D. Simberloff, 'Using island biogeographic distributions to determine if colonization is stochastic', *American Naturalist*, 112 (1978), 713–726; S. H. Faeth and E. F. Connor, 'Supersaturated and relaxing island faunas; a critique of the species–age relationship', *Journal of Biogeography*, 6 (1979), 311–316; M. E. Gilpin and J. M. Diamond, 'Immigration and extinction probabilities for individual species: relation to incidence functions and species colonization curves', *Proceedings of the National Academy of Science*, 78 (1981), 392–396; S. J. Wright and C. C. Biehl, 'Island biogeographic distributions: testing for random, regular, and aggregated patterns of species occurrence', *American Naturalist*, 119 (1982), 345–357.

[20] W. A. Niering, 'Terrestrial ecology of Kapingamarangi Atoll, Caroline Islands', *Ecological Monographs*, 33, 131–160.

[21] R. P. Seufert, 'Clumps of *Heliconia* inflorescences as ecological islands', *Ecology*, 56 (1975), 1416–1422.

[22] B. Banerjee, 'An analysis of the effect of latitude, age and area on the number of arthropod pest species of tea', *Journal of Applied Ecology*, 18 (1981), 339–342.

[23] B. N. K. Davis, 'The colonization of isolated patches of nettles (*Urtica dioica* L.) by insects', *Journal of Applied Ecology*, 12 (1975), 1–14.

[24] D. Culver, J. R. Holringer and R. Baroody, 'Toward a predictive cave biogeography: the Greenbrier Valley as a case study', *Evolution*, 27 (1973), 689–695.

[25] M. D. Hubbard, 'Experimental insular biogeography: ponds as islands', *Florida Scientist*, 36 (1973), 132–141.

[26] S. H. Faeth and T. C. Kane, 'Urban biogeography: city parks as islands for Diptera and Coleoptera', *Oecologia*, 32 (1978), 127–133.

[27] A. Schoener, 'Colonization curves for planar marine islands', *Evolution*, 55 (1974), 818–827; B. Wallace, 'Biogeography of laboratory islands', *Evolution*, 29 (1975), 622–635.

[28] D. R. Stoddart and F. R. Fosberg, 'Species–area relationships on small islands: floristic data from Belizean sand cays', *Smithsonian Contributions to the Marine Sciences*, 12 (1982), 527–539; D. R. Stoddart and F. R. Fosberg, 'Vegetation and floristics of western Indian Ocean coral islands', in D. R. Stoddart, ed., *Biogeography and ecology of the Seychelles islands* (The Hague: W. Junk, 1984), 221–238.

information[29] or through the realization that parameters such as area may themselves be variable over the time during which colonization has occurred.[30] Buckley[31] has drawn attention to the heterogeneity of islands and hence the difficulty of using them as homogeneous units of comparison in species–area analysis, and I have used the case of the Dry Tortugas, on which MacArthur and Wilson relied so heavily, to demonstrate the grave deficiencies of data available even for the best-studied islands.[32]

Perhaps the major support and extension of MacArthur and Wilson's theory, however, has come from Diamond's work on the California Channel Islands and on the satellite islands of New Guinea, in which he introduced the concept of relaxation time for the adjustment of a biota to a new equilibrium following environmental perturbations such as glacial changes of sea-level.[33] At the same time Simberloff and Wilson were demonstrating experimentally by defaunating and subdividing small mangrove islets in Florida that new equilibria in species numbers were rapidly established on 'virgin' islands independently of the taxa composing the fauna and were constrained simply by trophic considerations, distance and area.[34] Heatwole and Levins in a rare study of short-term turnover in the real world had reached similar conclusions on a Puerto Rican sand cay. The same authors also showed the reality of rafting on flotsam

29 M. P. Johnson and P. H. Raven, 'Species number and endemism: the Galapagos Archipelago revisited', *Science, N.Y.*, 179 (1973), 893–895.

30 B. B. Simpson, 'Glacial migrations of plants: island biogeographical evidence', *Science, N.Y.*, 185 (1974), 698–700.

31 R. C. Buckley, 'Scale-dependent equilibrium in highly heterogeneous islands: plant geography of the northern Great Barrier Reef sand cays and shingle islets', *Australian Journal of Ecology*, 6 (1981), 143–147.

32 D. R. Stoddart and F. R. Fosberg, 'Topographic and floristic change, Dry Tortugas, Florida, 1904–1977', *Atoll Research Bulletin*, 253 (1981), 1–55.

33 J. M. Diamond, 'Distributional ecology of New Guinea birds', *Science, N.Y.*, 179 (1973), 759–769; J. M. Diamond and E. Mayr, 'Species–area relation for birds of the Solomon Archipelago', *Proceedings of the National Academy of Science*, 73 (1976), 262–266; M. E. Gilpin and J. M. Diamond, 'Calculation of immigration and extinction curves for the species–area–distance relationship', *Proceedings of the National Academy of Science*, 73 (1976), 4130–4134.

34 Discussion by D. S. Simberloff, 'Equilibrium theory of island biogeography and ecology', *Annual Review of Ecology and Systematics*, 5 (1974), 161–182; D. S. Simberloff, 'Species turnover and equilibrium island biogeography', *Science, N.Y.*, 194 (1976), 572–578. For the classic colonization experiments, D. S. Simberloff and E. O. Wilson, 'Experimental zoogeography of islands: the colonization of empty islands', *Ecology*, 50 (1969), 278–296, and 'Experimental zoogeography of islands: a two year record of colonization', *Ecology*, 51 (1970), 934–937.

as a means of distribution of land animals across the sea.[35]

However, the support provided for the theory has not gone without challenge. On a rather trivial level, there has been concern over some of the computational procedures used, which give misleading results when extreme values are involved,[36] and also over the fact that the analyses may throw more light on levels of knowledge than on numbers of species.[37] More fundamentally there has been concern over two points. To what extent does the simple counting of species have any ecological meaning? Can a regular visitor, a vagrant and a breeding bird be counted in the same way? (Indeed, can they always be differentiated during rapid surveys?) How can 'colonization' be operationally defined? And then: what of extinction? Sighting a species is a fairly certain indication of its presence (though by no means an infallible one), but the reverse is by no means true, especially on rugged forested islands. What does extinction mean? Could one rank together as comparable events the extinction of the sparrow and the osprey on the British islands?

These doubts came to a head in a re-examination of Diamond's Channel Islands data, where high turnover (i.e. high colonization rates and high extinction rates) had apparently been demonstrated over the last 50 years. Lynch and Johnson showed that the available information, especially the 1916 baseline data, could not support the theoretical load placed on it.[38] They also argued that if apparent changes were to be accepted as evidence of 'equilibrium turnover' they had to be demonstrated to be stochastic in nature rather than the result of either long-term or short-term habitat change (palaeoturnover, successional turnover), inadequate information (pseudoturnover), or human activities. They set rigorous criteria for the recognition of immigration and extinction, while recognizing that their criteria could rarely be met, given the nature of past records of species and also the prevalence of artificial habitat change in the present century. As a result, in re-analysing Diamond's data, they found that only one out of 41

[35] H. Heatwole and R. Levins, 'Biogeography of the Puerto Rico Bank: flotsam transport of terrestrial animals', *Ecology*, 53 (1972), 112–117, and 'Biogeography of the Puerto Rico Bank: species turnover on a small cay, Cayo Ahogado', *Ecology*, 54 (1973), 1042–1055.

[36] E. D. McCoy and E. F. Connor, 'Environmental determinants of island species number in the British Isles: a reconsideration', *Journal of Biogeography*, 3 (1976), 381–382.

[37] See the amusing exchange on rat ectoparasite diversity by W. Dritschilo, H. Cornell, D. Nafus and B. O'Connor, 'Insular biogeography: of mice and mites', *Science, N.Y.*, 190 (1975), 467–469, and A. M. Kuris and A. Blaustein, 'Ectoparasitic mites on rodents: application of the island biogeography theory', *Science, N.Y.*, 195 (1977), 596–598.

[38] J. F. Lynch and N. K. Johnson, 'Turnover and equilibria in insular avifaunas, with special reference to the California Channel Islands', *Condor*, 76 (1974), 370–384.

extinctions recorded by him could be considered natural. Thus if the similarly high colonization rates proposed by Diamond were themselves accurate, the faunas could not in consequence be in equilibrium.

It has also been observed that numbers of species need to be disaggregated into trophic levels if they are to have ecological meaning,[39] and a rather similar point is made for the birds of the western Pacific by Abbott and Grant.[40] For example in the California Channel Islands raptors account for 38 per cent of all extinctions,[41] which is equivalent to saying that those species highest in the food chain are most prone to extinction, since they are fewer in number and require a larger area to support them.[42] Simberloff's reconsideration of the turnover issue leans heavily on his own mangrove islet data, and would be more attractive if his islets were not so minute.[43]

Doubt from a quite different quarter comes from Lack's final studies on West Indian birds.[44] He does not necessarily dispute the empirical regularities which form the basis of MacArthur and Wilson's theory, but he cannot accept that distance and area are sufficient conditions for the explanation of the presence or absence of birds on particular islands. Thus, although hummingbirds are frequently observed in the Cayman Islands, and are thus not excluded by distance, they have never become established.[45] The reason, Lack argues, must be ecological: to do with the nature of the habitat or the structure of the communities present, but in what manner this control acts, other than through competitive exclusion, Lack is uncertain. A similar observation has been made to account for low floristic diversity on south Pacific coral islands immediately adjacent to high volcanic island 'species reservoirs'.[46] Terborgh, on the other hand, has noted that at least in numbers of bird species in the West Indies, the faunas of high wet forested islands are identical to those of low dry scrubby ones, and he draws the

[39] G. L. Hunt Jr and M. W. Hunt, 'Trophic levels and turnover rates: the avifauna of Santa Barbara Island, California', *Condor*, 76 (1974), 363–369.

[40] I. Abbott and P. R. Grant, 'Nonequilibrial bird faunas on islands', *American Naturalist*, 110 (1976), 507–528.

[41] J. F. Lynch and N. K. Johnson, *op. cit.* (note 38).

[42] J. Terborgh, 'Preservation of natural diversity: the problem of extinction prone species', *BioScience*, 24 (1974), 715–722.

[43] D. S. Simberloff, *op. cit.* (1976) (note 34).

[44] D. Lack, *Island biogeography illustrated by the land birds of Jamaica* (Oxford: Basil Blackwell, 1976).

[45] D. Lack, 'The numbers of species of hummingbirds in the West Indies', *Evolution*, 27 (1973), 326–337.

[46] D. R. Stoddart, 'Vegetation and floristics of the Aitutaki motus', *Atoll Research Bulletin*, 190 (1975), 87–116.

conclusion that habitat differences cannot be of great importance.[47]

The major problem in such studies is that there are so few direct observations of either colonization or extinction, and it is difficult to see how in many cases these processes can be amenable to direct study. At present many analyses lean heavily on older and necessarily deficient surveys, and some of the questions raised will not be resolved until surveys comparable in quality to those now available are made in future decades.

In spite of these uncertainties, empirical studies continue to be of interest. Let me cite examples from the tropical seas. Hnatiuk studied the plant species of 100 small lagoon islands at Aldabra Atoll in the southern Seychelles.[48] She found that log island area accounted for 70 per cent of variance in plant species number on the islands; islet height a further 9 per cent. The increase in species numbers is largely accounted for by woody plants, whereas on larger tropical islands such as those of Kapingamarangi and Aitutaki the increase is accounted for by herbaceous weeds as a result of human interference.[49] Log area was also the main factor on the lagoon islands explaining the number of species of land birds (66.5 per cent) and wading birds, but to a much lesser degree (36.2 per cent) for seabirds. Conversely the main factor accounting for insect diversity is log number of plant species,and for land crustaceans it is the complexity of the vegetation structure. Values of z relating area and number of species range from 0.345 for plants, 0.205 for land birds, 0.174 for waders, to 0.136 for land crustaceans.[50]

A similarly large number of comparable limestone islets has been studied near Perth, Western Australia, with respect to land plants and land birds. For land birds, these studies show no simple relationship between habitat diversity and area, and the latter cannot thus be regarded as an easily measured surrogate for the former.[51] Sequential studies on these islands make possible statements about turnover on different time-scales,[52] as well

[47] J. Terborgh, 'Chance, habitat and dispersal in the distribution of birds in the West Indies', *Evolution*, 27 (1973), 338–349.

[48] S. H. Hnatiuk, 'Numbers of plant species on the islands of Aldabra Atoll', *Philosophical Transactions of the Royal Society of London*, B 286 (1979), 247–254.

[49] D. R. Stoddart, *op. cit.* (note 46).

[50] S. H. Hnatiuk, 'The numbers of land birds, waders, sea birds, land crustacea, and certain insects on the lagoon islands of Aldabra Atoll', *Biological Journal of the Linnean Society*, 14 (1980), 151–161.

[51] I. Abbott, 'Species richness, turnover and equilibrium in insular floras near Perth, Western Australia', *Oecologia*, 33 (1978), 221–233.

[52] I. Abbott and R. Black, 'Changes in species composition on floras on islets near Perth, Western Australia', *Journal of Biogeography*, 7 (1980), 399–410.

as about the relative importance of competitive exclusion and stochastic colonization processes.[53]

What is, however, apparent about most of these studies in island biogeography is that the data once collected are handled in a completely aspatial manner. Island area becomes a statistic divorced from location, as do other attributes of islands; likewise distance becomes a geometric notion divorced from the specifics of locality.

DISPERSAL AND DRIFT

Simultaneously with the development of island biogeographic studies, and on a quite different scale, botanists and zoologists were engaged in an increasingly acrimonious debate over the methods and techniques of 'classical' biogeography on the one hand and over the new taxonomic methods of cladistics, combined with the insights of plate tectonics, on the other. Let us examine this debate initially by focusing on the efficiency – or indeed existence – of dispersal mechanisms, since these lie at the heart both of island biogeography and of the so-called 'new biogeography' proposed with some vehemence by the followers of Leon Croizat.

For biogeographers such as Mayr and Simpson, Fosberg and Zimmerman, much of plant and animal distribution can be accounted for in terms of the movements of individuals or propagules. Dispersal has become a basic tenet of both island and continental biogeography, an emphasis traced back by Nelson to Darwin.[54] Nelson makes the distinction between what he terms 'ordinary' dispersal in the absence of a barrier, and 'improbable dispersal' across a barrier.[55] By the latter he means dispersal events of low probability, typified by the occasional colonizations of oceanic islands through overwater dispersal by air or sea. Because such events are infrequent and scattered they are rarely observed, but given sufficient time, as Simpson argued,[56] their occurrence at any point becomes increasingly probable.

[53] I. Abbott, 'The composition of landbird faunas of islands round south-western Australia; is there evidence for competitive exclusion?', *Journal of Biogeography*, 8 (1981), 135–144.

[54] G. Nelson, 'From Candolle to Croizat: comments on the history of biogeography', *Journal of the History of Biology*, 11 (1978), 269–305.

[55] The *Concise Oxford Dictionary* gives, as the first meaning of 'improbable', 'not likely to be true'. Such polemical use of pejorative vocabulary, while characteristic of some vicariance biogeographers, does not adequately convey the technical point required.

[56] G. G. Simpson, 'Probabilities of dispersal in geologic time', *Bulletin of the American Museum of Natural History*, 99 (1952), 163–176; D. I. Axelrod, 'Variables affecting the probabilities of dispersal in geologic time', *Bulletin of the American Museum of Natural History*, 99 (1952), 177–188.

Thus, it is argued, dispersal provided for Darwin 'a general explanation for all [*sic*] anomalous distributions', and Darwinian biogeography 'became a science of the improbable, the rare, the mysterious, and the miraculous'.[57] Why this 'misadventure in science', as Nelson calls it? He volunteers the 'explanation' that it was because Darwin himself 'worked in an age of colonialism' which led him naturally to find attraction in ideas of colonization, competition and dominance.[58] Rosen too believes that Darwin's theories were based on the 'biological application and affirmation of 19th-century axioms about human social values'.[59]

The criticism is often made that proponents of dispersal concentrate on highly vagile organisms such as birds: the great expansion of the Cattle Egret *Bubulcus ibis* to and through the western hemisphere in the last 50 years is repeatedly cited as an example. Similarly Burger *et al.*[60] describe the case of the sub-Antarctic Prince Edward Islands, which have but one species of breeding land bird, but where over a period of 64 months no less than 28 vagrant individuals in more than 17 species arrived but failed to establish themselves: in this instance dispersal is clearly demonstrated. In the case of ferns, Tryon[61] finds their means of dispersal by wind so efficient that their relative distributions in the Greater and Lesser Antilles are ecologically constrained rather than limited by distance. As a final example, Vagvolgyi has shown that the land Mollusca of Pacific islands have significantly smaller shells than those of continental areas, and he asks what other more plausible explanation could there be, if aerial dispersal does not account for this.[62]

There is, in fact, considerable interest in comparing the reviews of land

[57] G. Nelson, *op. cit.* (note 54), p. 289.

[58] *Ibid.*, p. 203.

[59] D. E. Rosen, 'Introduction', in G. Nelson and D. E. Rosen, eds, *Vicariance biogeography: a critique* (New York: Columbia University Press, 1981), 1–5; reference on p. 2. The suggestion inverts the usual relationship between Darwinism and nineteenth-century social thought, and as it applies to Darwin himself it is profoundly unhistorical. Doubtless it would have given Governor Eyre some satisfaction to think that Darwin would come to be accused of being a colonialist. Darwin's concepts of biogeography and tectonics were vastly more subtle: see D. R. Stoddart, 'Darwin, Lyell, and the geological significance of coral reefs', *British Journal for the History of Science*, 9 (1976), 199–218.

[60] A. E. Burger, A. J. Williams and J. C. Sinclair, 'Vagrants and the paucity of land bird species at the Prince Edward Islands', *Journal of Biogeography*, 7 (1980), 305–310.

[61] R. Tryon, 'Biogeography of the Antillean fern flora', in D. Bramwell, ed., *Plants and islands* (London: Academic Press, 1979), 55–68.

[62] J. Vagvolgyi, 'Body size, aerial dispersal, and origin of the Pacific land snail fauna', *Systematic Zoology*, 24 (1976), 465–488; J. Vagvolgyi, 'Why are so many minute land snails on the Pacific islands: a response to Leon Croizat', *Systematic Zoology*, 27 (1978), 213.

snails on islands contributed by Solem[63] and Peake.[64] Their data show clearly that snails have dispersed across water, but that their history is not a simple one. In general snails are poor dispersers but good colonists, as the record of historical introductions on islands shows; in this they contrast with, for example, land birds, which are good dispersers but less effective colonizers. Thus dispersability alone is but part of the story, which requires ecological understanding for its completion. Solem concludes that 'land snails show evidence that pleases neither the dispersalist nor the vicariant biogeographer. Unaware of such controversies, the snails remain happily munching on litter, oblivious to changing flora and fauna. They worry about water and shelter, not about systematist or biogeographer.'[65]

At the opposite end of the animal kingdom from the light-shelled land snails is the elephant. Dramatic evidence of dispersal comes from Johnson's investigation of pygmy proboscideans *Mammuthus exilis* on the California Channel Islands.[66] In contrast to existing ideas that elephants cannot swim and hence must have walked over land bridges to insular locations, Johnson clearly shows that elephants are excellent swimmers at speeds of up to 2.7 km/hr and distances of up to 48 km.[67]

The cases where insular biotas can be explained by dispersal are legion. Let us return to Aldabra Atoll in the western Indian Ocean. Here Wickens has analysed the dispersal mechanisms of the 176 native terrestrial angiosperms,[68] and has described in detail the dispersal adaptations of all 263 species in the flora.[69] He estimates that propagules would require to remain buoyant for 5–7.5 days to reach Aldabra by sea from Africa or Madagascar; and that bird dispersal can account for the introduction of one

[63] A. Solem, 'A theory of land snail biogeographic patterns through time', in S. van der Spoel, A. van Bruggen and J. Lever, eds, *Pathways in malacology* (Amsterdam: European Malacological Congress, 1979), 225–249; A. Solem, 'Biogeographic significance of land snails, Paleozoic to Recent', in J. Gray and A. J. Boucot, eds, *Historical biogeography, plate tectonics, and the changing environment* (Corvallis: Oregon State University Press, 1979), 277–287.

[64] J. F. Peake, 'The land snails of islands – a dispersalist's viewpoint', in P. L. Forey, ed., *The evolving biosphere* (Cambridge: Cambridge University Press, 1981), 247–263.

[65] A. Solem, 'Land-snail biogeography: a true snail's pace of change', in G. Nelson and D. E. Rosen, eds, *op. cit.* (note 59), 197–221; reference on p. 221.

[66] D. L. Johnson, 'The origin of island mammoths, and the Quaternary land bridge history of the Northern Channel Islands, California', *Quaternary Research*, 10 (1978), 204–225.

[67] D. L. Johnson, 'Problems in the land vertebrate zoogeography of certain islands and the swimming powers of elephants', *Journal of Biogeography*, 7 (1980), 383–398.

[68] G. E. Wickens, 'Speculations on seed dispersal and the flora of the Aldabra archipelago', *Philosophical Transactions of the Royal Society of London*, B 286 (1979), 85–97.

[69] G. E. Wickens, 'The propagules of the terrestrial flora of the Aldabra archipelago, western Indian Ocean', *Atoll Research Bulletin*, 229 (1979), 1–38.

taxon in each 400 years. Supporting evidence comes from Hnatiuk's finding that 56 of 69 species whose seeds were immersed in seawater for up to 56 days showed no adverse effects on germination.[70] She also showed that 28 species of grasses, herbs and woody plants could be germinated from the faeces of the Aldabra Giant Tortoise *Geochelone gigantea*. The average time of passage of food through the tortoise gut is 27 days and may reach 50 days. 'Assuming that neither current speeds and directions nor the digestive rates among tortoises have changed since tortoises became extinct on Madagascar, a tortoise floating to Aldabra from Madagascar could have reached the atoll and penetrated some distance inland in less time than it would have taken for its last Madagascan meal to be voided.'[71] There is, it might be noted, no doubt about the ability of tortoises to survive at sea and thus make possible their own dispersal and those of ingested seeds: the geological record at Aldabra shows successive colonizations and extinctions with the emergence and submergence of the land area during the Pleistocene.[72] Records also exist of both Aldabran and Galapagos tortoises floating at sea. Nevertheless, in the Pacific case, Melville still persists in welding the Galapagos Islands on to Peru so that 'tortoises, which soon drown in water, would have been able to cross unimpeded' – this in support of a vicariance rather than a dispersal explanation of the biota of the islands.[73]

All of these examples, it is true, refer to islands rather than to continents. But the point can be made that if present continental distributions result from the fragmentation of previously intact distributions by tectonic and other environmental events, as vicariance biogeographers demand, then these former more extensive distributions must have resulted from the dispersal and extension of range of the taxa concerned. Analysis of this prior dispersal ought thus to precede that of the fragmentation of the distribution, and logically must be as important in explaining present

[70] S. H. Hnatiuk, 'A survey of germination of seeds from some vascular plants found on Aldabra Atoll', *Journal of Biogeography*, 6 (1979), 105–114.

[71] S. H. Hnatiuk, 'Plant dispersal by the Aldabra Giant Tortoise *Geochelone gigantea* (Schweigger)', *Oecologia*, 36 (1978), 345–350.

[72] C. J. R. Braithwaite, J. D. Taylor and W. J. Kennedy, 'The evolution of an atoll: the depositional and erosional history of Aldabra', *Philosophical Transactions of the Royal Society of London*, B 226 (1973), 307–346; J. D. Taylor, C. J. R. Braithwaite, J. F. Peake and E. N. Arnold, 'Terrestrial faunas and habitats of Aldabra during the late Pleistocene', *Philosophical Transactions of the Royal Society of London*, B 286 (1979), 47–66.

[73] R. Melville, 'Vicarious plant distributions and paleogeography of the Pacific region', In G. Nelson and D. E. Rosen, eds, *op. cit.* (note 59), 238–274, with discussion by B. N. Haugh and G. F. Edmunds Jr, 275–302.

patterns. It is a point which Croizat dismisses as 'irrelevant'.[74] Nevertheless, dispersal (both 'ordinary' and 'improbable') remains a demonstrable fact: Croizat's attempt[75] to discredit it by drawing attention to the comparably extensive distributions of both earthworms and swallows, and inferring from this the unimportance of dispersal processes, is simply rhetoric, not science.

CROIZAT, CLADISTICS AND VICARIANCE BIOGEOGRAPHY

This discussion brings us to the central issues which have dominated the research frontier in biogeography for more than a decade. That these cannot be ignored is shown by two major symposia, at the American Museum of Natural History in 1979[76] and at the British Museum (Natural History) in 1981,[77] as well as by the first textbook of what might be termed (perhaps prematurely) a 'new biogeography'.[78] This 'new biogeography' finds its intellectual origins in Leon Croizat, a Venezuelan Italian who died at the age of 88 in 1982 after a turbulent scholarly and personal career.[79]

Croizat's major works – his *Panbiogeography* and *Space, time, form: the biological synthesis*[80] – initially attracted little attention. But they were sympathetically reviewed by Nelson in 1973,[81] summarized and extended by additional studies of Africa and Asia by Croizat himself,[82] and have been

[74] L. Croizat, G. Nelson and D. E. Rosen, 'Centers of origin and related concepts', *Systematic Zoology*, 23 (1974), 265–287; reference on p. 279.

[75] L. Croizat, 'Biogeography: past, present, and future', in G. Nelson and D. E. Rosen, eds, *op. cit.* (note 59), 501–523; reference on p. 505.

[76] G. Nelson and D. E. Rosen, eds, *op. cit.* (note 59).

[77] R. Sims and J. H. Price, eds, *Evolution, time and space: the evolution of the biosphere* (London: Academic Press, 1983).

[78] G. Nelson and N. I. Platnick, *Systematics and biogeography: cladistics and vicariance* (New York: Columbia University Press, 1981).

[79] R. C. Craw, 'Never a serious scientist: the life of Leon Croizat', *Tuatara*, 27 (1984), 5–7.

[80] L. Croizat, *Panbiogeography, or an introductory synthesis of zoogeography, phytogeography, and geology; with notes on evolution, systematics, ecology, anthropology, etc.* (Caracas: The Author, 2 volumes, 1958); L. Croizat, *Space, time, form: the biological synthesis* (Caracas: The Author, 1964).

[81] G. Nelson, 'Comments on Leon Croizat's Biogeography', *Systematic Zoology*, 22 (1973), 312–320.

[82] L. Croizat, 'The biogeography of the tropical lands and islands east of Suez-Madagascar: with particular reference to the dispersal and form-making of *Ficus* L., and different other vegetal and animal groups', *Atti dell'Istituto Botanica della Universita Laboratorio Crittogamico, Pavia*, (6) 4 (1968), 1–400; L. Croizat, 'Introduction raisonnée à la biogéographie de l'Afrique', *Memorias de la Sociedad Broteriana*, 20 (1968), 1–451.

the subject of many recent commentaries, including the symposia mentioned above. Simply stated, Croizat's method is to map distributions of whole varieties of taxa, many of them now disjunct; to establish by comparison concordances of pattern; thus to identify ancestral biotas; and hence to require explanations for the fragmentation of previously more extensive distributions. The method of argument is thus apparently the inverse of the Simpson–Mayr–Darlington approach based on the identification of centres of origin and subsequent dispersal or migration from them. What has given the Croizat revival its power, however, is that biogeographers such as Simpson and Darlington phrased their historical reconstructions in terms of continental stability, whereas their accepted mobility provides an immediate explanation of broad disjunction and of the fragmentation of biotas as well as of land masses. Characteristically, however, Croizat himself declines to utilize geological evidence, which is 'of no account the moment these theories contradict the evidence from biogeography as such'.[83]

The phenomenon of the fragmentation of ancestral biotas is termed vicariance, and monophyletic taxa so isolated are termed vicariant. The basic method used by Croizat is to derive 'generalized patterns of biotic distribution' from the correspondence of individual species distributions: 'If a given type of distribution (individual track) recurs in group after group of organisms, the region delineated by the coincident distributions (generalized track) becomes statistically and, therefore, geographically significant and invites explanation on a general level.'[84] Among others, Rosen in particular has used the method both for a group of salmoniform fishes and for a region, the Caribbean.[85]

Critics have not been slow to point out that the method is essentially inductive. McDowall states that there is never any problem in finding additional species whose distributions fit a 'generalized track', and that such records merely corroborate, never confirm, and cannot falsify a generalization.[86] What matters for Ball is not the number of corroborative instances provided, but their *quality*:[87] thus McDowall cites the case of the

[83] L. Croizat, *op. cit.*

[84] L. Croizat *et al.*, *op. cit.* (note 74); reference on p. 266.

[85] D. E. Rosen, 'The phylogeny and zoogeography of salmoniform fishes and the relationships of *Lepidogalaxias salamandroides*', *Bulletin of the American Museum of Natural History*, 153 (1974), 265–326; D. E. Rosen, 'A vicariance model of Caribbean biogeography', *Systematic Zoology*, 24 (1976), 431–464.

[86] R. B. McDowall, 'Generalized tracks and dispersal in biogeography', *Systematic Zoology*, 27 (1978), 88–104.

[87] I. R. Ball, 'Nature and formulation of biogeographical hypotheses', *Systematic Zoology*, 24 (1976), 407–430.

Whitefaced Heron *Ardea novaehollandiae* with an apparently Mesozoic distribution in Australia and New Zealand but which is known to have dispersed across the Tasman Sea during the present century. Platnick and Nelson[88] look to logic and Occam's Razor in generalizing tracks for three or more taxa; Croizat, Nelson and Rosen call for the formulation of 'explicit methods of statistical analysis (based on the concepts of generalized tracks) that yield unambiguous and repeatable results'.[89] The latter, at least, have so far been unsuccessful: 'I have been unable to find in their writings', says McDowall of the Croizat school, 'any statistical analysis of any sort.'[90] A geographer might add that a common scale of map used by Croizat and his followers for plotting distributions is 1:200,000,000, which scarcely allows any detailed analysis, statistical or otherwise.

This has undoubtedly led many systematists into grave suspicion of Croizat's methods and results (though it is fair to add that he himself is not unaware of ambiguities),[91] and this in its turn has led many to a renewed emphasis on prior phylogenetic analysis to establish the validity and meaning of distributional records: 'Generalized tracks become meaningful only when strict phylogenetic principles have been applied, which Croizat never did.'[92] This aim, in its turn, involves a second confrontation: between the methods of conventional evolutionary taxonomy and those of Hennig's phylogenetic systematics (now usually termed cladistics). Cladistics is a method of classification, resulting in graphs of relative affinity termed cladograms, which makes no *a priori* assumptions about the nature of the relationships, especially evolutionary relationships, which are involved. If the taxa represented in cladograms are then replaced by the locations they inhabit, cladograms of affinities of areas then result: the plotting of a phylogenetic tree on a map thus gives the distributional history of the organisms considered.

Cladists thus reject the methods employed by evolutionary taxonomists such as Simpson, Mayr and Darlington, and it is thus not surprising that their work has led to dissension. Rosen speaks of the 'open hostility' between respective 'battle lines', and of anger, incredulity and bitterness

[88] N. I. Platnick and G. Nelson, 'A method of analysis for historical biogeography', *Systematic Zoology*, 27 (1978), 1–16.

[89] L. Croizat *et al.*, *op. cit.* (note 74), p. 277.

[90] R. B. McDowall, *op. cit.* (note 86), p. 89. But see P. H. A. Sneath and K. G. McKenzie, 'Statistical methods for the study of biogeography', in N. F. Hughes, ed., *Organisms and continents through time* (London: Palaeontological Association, 1973), 45–59.

[91] L. Croizat, 'De la "pseudovicariance" et de la "disjonction illusoire"', *Anuario de la Sociedad Broteriana*, 37 (1971), 113–140.

[92] I. R. Ball, *op. cit.* (note 87), p. 421.

within the American Museum of Natural History.[93] The debate became public in November 1980 with Halstead's attack on the role of cladistics in the exhibitions and publications of the British Museum (Natural History), which for him symbolized not merely the negation of Darwinian explanations but also the acceptance of a 'fundamentally Marxist view of the history of life'.[94]

Even more extreme denunciation comes from Croizat himself, seen by Nelson as (with Hennig) one of the founders of the only true biogeography, which traces its descent from Lyell and De Candolle, though subverted for a century by Darwin and Wallace.[95] In his last sad declining years, Croizat saw Darwin's contributions as 'worthless as instruments of learning'; as a theorist the author of *On the origin of species* is disqualified as worthy of attention; and his writings consist of 'piffle' rather than 'legitimate knowledge'.[96] Poor Darwin, 'in no way an acute, logical thinker', not an 'authentic genius', responsible for the 'pseudoscience' of classical biogeography, is 'not worth being taken up seriously'.[97] Nor is Croizat alone in this assessment, for Nelson and Platnick manage simultaneously to disapprove deeply of the malign influence of the Darwinian revolution and to question whether in fact it ever occurred.[98] It would be charitable to draw a veil over Croizat's remarks on Mayr, a man who 'has no clever understanding, if any at all, of the subject he believes to master', 'no title to pose as a genuine biogeographer', 'dismissed by knockout from the ring of scientific biogeography', and so on. Mayr's science is 'childish', 'tomfoolery', consists of 'half truths', 'propaganda and make-believe'. Even Hennig's systematics is no more than 'sterile daydreaming', while Croizat's own chief disciples, Nelson and Platnick, are engaged in 'pointless chatter' and are 'not authentic biogeographers'.[99]

Faced with such bizarre accusations, and their ready acceptance within the ranks of the 'new biogeographers', it is perhaps difficult to subsume the terms of the debate within conventional categories. I shall first consider issues of theory and method, and then at greater length the more solid advances in knowledge which bear on them.

[93] D. E. Rosen, *op. cit.* (note 59), p. 3.
[94] L. B. Halstead, 'Museum of errors', *Nature, Lond.*, 288 (1980), 208.
[95] G. J. Nelson, *op. cit.* (note 54).
[96] L. Croizat, *op. cit.* (note 75).
[97] L. Croizat, 'Carlos Darwin y sus Teorias', *Boletin de la Academia de Ciencias Fisicas, Matematicas y Naturales, Caracas*, 37(113) (1977), 15–90, and partial translation as 'Charles Darwin and his theories', *Tuatara*, 27 (1984), 21–25, *passim*.
[98] G. J. Nelson and N. I. Platnick, *op. cit.* (note 78), p. 374.
[99] L. Croizat, *op. cit.* (note 2), pp. 52–55 and 60.

POPPER AND PROCEDURE

The fundamental claim of cladistics is that its method is objective, and that it yields predictions or conclusions susceptible to test: it thus fulfils Popper's requirements for scientific status. Darwinian methods, both in taxonomy and biogeography, seek simply to establish plausible 'narratives' about particular 'scenarios', and these can never be falsified. They thus lead to 'a metaphysical and not a true science because its basic statements about the natural world are framed in such a way that a given prediction will always be realized'.[100] Thus the new biogeographers must tread the now familiar path from idiographic to nomothetic methods, disdain the unique, search for laws and make predictions which can then be tested against reality; they must be deductive rather than inductive (though Croizat himself[101] has shown that he does not know what these terms mean). Though these particular trails were blazed in *History and theory* a quarter of a century ago, and broadly in geography thereafter, the cladists are opening them up anew, and not surprisingly falling into the same traps their forerunners had learned to avoid.[102]

Let us examine methods of procedure in historical biogeography. Ball draws firm distinctions between the empirical–descriptive, the narrative and the analytical.[103] The first of these is the class method of establishing distributions and their boundaries; its data are derived from the present rather than from the past; and its results are given in terms of regional schemes and identified disjunctions (e.g. Wallace's Line). The narrative method seeks to provide a reasonable explanation for the patterns so established; these explanations are usually inductively based and circumstantial, are not predictive and are not readily testable. Most of the classical biogeographic syntheses of workers such as Simpson, Mayr and Darlington have been of this kind.[104] Ball calls, therefore, for a third kind of

[100] D. E. Rosen, *op. cit.* (note 59), pp. 1–2.

[101] L. Croizat, 'Deduction, induction, and biogeography', *Systematic Zoology*, 27 (1978), 209–213.

[102] M. T. Ghiselin, 'Biogeographical units: more on radical solutions', *Systematic Zoology*, 29 (1980), 80–85.

[103] I. R. Ball, *op. cit.* (note 87).

[104] P. J. Darlington Jr, *Zoogeography: the geographical distribution of animals* (New York: John Wiley, 1957); P. J. Darlington Jr, *Biogeography of the southern end of the world* (Cambridge, Mass.: Harvard University Press, 1965); E. Mayr, *Animal species and their evolution* (Cambridge, Mass.: Belknap Press, 1963); G. G. Simpson, *Splendid isolation – the curious history of South American mammals* (New Haven: Yale University Press, 1980); G. G. Simpson, *Why and how: some problems and methods in historical biology* (Oxford: Pergamon Press, 1980).

biogeography, analytical, of hypothetico-deductive form, based on theoretical propositions which, following Popper, can be falsified. How this is to be done in a largely historical field has not only generated considerable debate but also recalls the intense controversy over generalization and prediction in geology itself during the 1960s.[105]

Ball's three modes broadly correspond with three dominant methods of working identified by Cracraft.[106] One of these concerns itself exclusively with distributions: localities are plotted over space, and linkages deduced over time are calibrated where possible from the palaeontological record. This concern with area and distribution[107] has been the basis of interpretations in terms of 'centres of origin' and migration routes of taxa, both highly inferential concepts which go beyond the distributional data they are introduced to explain. Such explanations in the past have often required land-bridges or stepping-stones, and thus reconstructions of earth history, to reconcile the facts.

Because of the paucity of the fossil record (e.g. of herbaceous as opposed to woody plants), and perhaps because those concerned with explaining individual distributions have been mainly systematists, the main support for the inferred development of known distribution patterns has been phylogenetic. Ball, who works on freshwater planarians, states that 'a phylogenetic systematic background is essential' in biogeographic reconstruction;[108] the fish specialist McDowall adds that 'if biogeographical hypotheses of any sort are to be rigorously formulated they must be established upon hypotheses of phylogenetic relationships of the taxa under study';[109] and Edmunds from his work on Ephemeroptera states that 'biogeography is only as meaningful as the accuracy of out interpretation of the phylogeny of the group.'[110] This, if true, automatically excludes non-systematists from the study of biogeography, but as a principle it raises two major problems: first, biogeography must become the study of

[105] G. G. Simpson, 'Historical science', in C. C. Albritton, ed., *The fabric of geology* (Reading, Mass.: Addison-Wesley, 1963), 24–48; D. B. Kitts, 'Historical explanation in geology', *Journal of Geology*, 71 (1963), 297–313; R. A. Watson, 'Explanation and prediction in geology', *Journal of Geology*, 77 (1969), 488–494.

[106] J. Cracraft, 'Historical biogeography and earth history: perspectives for a future synthesis', *Annals of the Missouri Botanical Garden*, 62 (1975), 227–250.

[107] G. Rotramel, 'The development and application of the area concept in biogeography', *Systematic Zoology*, 22 (1973), 227–232.

[108] I. R. Ball, *op. cit.* (note 87), p. 407.

[109] R. B. McDowall, *op. cit.* (note 83), p. 88.

[110] G. F. Edmunds Jr, 'Phylogenetic biogeography of mayflies', *Annals of the Missouri Botanical Garden*, 62 (1975), 251–263; reference on p. 252.

restricted taxonomic groups rather than biotas; and second, there are in many if not most groups alternative phylogenies available (Felsenstein,[111] aiming 'to frighten taxonomists', points out that the number of possible dendrograms in a system with only 20 tips is 8.87×10^{23}), the choice between which is a matter for rational argument rather than for proof. For this latter reason, in particular, many biogeographers have attempted to abandon phylogenetic reasoning altogether and to concentrate on establishing the 'facts' of distribution. In a sense this is Croizat's own justification for his methodology, but I wish to generalize the point and to argue that it is in the establishment of distributions and the identification of patterns that biogeography finds its primary rationale.

DISTRIBUTION AND PATTERN

The basic data of biogeography consist of distributional patterns, both in space and time. These patterns must be studied in themselves, and not with reference to pre-existing ideas of their genesis. Leaving aside the difficulty that what constitutes pattern and enables it to be recognized is a prior notion of structure and order, two points are remarkable about recent pattern studies. The first is that pattern is still established by plotting dots on maps to record specific occurrences (often still on Mercator world maps), by putting boundary lines round areas of distribution (suggesting a homogeneity and continuity which is usually illusory), or by plotting numbers of taxa found within arbitrary grid squares. The patterns are thus naively apprehended, and nothing more is done with them to establish degrees of similarity, overlap or discordance. The technical apparatus familiar to geographers is thus totally ignored by most biogeographers. This is of course a characteristic of the encyclopaedic compilations of Croizat, whose method was to accumulate many hundreds of distribution maps of organisms, and allow the weight of numbers of comparable distributions to supply his generalizations for him, irrespective of the attributes of the organisms concerned. It was surely for fact rather than for method that Corner spoke – it is often quoted – of Croizat's *Panbiogeography* as 'the most important contribution to plant and animal distribution that has appeared'; it was presumably for method rather than for fact that he also added – it is rarely mentioned – that the book also contains 'much tripe'.[112]

[111] J. Felsenstein, 'The number of evolutionary trees', *Systematic Zoology*, 27 (1978), 27–33.

[112] E. J. H. Corner, 'Panbiogeography [review]', *New Phytologist*, 58 (1959), 237–238.

Patterns must result from processes. Traditionally, processes have been used to *explain* patterns; they should, we are told, rather be used to predict them, and the predictions then compared with actuality.[113] Croizat and his followers take the naive view that only two processes require consideration, and that they are mutually exclusive: 'the contest is between the concepts of chance dispersal and vicariance. One of the two concepts is to die out, and the sooner the better';[114] Croizat gratuitously adds that 'either the Darwinians bury me, or I them.'[115] It is now a commonplace that for the new biogeography only vicariance (ultimately explained primarily by plate tectonics) is of any significance.

Before considering recent substantive work in vicariance biogeography, several crucial points must be made. The first is that pattern alone can say nothing of causes in the absence of understanding of the biological potentials of the taxa involved; and that in those cases where similar patterns could result from alternative causes then the patterns alone cannot be used to distinguish their relative importance. As Solem points out: 'Coincidence, repetition of historical events, simultaneous vicariances, isolated dispersal events, or simple misunderstanding of data all can produce similar maps.'[116] Second, whereas it would be possible to evaluate different hypotheses using statistical tests, no such tests have yet been employed. Third, appeal may be made to the principle of parsimony, in spite of Darlington's conclusion that 'I do not trust Occam's razor. The simplest explanations are not necessarily the right ones in biogeography.'[117]

Moreover, if one does admit the possibility that both vicariance and dispersal are important, grave problems arise in explaining patterns: there is then 'no necessary relationship between geography and phylogeny, and objectively testable historical biogeographical models are impossible to construct.' Tattersall[118] is thus led to the conclusion, anathema to the cladists, that 'biogeographical hypotheses must exist at the level of

[113] Discussion by N. Eldredge following M. D. F. Udvardy, '"The riddle of dispersal" dispersal theories and how they affect biogeography', in G. Nelson and D. E. Rosen, eds, *op. cit.* (note 59), 6–29, discussion by L. L. Short and N. Eldredge, 30–39; reference on pp. 35–36. The same argument was presented in a more general geographical context by W. Bunge, *Theoretical geography* (Lund: Gleerup, 1962), e.g. p. 195.

[114] L. Croizat, *op. cit.* (note 101), p. 210.

[115] L. Croizat, *op. cit.* (note 75), p. 517.

[116] A. Solem, *op. cit.* (note 65), p. 236.

[117] P. J. Darlington Jr, *op. cit.* (1965) (note 104), p. 216.

[118] Discussion by I. Tattersall following J. Haffer, 'Aspects of neotropical bird speciation during the Cenozoic', in G. Nelson and D. E. Rosen, eds, *op. cit.* (note 59), 395–412, discussion by G. T. Prance and I. Tattersall; reference on p. 409.

complexity of what has been called the evolutionary 'scenario' . . . testable only by the criterion of plausibility.' Instead of judging a theory by its rigour and logical structure, it is judged by its utility: 'The value of a theory', Mankiewicz argues, 'is measured by its beauty, its applicability, what it discloses to the practitioners of science, and its usefulness to the world at large.'[119]

BIOGEOGRAPHY AND EARTH HISTORY

The new biogeography implicitly accepts continental movement as the motor of vicariance, but often betrays uneasiness at the direction of the argument. McDowall denies that geological explanations are themselves unambiguous,[120] while Rosen insists that the sequence of distributional events must be derived from the taxa themselves before the geological evidence is introduced.[121] Ball feels no especial need to conform with what he rather disparagingly calls 'current geological opinion'.[122] Given the uncertainties inherent in the inductive method in biogeography, the interpretative nature of phylogenies, and the explanatory power of the new global tectonics, this scarcely seems an economical – or sensible – procedure. The point is made particularly explicitly, however, in Parenti's study of cyprinodontiform fishes. He derives cladograms of area from cladograms of taxa, and then from the coincidence of areal patterns for different taxa 'a pattern of earth history is suggested. This pattern is independent of geological hypotheses; however, a proposed geological model may fit such a pattern. If none is found, this does not necessarily suggest that the phylogeny of these organisms is incorrect, but that the geological models may be inappropriate.'[123] For some, this doubtless amounts to the welcome abandonment of what Straney termed the 'full glut of historical miscellany'[124] in Darwinian methodology in favour of more

[119] Discussion by P. Mankiewicz following J. A. Wolfe, 'Vicariance biogeography of angiosperms in relation to paleobotanical data', in G. Nelson and D. E. Rosen, eds, *op. cit.* (note 59), 413–427, with discussion by K. J. Niklas and P. Mankiewicz, 428–445; reference on p. 442.

[120] R. B. McDowall, *op. cit.* (note 86).

[121] D. E. Rosen, *op. cit.* (1976) (note 85).

[122] I. R. Ball, *op. cit.* (note 87), p. 422.

[123] L. R. Parenti, 'A phylogenetic and biogeographical analysis of cyprinodontiform fishes (Teleostei, Atherinomorpha)', *Bulletin of the American Museum of Natural History*, 168 (1981), 335–557.

[124] D. O. Straney, 'A framework for systematics', *Science, N.Y.*, 214 (1981), 788–789.

objective techniques; for others it will seem a rash abandonment of potent sources of information; while for others still it will raise profound questions of hypothesis testing and of the criteria – or truth? or appropriateness? – for accepting or rejecting biogeographical reconstructions. As it stands, Parenti's statement comes close to repeating W. M. Davis's exhortation 80 years ago, that an attempt should be made to match his cyclic reconstructions of landforms where possible with reality, though if appropriate real landforms could not be found this did not weaken the cyclic method, which was itself simply 'a scheme of the imagination'.

It is, furthermore, the case that the interpretation of vicariant distributions as the result of plate tectonics seems inherently less complex than as a consequence of dispersal, if only because the causative factors are independently calibrated and can be viewed deterministically. The successes of the new interpretations are of course beyond doubt. The transition from the Mesozoic world of reptiles and gymnosperms to that of the angiosperms during the Cretaceous and of mammals somewhat later, approximately coincidentally with the major continental rifting, supplies a key to otherwise puzzling distributions.[125] Classic examples include the southern cold temperate floras of all the southern continents except Africa, and of the relative distributions of marsupials and placental mammals.[126] Such explanations apply not only to land areas interrupted by sea, but also to sea fragmented by land,[127] and Valentine in particular has re-examined marine faunas in the light of plate tectonics.[128] There have been major new

[125] Tectonic events resulting from drift can have dramatic consequences for land and sea distribution on very short time scales, especially in mobile areas such as the East Indies. Here D. A. Hooijer ('Quaternary mammals east and west of Wallace's Line', *Netherlands Journal of Zoology*, 25 (1975), 45–56) has drawn attention to the existence of early Pleistocene pygmy elephants in southwestern Sulawesi, and of pgymy stegodonts in Sulawesi, Flores and Timor, all localities well outside Wallace's Line. This latter is thus of purely modern validity, if indeed it is the most appropriate boundary there is (see the review by G. G. Simpson, *op. cit.* (note 7)). M. G. Audley-Charles and D. A. Hooijer ('Relation of Pleistocene migrations of pygmy stegodonts to island arc tectonics in eastern Indonesia', *Nature, Lond.*, 241 (1973), 197–198) have suggested that only extensive earth movements can explain these patterns. But see D. L. Johnson's quite different explanation of a similar problem on the other side of the Pacific (*op. cit.*, notes 66 and 67).

[126] J. Fooden, 'Breakup of Pangaea and isolation of relict mammals in Australia, South America, and Madagascar', *Science, N.Y.*, 175 (1972), 894–898; C. B. Cox, *op. cit.* (note 10); A. Keast, 'Contemporary biota and the separation sequence of the southern continents', in D. H. Tarling and S. K. Runcorn, eds, *Implications of continental drift to the earth sciences* (London: Academic Press, 1964), volume 1, 309–343.

[127] J. C. Briggs, 'Operation of zoogeographic barriers', *Systematic Zoology*, 23 (1973), 248–256.

interpretations in terms of the fragmentation of a previously extensive Tethys biota for marine angiosperms, mangroves and scleractinian corals.[129] Many, if not most, of these studies have dealt with particular taxonomic groups at various systematic levels: an eclectic list would include Schuster and Whitmore on vascular plants,[130] Demoulin on Fungi,[131] Ferris *et al.* on soil nematodes,[132] Sterrer on the meiofauna,[133] Foin on the Cypraeidae,[134] Platnick on arachnids,[135] Rosen on fishes,[136] Cracraft on Rich on birds,[137] and Cracraft and McKenna on mammals.[138] Some of the applications have been of great ingenuity. Thus Carr and Coleman[139] suggest that the migration of the Green Turtle *Chelonia mydas* for 2000 km from the coast of Brazil to Ascension Island in order to breed reflects

[128] J. W. Valentine, 'Plate tectonics and shallow marine diversity and endemism, an actualistic model', *Systematic Zoology*, 20 (1971), 253–264; J. W. Valentine, *Evolutionary biogeography of the marine biosphere* (Englewood Cliffs: Prentice-Hall, 1973).

[129] M. D. Brasier, 'An outline history of seagrass communities', *Palaeontology*, 18 (1975), 681–702; E. Hadac, 'Species diversity of mangrove and continental drift', *Folia Geobotanica et Phytotaxonomica*, 11 (1976), 213–216; E. D. McCoy and K. L. Heck Jr, 'Biogeography of corals, seagrasses, and mangroves: an alternative to the center of origin concept', *Systematic Zoology*, 25 (1976), 201–210. See also my *Biogeographical atlas of tropical corals, seagrasses and mangroves* (Oxford: Basil Blackwell, forthcoming).

[130] R. M. Schuster, 'Continental movements, "Wallace's Line" and Indo-Malayan dispesal of land plants: some eclectic concepts', *Botanical Review*, 38 (1972), 1–86; T. C. Whitmore, *op. cit.* (note 10).

[131] V. Demoulin, 'Phytogeography of the fungal genus *Lycoperdon* in relation to the opening of the Atlantic', *Nature, Lond.*, 242 (1973), 123–125.

[132] V. R. Ferris, C. G. Goseco and J. M. Ferris, 'Biogeography of freeliving soil nematodes from the perspective of plate tectonics', *Science, N.Y.*, 193 (1976), 508–510.

[133] W. Sterrer, 'Plate tectonics as a mechanism for dispersal and speciation in interstitial sand fauna', *Netherlands Journal for Sea Research*, 7 (1973), 200–222.

[134] T. C. Foin, 'Plate tectonics and the biogeography of the Cypraeidae (Mollusca: Gastropoda)', *Journal of Biogeography*, 3 (1976), 19–34.

[135] N. I. Platnick, 'Drifting spiders or continents? Vicariance biogeography of the spider subfamily Lavoniinae (Araneae: Gnaphosidae)', *Systematic Zoology*, 25 (1976), 101–109.

[136] D. E. Rosen, *op. cit.* (note 85).

[137] J. Cracraft, 'Continental drift, paleoclimatology, and the evolution and biogeography of birds', *Journal of Zoology*, 169 (1973), 455–545; P. V. Rich, 'Antarctic dispersal routes, wandering continents, and the origin of Australia's non-passeriform avifauna', *Memoirs of the National Museum of Victoria*, 36 (1975), 63–125.

[138] J. Cracraft, 'Continental drift and vertebrate distribution', *Annual Review of Ecology and Systematics*, 5 (1974), 215–261; M. C. McKenna, 'Fossil mammals and early Eocene North Atlantic land continuity', *Annals of the Missouri Botanical Garden*, 62 (1975), 335–353.

[139] A. Carr and P. J. Coleman, 'Seafloor spreading theory and the odyssey of the Green Turtle', *Nature, Lond.*, 249 (1974), 128–130.

behaviour continuous since the early Cenozoic when the two localities were still closely adjacent: if this explanation has generality, then it may be possible to use Green Turtle migration routes in other parts of the world to infer drift histories.

Some distributional anomalies can be readily explained by the vicariance hypothesis. One such is Geister's spectacular discovery of the genus *Pocillopora* in last interglacial coral reefs of San Andres on the Nicaraguan Shelf,[140] and subsequently in Aruba, Barbados and Grand Cayman: this genus is now restricted to the Indo-Pacific, was hitherto known in the Caribbean not later than the Miocene of Cuba and Dominica, and has always been considered one of the many extinctions caused by the closure of the Panama seaway and the growing Cenozoic provinciality of the Caribbean. Other distributions remain enigmatic. Thus *Typhlatya* is a genus of mainly subterranean and freshwater caridean shrimp known from Barbuda, Mona, Cuba and Yucatan, but recently discovered in localities as far apart as the Galapagos and Ascension.[141]

In the Caribbean, several workers have developed Rosen's vicariance analysis.[142] Thus MacFadden[143] has interpreted the distribution of the insectivores *Nesophontes* and *Solenodon* in the Greater Antilles in terms of plate movement in the lat Cretaceous and early Cenozoic, though he recognizes that subsequent modification of the fauna has taken place by dispersal and extinction. Shields and Dvorak had similarly accounted for the distribution of Meso-American butterfly faunas, using a reconstruction that emphasizes the linkages between Yucatan, Cuba and Florida.[144] It is, however, noticeable that the criterion they advance for the acceptance of their model is one of plausibility, and that in no sense is it subject to rigorous test or even comparison with other possible reconstructions.

Pregill's critique[145] of Rosen's vicariance model as it refers to vertebrate

[140] J. Geister, 'Occurrence of Pocillopora in late Pleistocene Caribbean coral reefs', *Mémoires du Bureau de Recherche géologiques et minières*, 89 (1977), 378–388.

[141] F. A. Chase Jr and R. B. Manning, 'Two new caridean shrimps, one representing a new family, from marine pools on Ascension Island (Crustacea: Decapoda: Natantia)', *Smithsonian Contributions to Zoology*, 131 (1972), 1–18.

[142] D. E. Rosen, *op. cit.* (1976) (note 85).

[143] B. J. MacFadden, 'Rafting mammals or drifting islands? Biogeography of the Greater Antillean insectivores *Nesophontes* and *Solenodon*', *Journal of Biogeography*, 7 (1980), 11–22.

[144] O. Shields and S. K. Dvorak, 'Butterfly distribution and continental drift between the Americas, the Caribbean and Africa', *Journal of Natural History*, 13 (1979), 221–250.

[145] G. K. Pregill, 'An appraisal of the vicariance hypothesis of Caribbean biogeography and its application to West Indian terrestrial vertebrates', *Systematic Zoology*, 30 (1981), 147–155.

biogeography is of particular interest. On the one hand, Pregill found it difficult to distinguish Rosen's results from those of classical biogeography (i.e. to differentiate generalized tracks from centres of origin), or to specify how his explanations were more testable than those they replaced. Second, he shows that the geological evidence does not support Rosen's inferred geological history (though it is fair to add that there is much disagreement over palaeogeographical reconstructions in the Caribbean area). And third, he shows that known faunistic distribution patterns and fossil evidence are themselves inconsistent with Rosen's reconstructions (most especially in the critical absence from the Greater Antilles of marsupials, carnivores and ungulates). While Pregill's paper lacks the presumed elegance of statistical tests and null hypotheses, the points emerging from his 'glut of historical miscellany' approach cannot simply be dismissed by the cladists as 'inappropriate' to their models.

There is, of course, room for a quite extraordinary amount of misunderstanding and misinterpretation in the application of plate-tectonic ideas to vicariance biogeography. This is shown by the anachronistic pursuit of a lost 'Pacifica' continent by Melville and by Nur and Ben-Avraham.[147] Unfortunately for Melville, the movements he requires to fragment his angiosperm distributions predated the emergence of the angiosperms, while his geological and geophysical evidence is stigmatized as 'astounding', 'impossible', 'naive', 'ludicrous', 'fictional' and a 'geofantasy' by Haugh.[148]

Not all vicariant distributions result from plate tectonics: of equal interest are those resulting from Pleistocene climatic events and sea-level changes, especially in the humid tropical lowlands and shallow seas. In the case of the Amazon lowland, for example, Simpson and Haffer review the history of the biota as a whole,[149] Haffer that of the birds,[150] and Prance and Steyermark the plants.[151] Similarly Vuillemier and Simberloff discuss the

[146] R. Melville, *op. cit.* (note 73).

[147] A. Nur and Z. Ben-Avraham, 'Lost Pacifica continent: a mobilistic speculation', in G. Nelson and D. E. Rosen, eds, *op. cit.* (note 59), 341–366.

[148] Discussion by B. N. Haugh following R. Melville, *op. cit.* (note 73); reference on p. 275.

[149] B. B. Simpson and J. Haffer, 'Speciation patterns in the Amazonian forest biota', *Annual Review of Ecology and Systematics*, 9 (1978), 497–518.

[150] J. Haffer, 'Distribution of Amazonian forest birds', *Bonner Zoologische Beiträge*, 29 (1978), 38–78.

[151] G. T. Prance, 'Distribution patterns of lowland neotropical species with relation to history, dispersal and ecology, with special reference to Chrysobalanaceae, Caryocacaceae and Lecythidaceae', in K. Larsen and L. B. Holm-Nielsen, eds, *Tropical botany* (London: Academic Press, 1979), 59–87; J. A. Steyermark, 'Plant refuge and dispersal centres in Venezuela: their relict and endemic elements', *ibid.*, 185–221.

adjacent high Andes.[152] Work has also concentrated on the classical problems of Wallace's Line and other biogeographic limits in southeast Asia, a set of problems recently reviewed by Whitmore.[153] In particular Musser has reviewed the systematics of Indo-Malayan murid rodents in the area, and has given an account of the Giant Rat and other mammals of the island of Flores in Indonesia.[154] Here the total mammal fauna consists of six native rats and six other murids, the Javan porcupine, two stegodonts, one nisu, two pigs, a palm civet, a crab-eating macaque, nine bats and two shrews. The stegodonts and one native rat are represented by Pleistocene fossil remains, and four of the native rats by fossils 3000–4000 years old; 13 of the mammal species (52 per cent) are introduced. Of the total mammal fauna five rats and one shrew are endemic to Flores (19 per cent of the total), but of these four are extinct. Musser's study is a fascinating reminder of the need for detailed studies, locality by locality, making the fullest use of the palaeontological evidence, in areas where not only plate tectonics but also sea-level shifts and Pleistocene environmental changes have been major determinants of faunal distributions.[155] Also in the southeast Asian area, the dilemma posed by the apparent fossil occurrence of a pygmy rhinoceros in New Caledonia has been resolved by the demonstration that the tooth from which it was identified is in fact that of a marsupial:[156] as a rhinoceros it posed insuperable dilemmas both to vicariance biogeographers and to dispersalists, notwithstanding Johnson's demonstration of the swimming abilities of proboscideans.[157]

It must be a matter for regret that, following Croizat's lead, many of those who pursue vicariant explanations simultaneously reject the reality of dispersal in accounting for distributions. Croizat *et al.* castigate those

[152] F. Vuillemier and D. Simberloff, 'Ecology vs. history as determinants of patchy and insular distributions in high Andean birds', in M. K. Hecht, W. C. Steere and B. Wallace, eds, *Evolutionary biology*, volume 12 (New York: Plenum, 1980), 235–379.

[153] T. C. Whitmore, ed., *Wallace's Line and plate tectonics* (Oxford: Clarendon Press, 1981).

[154] G. G. Musser, 'Notes on systematics of Indo-Malayan murid rodents and descriptions of new genera and species from Ceylon, Sulawesi, and the Philippines', *Bulletin of the American Museum of Natural History*, 168 (1981), 225–234; G. G. Musser, 'The Giant Rat of Flores and its relatives east of Borneo and Pali', *Bulletin of the American Museum of Natural History*, 169 (1981), 67–176.

[155] H. T. Verstappen, 'Quaternary climatic changes and natural environments in southeast Asia', *GeoJournal*, 4 (1980), 45–54.

[156] C. Guérin, J. H. Winslow, M. Piboule and M. Faure, 'Le prétendu rhinocéros de Nouvelle Caledonie est un marsupial (*Zygomaturus diahotensis* nov.sp.): solution d'une énigme et conséquences paléogéographiques', *Geobios*, 14 (1981), 201–217.

[157] D. L. Johnson, *op. cit.* (notes 66 and 67).

biogeographers who 'founder in a self-created morass of chance hops; great capacities for, or mysterious means of, dispersal; rare accidents of over-sea transportation; small probabilities that with time become certainties; and other pseudo-explanations'.[158] Ball goes so far as to say that 'groups which are known to be distributed by passive dispersal make poor subjects for biogeographical enquiry . . . and should not form part of general hypotheses';[159] he cites arachnids, Protozoa and Malacostraca, but not the plants and land birds on which MacArthur and Wilson relied so heavily in their theory of island biogeography.

The attack is a double one: first on calculations such as those of Simpson on probabilities of dispersal in geological time;[160] second, on reconstructions such as those of Simpson and Darlington which require extensive migration. There is ambiguity in the arguments, however. Croizat's use of the term 'generalized (or individual) track' is itself curious, since 'track' implies mobility and movement rather than simply extent, which in any case must itself have resulted from some translocation: 'tract' would be preferable and more simply descriptive. But, second, those applying the method of vicariance analysis do not rule out secondary or limited dispersal following fragmentation (indeed this is seen as a cause of sympatric speciation). And, third, as we have seen, there is no difficulty whatever in demonstrating the reality of long-distance dispersal. As so often, complementary hypotheses are needlessly set in spurious opposition to each other by those seeking to defend or promote alternative viewpoints.

It is worth noting one particular instance where both plate tectonics and the dispersalist tenets of conventional island biogeography have been used to elucidate particularly difficult distribution patterns. Rotondo *et al.*[161] introduce the concept of island integration for the incorporation of islands moving on oceanic plates into pre-existing island groups through lateral movement, a concept derived partly from the close juxtaposition of islands or seamounts with substantially different ages. Thus Wentworth Seamount, close to the inflection of the Hawaiian-Emperor Chain, has an age of 71 million years (compared with 27.3 million years for an adjacent seamount),

158 L. Croizat *et al.*, *op cit.* (note 74), p. 276.

159 I. R. Ball, *op. cit.* (note 87), p. 420.

160 G. G. Simpson, *op. cit.* (note 56); also M. C. McKenna, 'Sweepstakes, filters, corridors, Noah's Arks, and beached Viking funeral ships in palaeogeography', in D. H. Tarling and S. K. Runcorn, eds, *op. cit.* (note 126), volume 1, 295–308.

161 G. M. Rotondo, V. G. Springer, G. A. J. Scott and S. O. Schlanger, 'Plate movement and island integration – a possible mechanism in the formation of endemic biotas, with special reference to the Hawaiian Islands', *Systematic Zoology*, 30 (1981), 12–21.

and Necker Island, halfway along the Hawaiian Chain, an age of 77.6 million years (compared with an adjacent island date of 10.1 million years). The topographic features are hypothesized to have been located at 4°S 144°W and 9°S 131°W at 40 million years B.P.; their present positions are 29°N 178°W and 23.5°N 164.5°W respectively. If, on reaching the islands of the Hawaiian Chain (generated over the Hawaiian hot spot), Wentworth and Necker had still been high volcanic islands (over which there is, however, room for doubt), this would give an opportunity for the admixture of non- and pre-Hawaiian elements into the Hawaiian biota, thus supplementing the dispersal mechanisms classically proposed by Zimmerman, Fosberg and Carlquist. Rotondo *et al.* suggest that this mechanism explains existing extra-Hawaiian faunal relationships in, for example, marine fish, marine molluscs and land snails, better than dispersal mechanisms alone. Their hypothesis makes it possible to re-examine the extraordinary biota of Laysan Island, Leeward Hawaiian Islands, which in spite of being a low (but large) coral island possesses essentially high-island biotic elements.[162] These include a rail *Porzana*, honeycreepers *Psittirostra* and *Himatione*, a warbler *Acrocephalus*, land snails Tornatellidae, and unexpected angiosperms (*Santalum*, *Schiedea*, *Phyllostegia*, *Lipochaete* and the palm *Pritchardia*).

GEOGRAPHY IN BIOGEOGRAPHY

Analytical biogeography at present spans extraordinary – and probably quite unnecessary – divergence of views, compounded by major controversies over taxonomic practice, the status of evolutionary theory and the role and significance of Darwin. At the root of these disputes there is the patently unnecessary opposition of dispersal and vicariance as explanations of distributions. As Green pointed out in his analysis of New Hebridean and New Caledonian plants, 'no single factor may explain present-day, disjunct distribution patterns. In some cases dispersal, even long distant dispersal, may be the explanation, in others the disappearance of land once available for plant growth and in yet others extinction from intermediate areas';[163] many others, perhaps especially fieldworkers rather than museum special-

[162] S. O. Schlanger and G. W. Gillett, 'A geological perspective of the upland biota of Laysan Atoll (Hawaiian Islands)', *Biological Journal of the Linnean Society*, 8 (1976), 205–216.

[163] P. S. Green, 'Observations on the phytogeography of the New Hebrides, Lord Howe Island and Norfolk Island', in D. Bramwell, ed., *op. cit.* (note 61), 41–53; reference on p. 52.

ists, have echoed these statements in diverse situations, yet the vicariance biogeographers on the whole reject them.

Croizat was undoubtedly right in his basic proposition: that it is distribution, pattern and coincidence of pattern that supply the fundamental data of biogeography, and indeed which make it geographical at all. Croizat's weakness, of course, was that he established pattern and coincidence in a mechanical manner, without considering the functional relationships of the taxa involved. This could indeed be considered a general criticism of much biogeographical work by systematists, typified by their reluctance to call on geological evidence at the outset of their enquiry. For to do so is not to prejudge the conclusions, but to establish the nature of the stage on which, in Hutchinson's analogy, the evolutionary play is acted out. To adopt a perhaps familiar criticism, the overriding weakness of much systematic biogeography is that it is set on featureless plains distinguished only by geographical coordinates and sharp disjunctions from other comparable areas. The sense of geography is lacking.

The force of this criticism is best realized by taking as examples studies to which it is *not* applicable. This is particularly so of the work of Axelrod and Raven, especially in their reviews of Australasian, South American and African historical biogeography.[164] The Australian case is particularly instructive.[165] Here a Cretaceous landmass covered with a temperate woodland of gymnosperms and evergreen angiosperms, inhabited by marsupials, began moving northwards and by the Eocene was entering low rainfall latitudes. The great development of xeric and broadleaf sclerophyll plants (*Acacia*, *Eucalyptus*) which resulted has given Australian vegetation its unique physiognomy. As northward drift continued, the northern coastlands emerged from the desert belt into humid tropical latitudes and

[164] P. H. Raven and D. I. Axelrod, 'Plate tectonics and Australasian paleobiogeography', *Science, N.Y.*, 176 (1972), 1379–1386; P. H. Raven and D. I. Axelrod, 'History of the flora and fauna of Latin America', *American Scientist*, 63 (1975), 420–429; D. I. Axelrod and P. H. Raven, 'Late Mesozoic and Tertiary vegetation history of Africa', in M. J. A. Werger, ed., *Biogeography and ecology of southern Africa* (The Hague: W. Junk, 1978), volume 1, 77–130. For a similarly wide-ranging synthesis for a largely oceanic realm, see E. A. Kay, 'Little worlds of the Pacific: an essay on Pacific basin biogeography', *University of Hawaii Harold L. Lyon Arboretum Lectures*, 9 (1980), 1–40.

[165] E. D. Gill, 'Evolution of Australia's unique flora and fauna in relation to the plate tectonic theory', *Proceedings of the Royal Society of Victoria*, 87 (1975), 215–234; P. V. Rich, *op. cit.* (note 137); J. S. Beard, 'Tertiary evolution of the Australian flora in the light of latitudinal movements of the continent', *Journal of Biogeography*, 4 (1977), 111–118; E. M. Kemp, 'Tertiary climatic evolution and vegetation history in the southeast Indian Ocean region', *Palaeogeography, Palaeoclimatology, Palaeoecology*, 24 (1978), 169–208.

collided with New Guinea during the Oligocene and Miocene. Biotic interactions between the two have led to the present pattern of outliers of rainforest in a largely grass-covered northern Australia, and outliers of savanna in a largely forested New Guinea. It is also no coincidence that the present coastlands of Queensland reached tropical latitudes in the Miocene[166] and that the earliest reefs of the Great Barrier Reef are of the same age.[167] New Zealand largely escaped these climatic tribulations; its biota is thus more closely allied to that of temperate Gondwanaland, and drift dispersal has played a greater part in its subsequent modification.[168] New Caledonia too has a relict fauna and flora of Cretaceous age.[169] India moved northwards through some 50° of latitude before colliding with the Asian plate and thrusting up the Himalayas; its biotic history has been considered by Axelrod, Lakhanpal, Prakash, and Sahni and Kumar.[170]

Nor have synthetic studies on this scale been restricted to particular geographical areas. Following a general survey of angiosperm biogeography by Raven and Axelrod,[171] Axelrod has analysed the development and distribution of North American and Eurasian (Madrean-Tethyan) broadleaf sclerophyll vegetation in a paper remarkable for the attention it gives to palaeoclimate, palaeooceanography, geology and even geomorphology in establishing the context for plant distribution and vegetation growth.[172]

Many ancillary questions are raised by these essentially geographical analyses. One concerns the reasons for massive extinctions at the end of the

[166] K. A. W. Crook and L. Belbin, 'The southwest Pacific during the last 90 million years', *Journal of the Geological Society of Australia*, 25 (1978), 23–40.

[167] A. R. Lloyd, 'Foraminifera of the Great Barrier Reef bores', in O. A. Jones and R. Endean, eds, *Biology and geology of coral reefs* (New York: Academic Press, 1973), volume 1, 347–366.

[168] J. P. Skipworth, 'Continental drift and the New Zealand biota', *New Zealand Journal of Geography*, 57 (1974), 1–13.

[169] R. F. Thorne, 'Floristic relationships of New Caledonia', *State University of Iowa Studies in Natural History*, 20(7) (1965), 1–14.

[170] D. I. Axelrod, 'Plate tectonics in relation to the history of angiosperm vegetation in India', *Special Publications of the Birbal Sahni Institute of Palaeobotany*, 1 (1974), 5–18; R. N. Lakhanpal, 'Tertiary floras of India and their bearing on the historical geology of the region', *Taxon*, 19 (1970), 675–694; U. Prakash, 'Palaeoenvironmental analysis of Indian tertiary floras', *Geophytology*, 2 (1972), 178–205; A. Sahni and V. Kumar, 'Palaeogene palaeobiogeography of the Indian subcontinent', *Palaeogeography, Palaeoclimatology, Palaeoecology*, 15 (1974), 209–226.

[171] P. H. Raven and D. I. Axelrod, 'Angiosperm biogeography and past continental movements', *Annals of the Missouri Botanical Garden*, 61 (1974), 539–673.

[172] D. I. Axelrod, 'Evolution and biogeography of Madrean-Tethyan sclerophyll vegetation', *Annals of the Missouri Botanical Garden*, 62 (1975), 280–334.

Mesozoic and again in the Pleistocene. Recent contributions have stressed both extraterrestrial mechanisms[173] as well as interrelated changes of sea-level and climate;[174] there also continues to be debate over the relative role of man as predator and of environmental change in accounting for Pleistocene extinctions.[175] A second set of problems concerns interrelationships between plants and animals in evolving systems. On a general level this includes the development of defence mechanisms in plants in response to herbivory[176] and the development of unusual fertilization procedures.[177] A particularly attractive example of such interdependence is given by Temple's discovery[178] of the imminent extinction of the tree *Calvaria major* in Mauritius. The species is now represented only by old individuals. Its seed has an unusually thick pericarp, and Temple hypothesized that it only germinated after passing through the digestive system of the Dodo *Rhaphus cucullatus*, a giant flightless pigeon now extinct, in much the same fashion as the Galapagos Tomato *Lycopersicon* species is related to the Galapagos tortoise *Geochelone elephantopus*.[179] Temple tested this supposition by feeding seeds of *Calvaria* to turkeys as a Dodo substitute, and found that germination occurred when the seeds were excreted. A third implication concerns the biogeochemical significance of the evolution of the land biota for the global carbon cycle and the implications of this for world radiation balance and hence for the nature of terrestrial environments.[180]

[173] D. H. Clark, W. H. McCrea and F. R. Stephenson, 'Frequency of nearby supernovae and climatic and biological catastrophes', *Nature, Lond.*, 265 (1977), 318–319; G. E. Hunt, 'Possible climatic and biological impact of nearby supernovae', *Nature, Lond.*, 271 (1978), 430–431.

[174] N.-A. Mörner, 'Low sea levels, droughts and mammalian extinctions', *Nature, Lond.*, 271 (1978), 738–739.

[175] D. K. Grayson, 'Pleistocene avifaunas and the overkill hypothesis', *Science, N.Y.*, 195 (1977), 691–693; R. Gillespie, D. R. Horton, P. Ladd, P. G. Macumber, T. H. Rich, T. Thorne and R. V. S. Wright, 'Lancefield Swamp and the extinction of the Australian megafauna', *Science, N.Y.*, 200 (1978), 1044–1048; D. A. Russell, 'The dilemma of the extinction of the dinosaurs', *Annual Review of Earth and Planetary Science*, 7 (1979), 163–182.

[176] P. R. Atsatt and D. J. O'Dowd, 'Plant defense guilds: many plants are functionally interdependent with respect to their herbivores', *Science, N.Y.*, 193 (1976), 24–29; R. T. Bakker, 'Dinosaur feeding behaviour and the origin of flowering plants', *Nature, Lond.*, 274 (1978), 661–663.

[177] R. W. Sussman and P. H. Raven, 'Pollination by lemurs and marsupials: an archaic coevolutionary system', *Science, N.Y.*, 200 (1978), 731–736.

[178] S. A. Temple, 'Plant–animal mutualism: coevolution with Dodo leads to neat extinction of plant', *Science, N.Y.*, 197 (1977), 885–886.

[179] C. M. Rick and R. I. Bowman, 'Galapagos tomatoes and tortoises', *Evolution*, 15 (1961), 407–417.

[180] D. M. McLean, 'Land floras: the major late Phanerozoic atmospheric carbon

The potentials of this kind of comprehensive analysis and reconstruction of integrated distribution patterns are remarkable. Raven and Axelrod[181] realized some of the implications of the northward drift of Australia for the dynamics of the global climate, notably for the strength and continuity of the high-latitude westerlies. As Australia moved away from Antarctica, however, the westerly ocean circulation also became established, and thus the capacity of ocean currents for meridional heat transfer was materially reduced. Antarctica also itself became climatically isolated by the West Wind Drift and Roaring Forties, and widespread glaciation resulted as early as the Oligocene.[182] The increasingly zonal nature of the world climate which resulted could be seen as leading to the replacement of forests by savannas and deserts in subtropical latitudes, to the expansion of grasslands at the expense of trees, to the consequent descent of arboreal mammals to the ground, and in consequence to the evolution of bipedalism and indeed of man himself. This is a narrative and scenario which adds considerably more to our understanding of the biogeography of the planet than either Croizat's naive comparison of distributions or any multitude of cladograms: Darwin would undoubtedly have classed such insights among the 'grand facts' of nature.

CONCLUSION

It seems to me clear that both vicariance biogeography and dispersal are appropriate tools in biogeographical analysis at particular levels of enquiry, and that biogeography is weakened by the stubborn neglect of either. The heat generated by the recent disputes over the 'new biogeography' cannot detract from its achievements. On the other hand there is nothing to be gained by attaching to so massive an achievement as Darlington's such pejorative labels as 'colonialistic' or 'vacuum biogeography', as do Platnick

---

dioxide/oxygen control', *Science, N.Y.*, 200 (1978), 1060–1062; D. M. McLean, 'A terminal Mesozoic "greenhouse": lessons from the past', *Science, N.Y.*, 201 (1978), 401–406; B. Bolin, 'Changes of land biota and their importance for the carbon cycle', *Science, N.Y.*, 196 (1977), 613–615; G. M. Woodwell, R. H. Whittaker, W. A. Reiners, G. E. Likens, C. C. Delwiche and D. B. Botkin, 'The biota and the world carbon budget', *Science, N.Y.*, 199 (1978), 141–146.

[181] P. H. Raven and D. I. Axelrod, *op. cit.* (1972) (note 164).

[182] J. P. Kennett, 'Cenozoic evolution of Antarctic glaciation, the Circum-Antarctic Ocean, and their impact on global palaeoceanography', *Journal of Geophysical Research*, 82 (1977), 3843–3860.

and Nelson.[183] Nor do attacks on Darwin for his 'unfortunate' and 'tragic' interest in dispersal[184] materially advance the argument. I believe that geographers can supply a broader vision, capable of providing a technical capability in the analysis of distributions and the recognition of patterns that Croizat so sadly lacked, and uniting it with a holistic palaeoenvironmental reconstruction that goes beyond the simple relocation of land and sea. In this more comprehensive vision, ecosystems rather than species, occupying landscapes rather than areas, become the subjects of study.

Simberloff and his colleagues have already recognized that the only real differences in the appropriateness or otherwise of calling on particular factors in explanation of distributions are in 'time frames' and 'geographical scales'.[185] It is remarkable that only Udvardy has attempted to specify the scale factors which might structure our investigations.[186] He distinguishes:

1 The secular scale, with spatial dimensions of about 100 km and time dimensions of about 100 years, 'wherein the dynamics of distribution, spreading, and dispersal is actually studied as a biological phenomenon'. This is presumably equivalent to the 'ecological time' of Simberloff *et al.*[187]
2 The millenial scale. In a diagram Udvardy suggests that this covers post-Pleistocene time (*ca.* 12,000 years) and spatial scales of up to 1000 km, though in his text he extends its scope to cover the Pleistocene (an order-of-magnitude extension in the time scale). The major factors operating here are those of climatic and sea-level change.
3 The phylogenetic scale (or evolutionary time). Here the time scale may go up to 500 million years and the spatial extent to 40,000 km; and it is on these scales that continental displacement becomes important.

Though clearly capable of refinement, this scheme has a familiar ring. Schumm and Lichty[188] said much the same 20 years ago in an influential paper of which the systematists are uniformly ignorant. Perhaps the terms of

[183] N. I. Platnick and G. Nelson, *op. cit.* (note 88).
[184] L. Croizat *et al.*, *op. cit.* (note 74).
[185] D. Simberloff, K. L. Heck, E. D. McCoy and E. F. Connor, 'There have been no statistical tests of cladistic biogeographical hypotheses', in G. Nelson and D. E. Rosen, eds, *op. cit.* (note 59), 40–63.
[186] M. D. F. Udvardy, *op. cit.* (note 113).
[187] D. Simberloff *et al.*, *op. cit.* (note 185), p. 41.
[188] S. A. Schumm and R. W. Lichty, 'Time, space and causality in geomorphology', *American Journal of Science*, 263 (1965), 110–119.

the debate would be considerably sharpened if taxonomists working on biogeographical problems were familiar not only with the techniques of pattern analysis which they do not use, but also with the common concepts of modern geography of which they are apparently unaware.

Over a century ago, Thomas Henry Huxley argued that science was simply common sense.[189] While it is true that this can no longer serve as a sufficient definition, many of those involved in the biogeographical controversies of the past several years might reflect that it is at least a necessary one. Putting the *geography* back in the *bio-* means providing once more that comprehensive view of the distribution of life on earth, pioneered by Humboldt and established by Darwin. That this has in large degree been lost sight of in our increasingly technical and specialist concerns accounts for the wide array of seemingly disparate studies which currently shelter under the title of biogeography.

[189] T. H. Huxley, 'On the educational value of the natural history sciences', in T. H. Huxley, *Science and education* (London: Macmillan, 1893), reference on p. 45.

General Index

Index of Persons